BIOLOGICAL REGULATION
of the
CHONDROCYTES

Edited By

Monique Adolphe
Laboratory of Cellular Pharmacology
Ecole Pratique des Hautes Etudes
Paris, France

CRC Press
Boca Raton Ann Arbor London Tokyo

Library of Congress Cataloging-in-Publication Data
Biological regulation of the chondrocytes / Monique Adolphe, [editor].
    p. cm.
    Includes bibliographical references and index.
    ISBN 0-8493-6733-6
    1. Cartilage — Growth — Regulation. 2. Cartilage — Growth — Molecular
aspects. 3. Cartilage cells. 4. Growth factors. I. Adolphe, Monique.
    [DNLM: 1. Cartilage — cytology. 2. Cartilage — physiology. WE 300
B614]
QP88.23.B57   1992
611'.0183 — dc20
DNLM/DLC   91-45369
for Library of Congress   CIP

PREFACE

This book was born from the suggestions of Dr. Sokoloff who was one of the most important pioneers in cartilage research. However, as he declined to be editor of a new book in this field, he suggested my name to CRC Press and I am grateful to him for this confidence. With his encouragement and the help of Paul Benya, I have tried to contribute to an up-to-date book on "*biological regulation of chondrocytes*".

The chondrocyte is a fascinating cell when considering the wide range of its functions. Furthermore, it has been recently demonstrated that this cell produces and is sensitive to many growth factors and cytokines which appear to be regulators of the cartilaginous equilibrium.

The introductory chapter of this book is an overview of the different types of chondrocyte. Each subsequent chapter discusses the various biological regulations implicated in chondrocyte functions, especially the modulation of the differentiative properties. Chapters 2 and 3 deal with the extracellular matrix components, and Chapter 4 discusses the special case of cultured chondrocytes and the usefulness of *in vitro* approaches. Chapters 5, 6, and 7 focus on the complex role of cartilaginous growth factors and cytokines (FGF, TGFß, IGF, IL-1...) on the modulation of the chondrocyte properties. The interactions of chondrocytes with a very important neighboring cell, the synoviocyte, is treated in Chapter 9. The last four chapters focus on the biochemical and molecular pertubations appearing during repair, ageing, rheumatic diseases, and cancer.

Distinguished scientists coming from different parts of the world took the responsibility for one chapter each. It is important to note that overlappings occur among the chapters: for example, cytokines are implicated in nearly all the subjects since they act in the regulation of matrix components of cultured chondrocytes and in the disregulations occurring in cartilaginous diseases. This intentional redundancy should permit the highlighting of the main biologic regulations for which the chondrocyte is responsible. Finally, I hope this book will help biologists and researchers obtain a better understanding of the cartilaginous tissue physiopathology that may lead to new research directions in pharmacology and therapy.

I wish to thank the colleagues from my laboratory who helped me with the content as well as the form of this book, particularly S. Demignot, B. Benoit, S. Thenet, F. Vincent, and my efficient secretary R. Frolleau. I sincerely hope that all the contributors to this work have helped the scientific community. I also thank CRC Press for having given me the opportunity to work towards the completion of this fascinating up-to-date information on a cell which has been the center of my interest for the past 15 years.

Monique Adolphe

THE EDITOR

Monique Adolphe is the Director of the Laboratory of Cellular Pharmacology of the Ecole Pratique des Hautes Etudes in Paris, France, and President of the "Ecole Pratique des Hautes Etudes".

Monique Adolphe graduated in 1955 from the Faculty of Pharmacy, Paris and obtained her Ph.D. degree in 1969 from the University of Paris V. She successively interned in hospitals, elevating herself to Assistant at the faculty of Medicine, and later becoming Director of Laboratory in 1974. Monique Adolphe is member of Société Française de Thérapeutique et Pharmacodynamie, Association des Pharmacologistes, European Study Group for Cell Proliferation, European Cell Biology Organization, American Tissue Culture Association, European Tissue Culture Society, Association Française pour l'Avancement des Sciences, Cercle Français du Tissu Conjonctif, Société de Biologie, New York Academy of Sciences, Académie Nationale de Pharmacie, Society for Analytical Cytology, OPAL, Chimie Ecologie, Société de Pharmaco-Toxicologie Cellulaire, and Correspondant member de l'Académie Nationale de Médecine. She serves as President of Société de Pharmaco-Toxicologie Cellulaire (SPTC), and OPAL. She received awards from the Academy of Medicine, the Academy of Pharmacy, Society of Biology and OPAL. She became "Officier des Palmes Académiques" in 1988 and "Chevalier de la Légion d'Honneur" in 1991. She has been recipient of many research grants from the Centre National de la Recherche Scientifique (CNRS), the Institut National de la Recherche Médicale (INSERM), Ministry of Research, and ECC.

Monique Adolphe has published 147 research papers and presented many lectures at international meetings. Her current major research interests include the development of various models of cellular cultures in pharmaco-toxicology and more specifically division, differentiation, and aging of cultured chondrocytes.

CONTRIBUTORS

Monique Adolphe, Ph.D.
Director
Laboratoire de Pharmacologie
  Cellulaire de l'Ecole Pratique des
  Hautes Etudes
Centre de Recherches
  Biomédicales des Cordeliers
Paris, France

Geethani Bandara, Ph.D.
Research Associate
Ferguson Laboratory for
  Orthopaedic Research
Department of Orthopaedic
  Surgery
University of Pittsburgh School of
  Medicine
Pittsburgh, Pennsylvania

Paul Benya, Ph.D.
Associate Professor
Department of Orthopaedics
University of Southern California
  and
Laboratory of Cartilage Cell
  Biology
Orthopaedic Hospital of
  Los Angeles
Los Angeles, California

**Motomi I. Enomoto, Ph.D.,
  D.D.S.**
Lecturer
Department of Biochemistry and
  Calcified-Tissue Metabolism
Faculty of Dentistry
Osaka University
Suita, Osaka, Japan

C. H. Evans, Ph.D.
Associate Professor
Department of Orthopaedic
  Surgery
University of Pittsburgh School of
  Medicine
  and
Molecular Genetics and
  Biochemistry
Pittsburgh, Pennsylvania

Anne-Marie Freyria
Head
Institut Biologie Chimie des
  Protéines
Villeurbanne, France

**Christopher Handley, B.A.
  (Hons), Ph.D.**
Doctor
Department of Biochemistry
Monash University
Clayton, Victoria, Australia

Daniel Herbage, D.Sc.
Research Directory
Cytologie Moleculaire
Institut Biologie Chimie des
  Protéines
Villeurbanne, France

Thomas M. Hering, Ph.D.
Assistant Professor
Department of Medicine
Division of Rheumatic Diseases
Case Western University
Cleveland, Ohio

Ernst B. Hunziker, M.D.
Professor, Director
M.E. Müller Institute for
  Biomechanics
University of Bern
Bern, Switzerland

Yukio Kato, Ph.D., D.D.S.
Professor
Department of Biochemistry
School of Dentistry
Hiroshima University
Hiroshima City, Japan

Martin Lotz, M.D.
Associate Professor in Residence
Department of Medicine
Sam and Rose Stein Institute for
  Research on Aging
University of California,
  San Diego
La Jolla, California

Frank P. Luyten, M.D.
Bone Cell Biology Section
Laboratory of Cellular
  Development and Oncology
National Institute of Dental
  Research
National Institutes of Health
Bethesda, Maryland

Charles J. Malemud, Ph.D.
Professor
Department of Medicine and
  Anatomy
Division of Rheumatic Diseases
Case Western Reserve University
Cleveland, Ohio

Chee Keng Ng, Ph.D.
Doctor
Department of Biochemistry
Monash University
Clayton, Victoria, Australia

Beatrice Petit, Ph.D.
Research Assistant
Institut Biologie Chimie des
  Protéines
Laboratoire de Cytologie
  Moléculaire
Villeurbanne, France

A. H. Reddi, Ph.D.
Chief
Laboratory of Cellular
  Development and Oncology
National Institute of Dental
  Research
Bethesda, Maryland

Michel van der Rest, Ph.D.
Professor
Cytologie Moleculaire
Institute Biologie Chimie des
  Protéines
Villeurbanne, France

Jeremy Saklatvala, Ph.D., M.D.
Doctor
Strangeways Research Laboratory
Cambridge, England

**Masaharu Takigawa, Ph.D.,
  D.D.S.**
Associate Professor
Department of Biochemistry and
  Calcified Tissue Metabolism
Osaka University Faculty of
  Dentistry
Suita, Osaka, Japan

Stephen B. Trippel, M.D.
Assistant Professor
Department of Orthopaedic
  Surgery
Massachusetts General Hospital
  and
Harvard Medical School
Boston, Massachusetts

Gust Verbruggen, M.D.
Professor
Department of Rheumatology
University Hospital
Ghent, Belgium

Eric M. Veys, M.D.
Professor
Department of Rheumatology
University Hospital
Ghent, Belgium

TABLE OF CONTENTS

Chapter 1

THE DIFFERENT TYPES OF CHONDROCYTES AND THEIR FUNCTION *IN VIVO*

Ernst B. Hunziker

TABLE OF CONTENTS

I. INTRODUCTION

A comprehensive survey of chondrocyte diversification cannot be ventured upon without the necessary preliminary of briefly delineating the basic structure and function of the native tissue which hosts these cells, namely, cartilage. Most cartilages present in mammalian organisms are of the so-called "hyaline" type, this designation deriving from its "glass-like" (Greek, *hyalos,* glass) or transparent appearance in the native state. Quantitatively minor types include fibrous (e.g., intervertebral discs, menisci, and repair cartilage) and elastic (e.g., ear and nose) forms. Each of these cartilages fulfills a principally skeletal function, and together they constitute a significant portion of the skeletal system.

During fetal and postnatal development, hyaline cartilage serves as an intermediary tissue in bone formation processes (bone anlage), functions as the growth organ for long bones [epiphyseal (or growth) plate cartilage], and fulfills permanent functions within the skeleton, as represented by the articular cartilage of synovial joints or rib cartilage in the thoracic wall.

The skeletal role played by cartilage tissue may be attributed to its specific biomechanical properties, these being afforded by the intercellular substance, i.e., the matrix. Its combined properties of stiffness and elasticity provide cartilage tissue with the potential to absorb compressive, tensional, and shearing forces applied externally.[1-3] A feature unique to cartilage is that it may be formed even when such forces are being exerted, which bone cannot.[4-7] For this reason, cartilage is particularly useful in bone formation or fracture healing processes, in which cases it serves as a primary (and intermediary) stabilizing tissue.[4]

At first glance, cartilage tissue appears to have a relatively simple structure, consisting merely of chondrocytes and intercellular matrix substance, and usually possessing no nerve fibers or blood and lymphatic systems (except for those tissues involved in ontogenic processes).

In hyaline cartilages, the cellular and matrix components generally constitute about 10 and 90%, respectively, of the tissue volume.[8] The matrix is composed of water ($\sim$70%),[9] proteoglycans, and collagen, and it is the unique relationship existing between these components which is responsible for the biomechanical properties of this tissue.[1,10]

The proteoglycans are trapped within the fibrous collagen network in a highly underhydrated state,[11-16] with a swelling pressure in the order of 2 to 3 atmospheres (atm). Under these conditions, the collagen fibrillar network is maintained under a permanently high tensional load which, at the same time, determines tissue shape and accounts for its high tensile strength.

In some specialized forms of cartilage which meet specific biomechanical requirements, tissue composition is correspondingly modified.[17] In fibrous cartilage,[18,19] for example, collagen fibrils and fibers constitute the principal component, and this tissue is thereby able to withstand very high tension and

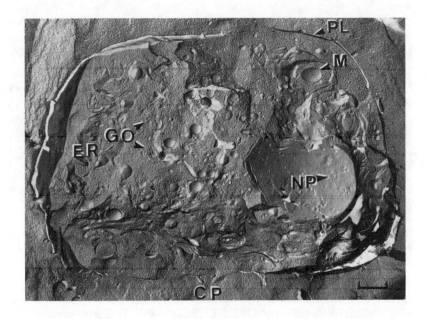

FIGURE 1. Electron micrograph of a replicated hypertrophic chondrocyte from rat growth plate cartilage (35-day-old; proximal tibia). Tissue was processed by high pressure freezing and freeze etching. The smooth plasmalemmal surface (PL) is interrupted at intervals by the protrusion of tiny cytoplasmic processes (CP). Abbreviations: rough endoplasmic reticulum (ER); Golgi-apparatus (GO); mitochondrion (M); nuclear pore (NP). Bar = 1 μm. (From Hunziker, E. B., Herrmann, W., Schenk, R. K., Mueller, M. and Moor, H., *J. Cell. Biol.*, 98, 267, 1984. With permission.)

shearing forces (menisci). The abundance of elastic fibrils in the matrix of elastic cartilage (outer ear) confers upon this tissue the capacity to meet deformational forces in all directions.[20-23]

This chapter will focus upon the various forms of hyaline cartilage and its chondrocytes, since it is this type which is most frequently used for experimental studies and is of greatest significance in pathology.

II. BASIC COMPOSITION OF CARTILAGE TISSUE

The morphological appearance of chondrocytes in histologic sections of cartilage tissue is critically dependent upon the preservation technique adopted.[24-27] In native postnatal tissue (observed using Normarsky optics), they usually have a rounded profile, with numerous tiny cytoplasmic processes projecting from an otherwise smooth surface.[28,29] These characteristic features are also preserved when tissue is processed by purely physical means, such as rapid high pressure freezing and freeze etching (Figure 1),[24] or by a unique combination of physical and chemical means, namely, rapid high pressure freezing, freeze substitution, and low temperature

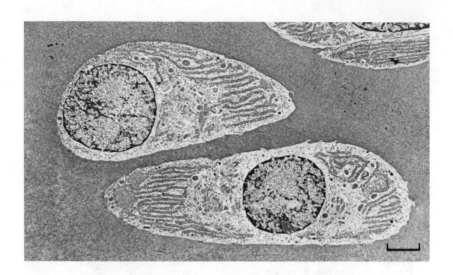

FIGURE 2. Electron micrograph of proliferating chondrocytes from rat growth plate cartilage (35-day-old; proximal tibia). Tissue was processed by high pressure freezing, freeze substitution, and low temperature embedding. Bar = 2 μm. (From Hunziker, E. B. and Herrmann, W., in *Ultrastructure of Skeletal Tissues,* Bonucci, E. and Motta, P. M., Eds., Kluwer Academic, New York, 1990, 79. With permission.)

embedding (Figure 2).[24,30] Since this latter procedure yields embedded tissue, it offers the potential of reinforcing morphological analysis, particularly of matrix components, by (immuno)histo-chemical means.[31] One considerable advantage of this processing technique lies in its capacity for preserving macromolecular components of the extracellular space in their native, expanded state and *in situ*. It is this site which is particularly prone to alterations under purely chemical processing conditions.[25,29] This situation may be largely overcome by introducing an appropriate cationic dye into the fixation media, by which means precipitation of cartilage proteoglycan molecules is effected (Figure 3).[25,28,32] If this measure is not taken, proteoglycan extraction in the order of 70% may be incurred.[33] Such losses lead to various artifact phenomena, such as cell shrinkage, lacuna formation, and alteration of matrix staining profile (see Figure 3).[24,25] Although the use of cationic dyes permits preservation of cell shape, the precipitation of proteoglycans (Figure 4) leads to a minor (compensatory) increase in cell volume with the resultant formation of artifactual intracellular vacuoles (Figure 3).[28,34] The choice of dye is also important. A large variety of these substances is available, and their tissue penetration properties and power for precipitating proteoglycans vary tremendously.[35] Moreover, most dye-proteoglycan complexes formed during aldehyde fixation are disrupted by osmium tetroxide during postfixation. The only exceptions to this phenomenon are ruthenium hexaammine trichloride (RHT) and its chemical analogs.[32]

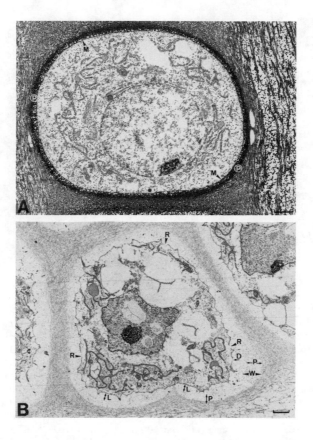

FIGURE 3. Electron micrographs of hypertrophic chondrocytes from rat growth plate cartilage (35-day-old; proximal tibia) chemically fixed in the presence (A) or absence (B) of ruthenium hexaammine trichloride (RHT). In A, the plasma membrane (M) is preserved intact and adherent to the pericellular matrix (P). Chemical fixation in the absence of RHT (B) leads to multiple rupturing of the plasma membrane (R) and its almost complete detachment (D) from the pericellular matrix (P), which is depleted of proteoglycans and artificially enlarged (W), thereby giving rise to a lacuna (L). Bar = 2 μm (A and B). (From Hunziker, E. B., Herrmann, W., and Schenk, R. K., *J. Histochem. Cytochem.*, 31, 717, 1983. With permission.)

As described above, the preservation quality of chondrocytes depends critically upon conservation of the cartilage matrix and, particularly, of its proteoglycan components. The matrix of mature cartilage tissue is structured into defined compartments, the morphological basis for which is the fibrillar collagen architecture.[36] Reproducible delineation of this organization is possible only at the ultrastructural level (Figure 5). The various regions are generally designated as the pericellular, territorial, and interterritorial compartments.[29,36] In some cartilage types, the interterritorium may also mineralize, a process which is believed to be under cellular control (Figure 6).[37-43]

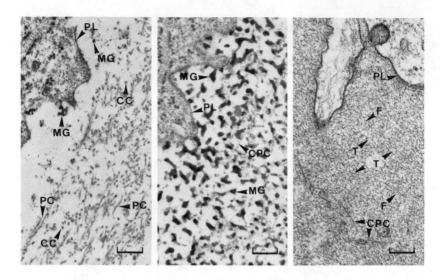

FIGURE 4. Electron micrographs of the pericellular region surrounding hypertrophic chondrocytes from rat growth plate cartilage (35-day-old; proximal tibia), following chemical fixation in the absence (left) or presence (middle) of ruthenium hexaammine trichloride, or high pressure freezing, freeze substitution, and low temperature embedding (right). Left: most proteoglycan molecules have been extracted, the few remaining being precipitated as matrix granules (MG). Abbreviations: cross-sectioned collagen fibrils (CC); collagen fibrils running parallel or obliquely to the section plane (PC); plasmalemma (PL). Middle: Proteoglycans are preserved in a precipitated form as matrix granules (MG). Abbreviations: longitudinally-sectioned collagen fibrils (CRC); plasmalemma (PL). Right: Proteoglycans are preserved in their expanded, native state, with thick (T) and fine (F) filamentous components. Abbreviations: collagen fibrils running parallel or obliquely to the section plane (CPC); plasmalemma (PL). Bar = 0.2 μm. (From Hunziker, E. B. and Schenk, R. K., *J. Cell Biol.*, 98, 277, 1984. With permission.)

However, as will be discovered in the following sections,[44-52] a universal definition for cartilage tissue cannot be given based upon the simple and general criteria detailed above, since no single component (marker) exists which is common and unique to all cartilage types.

III. AVIAN CARTILAGES

A. FETAL CARTILAGE

Fetal chick cartilage tissue is frequently used as a source of chondrocytes for studies *in vitro*. One of the preferred loci is the developing sternum in the 17-day-old fetus. At this stage, the sternum is still largely present as a cartilage anlage, but one which is no longer homogeneous in structure, since secondary centers of ossification have begun to appear. As a consequence of this, the anlage becomes invaded with blood vessel canals,[53] and in the actual centers, chondrocytes are undergoing differentiation and transformation into hypertrophic cells prior to induction of matrix mineralization (of the interterritorial

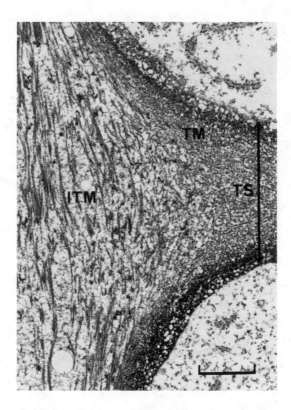

FIGURE 5. Electron micrograph of matrix compartments around hypertrophic chondrocytes in rat growth plate cartilage (35-day-old; proximal tibia). The cells are surrounded by an intensely stained pericellular ring which is devoid of fibrillar collagen. The territorial matrix (TM) is characterized by the presence of a fine network of fibrillar collagen and, in the interterritorial compartment (ITM), fibrils exhibit a parallel orientation. Abbreviation: transverse septum between cells (TS). Bar = 2 μm. (From Eggli, P. S., Herrmann, W., Hunziker, E. B., and Schenk, R. K., *Anat. Rec.,* 211, 246, 1985. With permission.)

compartment) and the formation of a mechanically stable framework. Following this process, the calcified hypertrophic cartilage is destroyed by invading monocytes/macrophages and chondroclasts (osteoclasts).[54,55] Accompanying perivascular mesenchymal cells differentiate and transform into osteoblasts which deposit bone matrix on the calcified cartilage matrix scaffold. Hence, within this anlage, there is considerable regional variation in the stage of development (Figures 7 and 8).

A typical feature of immature, fetal cartilage is the almost random distribution of cells within it (Figures 7 to 9). A change in the degree of anisotropy occurs only when differentiating processes are triggered, such as the formation of secondary centers of ossification. The randomness of cell organization also finds its counterpart in the matrix, which is not structured into compartments.

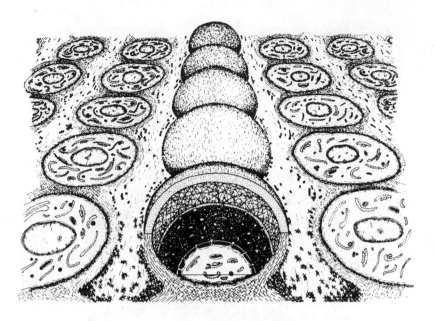

FIGURE 6. Schematic representation of matrix organization in that part of the growth plate where mineralization commences (lower hypertrophic zone). The pericellular matrix compartment is rich in proteoglycans and devoid of fibrillar collagen. The territorial matrix forms a basket-like network of fibrillar collagen around individual cells and cell groups (columns or chondrones); tiny cytoplasmic processes extend to this zone. In the interterritorial matrix, collagen fibrils run in parallel and are less densely packed than in the territorial matrix; the density of proteoglycan molecules is also lower here. (From Hunziker, E. B. and Herrmann, W., in *Ultrastructure of Skeletal Tissues,* Bonucci, E. and Motta, P. M., Eds., Kluwer Academic, New York, 1990, 79. With permission.)

Collagen fibrils are randomly orientated as a fine reticular network throughout (Figure 8). The formation of defined compartments occurs upon the triggering of differentiation processes.

B. POSTNATAL CARTILAGE

During the development of bones, various topographical regions become defined. In the central part of the epiphysis (distal and proximal ends of bone), fetal-type chondrocytes persist for a considerable period of time into the postnatal growth phase (Figure 9). This region fulfills a skeletal function and, in appearance, is almost identical to the 17-day-old fetal cartilage depicted in Figures 7 and 8. In mammalian systems, epiphyseal cartilage (see Figure 16) is usually replaced by bone tissue (via the formation of secondary centers of ossification).[4] This tissue thus needs to be well differentiated from growth (epiphyseal) plate cartilage in the regions containing resting, proliferating, and hypertrophic chondrocytes (Figure 10).[55,56] In contrast to epiphyseal cartilage chondrocytes (Figure 9), those in the growth plate (resting and proliferating zones) (Figure 10) represent a more differentiated state, and are surrounded by well-defined matrix compartments. Their responsiveness to various signaling substances (e.g., growth factors, cytokines, and hormones)[57-59] is also quite

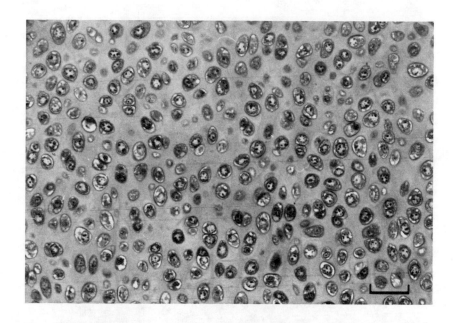

FIGURE 7. Light micrograph of avian fetal cartilage (17-day-old chick). Chondrocytes exhibit a predominantly ellipsoidal shape, and are randomly distributed within a homogeneously stained matrix. 1-μm section stained with Toluidine Blue 0. Bar = 20 μm.

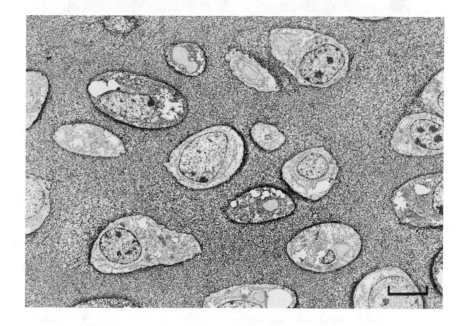

FIGURE 8. Electron micrograph of avian fetal cartilage (17-day-old chick). Collagen fibrils within the matrix are randomly orientated in a network-like fashion and extend to the cell surface. Compartmentalization of the matrix substance has not yet begun. Bar = 4 μm.

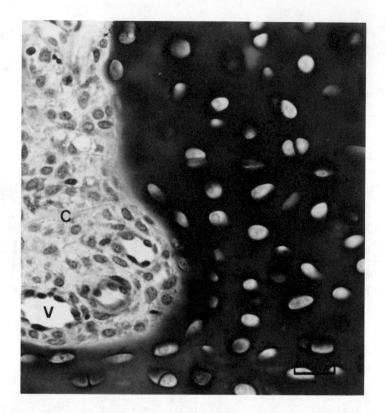

FIGURE 9. Light micrograph of avian epiphysis (5-week-old chick). At this stage, the epiphysis is still cartilaginous and blood vessel canals are present. Chondrocytes exist as single entities and exhibit no preferential orientation pattern. The matrix stains homogeneously (Toluidine Blue 0). Bar = 20 μm.

different. Hence, epiphyseal cartilage cells should not be used as a model system (either *in vitro* or *in vivo*) for resting or proliferating chondrocytes of growth plate cartilage (see below), even though they bear a superficial similarity to these cells based upon morphological criteria alone. In mammalian systems, the correspondence is much closer (cf. Figures 11, 16, and 17).

IV. MAMMALIAN CARTILAGES

A. IMMATURE FORMS
1. Rabbit Articular Cartilage

During the early postnatal development and growth phases of mammalian articular cartilage, a cell development cascade similar to that taking place in primary centers of ossification or in growth plate cartilage tissue may be seen (Figure 11A).[44] The resemblance is, however, superficial. Moving inward from

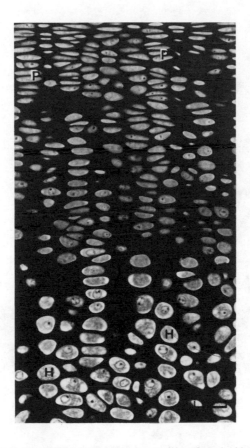

FIGURE 10. Light micrograph of avian growth plate cartilage (5-week-old chick) photographed at the junction between proliferating and hypertrophic zones. Proliferating cells within the chick growth plate are extremely flat, almost "spindle-like" in shape; but in the hypertrophic phase, they rapidly adopt the usual "superegg-like" (prolate spheroid) form. At all stages, the chondrocytes are surrounded by a matrix substance structured into three compartments (i.e., pericellular, territorial, and interterritorial). Thick (6 μm) section, stained by McNeil's Tetrachrome. Bar = 25 μm.

the joint surface, a gradient in cell size is identifiable, large (hypertrophic) chondrocytes being situated in deeper zones. The degree of anisotropy is lower than that apparent in growth plate cartilage, but higher than that occurring in fetal (bone anlage) cartilage. At maturity, the tissue is transformed into a highly organized structural unit consisting of axial cell columns and various horizontal strata (Figure 11B).[44] The total thickness of the tissue is, in addition, dramatically reduced. At this stage, the tissue functions uniquely as articular cartilage proper, whereas during the immature postnatal phase it serves dual functions, both as articular cartilage and as a superficial growth plate for the shaping and growth of the whole epiphysis.[60] Its functioning in this latter capacity accounts for the transformation and elimination processes occurring in

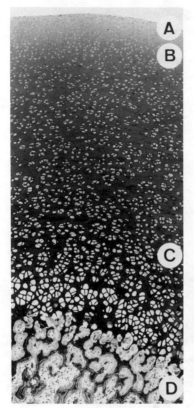
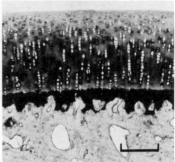

FIGURE 11. Light micrographs of immature (left: 1-month-old) and mature (right: 8-month-old) rabbit articular cartilage (femoral condyle). Note the difference in cartilage thickness between each developmental state (both illustrations are represented at identical magnification). In the immature condition (left), this tissue functions both as articular cartilage and as a superficial growth plate. The degree of structural anisotropy is still low, whereas in the mature form (right) it is very high, chondrocytes being organized into axial columns. A: superficial (tangential) zone; B: transitional zone; C: "radial zone" (cell hypertrophy); D: trabeculae of subchondral bone, including vascular invasion front. Bar = 200 μm. (From Hunziker, E. B., in *Arthroskopie bei Knorpelschaden und bei Arthrose,* Glinz, W., Kieser, C. and Munzinger, U., Eds., Ferdinand Enke Verlag, Stuttgart, 1990, 1. With permission.)

the deeper zones at the junction between cartilage and metaphyseal bone tissue (Figure 11A) and, on this basis, the structural similarities between immature postnatal articular cartilage (Figures 11 (left), 12, and 13) and the growth plate (Figures 14 and 15) are not so surprising.[44]

The significant structural and functional differences between immature and mature articular cartilage are also reflected in their biomechanical properties. In the adult state, when purely articular functions are being undertaken, the tissue is, for example, much more resistant to shearing and compressive forces than it is during the immature postnatal phase.[10]

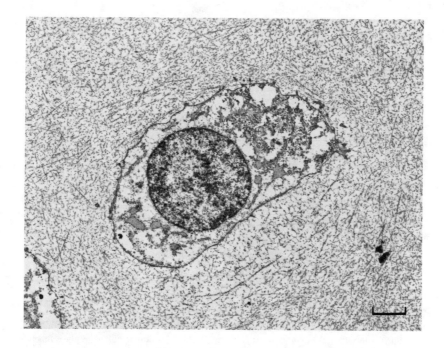

FIGURE 12. Electron micrograph of a chondrocyte within the transitional zone of immature (1-month-old) rabbit articular cartilage. The cell corresponds functionally to a proliferating chondrocyte in the growth plate. Bar = 1 μm.

2. Rat Growth Plate Cartilage

For its degree of structural organization and anisotropy, the growth plate holds a unique place amongst cartilages.[34,61-64] It is organized into horizontal layers and axial cell columns, the latter representing the functional units of longitudinal bone growth. The structural organization is determined by the chondrocytes themselves which, nonetheless, require the support of a functionally, structurally, and biomechanically intact matrix. In instances where matrix composition is impaired, severe disturbances in cell arrangement and postnatal growth may be incurred (cf. chondrodystrophy syndromes).[65] Already at birth, this intercellular space is structurally organized into compartments.[64]

The chondrocytes themselves pass through a well-defined series of activities during their life span,[34,61] these being recapitulated in sequence within a cell column from the epiphyseal to the metaphyseal aspect of the growth plate (Figure 14). The synchronization of each cell activity phase between lateral neighbors accounts for the horizontal stratification. The various cell activity phases are recognized in the stem (resting) cell, proliferating and hypertrophic zones, respectively (Figure 14). Towards the end of hypertrophy, which represents a controlled cell phenotype modulation step promoting growth by a very efficient means, the cells initiate mineralization of the interterritorial matrix compartment (Figure 15).[66-69] On this basis, some authors define this

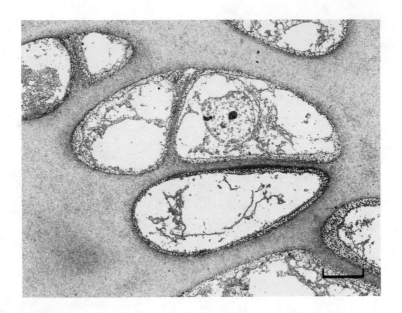

FIGURE 13. Electron micrograph of a group of hypertrophic chondrocytes within the radial zone of immature (1-month-old) rabbit articular cartilage. At this developmental stage, their function corresponds to that of the hypertrophic activity phase in growth plate cartilage. Bar = 4 μm. (From Hunziker, E. B., in *Arthroskopie bei Knorpelschaden und bei Arthrose,* Glinz, W., Kieser, C., and Munzinger, U., Eds., Ferdinand Enke Verlag, Stuttgart, 1990, 1. With permission.)

deepest layer as the zone of mineralization. The region bordering on the metaphyseal bone is invaded by blood vessels and monocytes/macrophages; in this location, cartilage tissue is continually eliminated and replaced by bone tissue.[70]

3. Bovine Epiphyseal Cartilage

Cartilage tissue which is formed as a cartilage anlage during fetal life may persist in certain bones well into the postnatal phase. An example of this is the distal ulnar epiphyseal cartilage interposed between the growing secondary (epiphyseal) center of ossification and the metaphyseal vascular invasion front. This tissue contains a large number of blood vessel canals[71,72] and immature, spheroid-shaped chondrocytes (Figure 16). Owing to its ready availability, it is frequently used for experimental purposes.[73] The same holds true for "young" (i.e., calf[74]) articular cartilage which forms a similar type of tissue beneath the joint surfaces in large mammals.

B. MATURE FORMS
1. Rabbit Articular Cartilage

Mature articular cartilage tissue fulfills several functions: it serves to transmit load between skeletal elements, acts as a shock absorber, and provides a practically friction-free gliding surface. In structure, the tissue bears a similar-

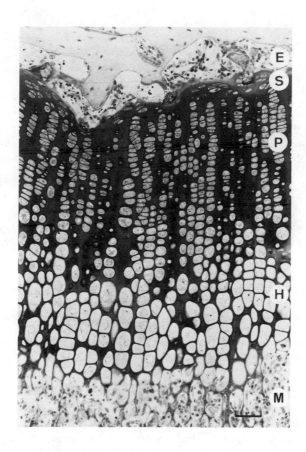

FIGURE 14. Light micrograph of rat growth plate cartilage (35-day-old; proximal tibia). The growth plate is a disc-like structure situated between the epi- (E) and metaphyses (M). Chondrocytes are organized into axial cell columns which represent the functional units of longitudinal bone growth. The coordination of cell activity between lateral neighbors leads to a horizontal stratification: stem cell zone (S), proliferating zone (P), and hypertrophic zone (H). During the final stages of hypertrophy, cells induce matrix mineralization; on this basis, some authors designate the lower hypertrophic zone as the zone of mineralization. Its boundary with the metaphysis is marked by the ingrowth of blood vessels (zone of vascular invasion). 1-µm section stained with Toluidine Blue 0. Bar = 50 µm. (From Hunziker, E. B., in *Cartilage Degradation: Basic and Clinical Aspects*, Woessner, J. F., Jr. and Howell, D. S., Eds., Marcel Dekker, New York, 1992. With permission.)

ity to the growth plate, being organized into horizontal layers (superficial, transitional, and radial zones) and axial cell columns (Figure 17).[35,44,75] Cells in the superficial zone are flat, almost spindle-like in shape. In the transitional zone, chondrocytes usually have a more rounded or circular profile. In the radial zone, which extends down to the zone of mineralized cartilage, cells generally appear as prolate spheroids (i.e., "superegg-like" in shape), there being a gradual reduction in size towards the tide mark; these chondrocytes are organized into either axial columns (Figure 17) or groups (chondrones) (Figure 18),[75] which may be isolated as discrete entities for experimental purposes.[76,77]

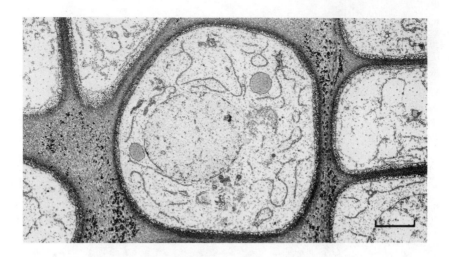

FIGURE 15. Electron micrograph of a late hypertrophic chondrocyte (zone of mineralization) within rat growth plate cartilage (35-day-old; proximal tibia). Mineralization of the interterritorial matrix compartment has commenced, and yet the cell is still structurally (and functionally) intact. The membrane systems of late hypertrophic chondrocytes are particularly susceptible to fixation conditions, and hence are frequently ruptured during processing. On this basis, many authors have previously believed these cells to be degenerate and have designated the region containing them as the "zone of degenerating chondrocytes". Bar = 5 μm.

Cells of this zone are particularly rich in microfilaments and glycogen particles.[75]

The matrix exhibits a unique architecture, with collagen fibrils assuming an arcade-like configuration.[78,79] The tips of these structures, which reside in the superficial zone, are small in diameter and supplemented by an additional tangentially-orientated fibrillar collagen network. In deeper zones, the fibrils run perpendicular to the surface, become increasingly thicker, and continue into the mineralized (calcified) cartilage layer (Figure 19).[44] They do not, however, penetrate beyond the junction between mineralized cartilage and bone tissue, where they end abruptly. Matrix metabolism is actively controlled by chondrocytes (Figure 20), the structural and functional integrity of which is manifested throughout the entire depth of articular cartilage tissue.[44]

2. Adult Human Articular Cartilage

The structure of adult human articular cartilage (Figure 21) is basically similar to that in the mature rabbit described above.[8,44] The relative thicknesses of the individual horizontal zones are, however, quite different, with the radial zone (constituting approximately 85% of the total tissue height) clearly dominating. The degree to which chondrocytes are able to remodel calcified cartilage tissue is somewhat less than is possible in unmineralized cartilage, this role being undertaken principally by osteoclasts.[44] Nonetheless, the chondrocytes within mineralized cartilage and, in particular, those bordering on the unmineralized tissue (i.e., at the tide mark) still maintain control over the

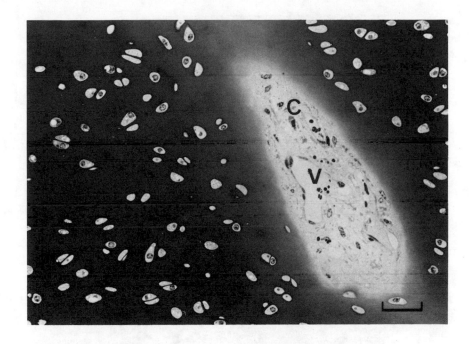

FIGURE 16. Light micrograph of immature mammalian epiphyseal cartilage (bovine; distal ulna). The cartilage contains vascular canals (V) which consist of blood vessels and perivascular connective tissue cells. The chondrocytes are predominantly ellipsoidal in shape and randomly distributed. Bar = 40 μm. (Tissue was kindly provided by Dr. M. Gray, Cambridge, MA.)

induction and progression of mineralization.[66-69] Their role in this capacity (Figure 22) appears to be one of continuous controlled suppression rather than one of active induction, as occurs during the postnatal growth phase.

3. Mammalian Fibrocartilage

Fibrocartilage (Figure 23) occurs at various locations in the mammalian organism (e.g., intervertebral discs and menisci).[19,21,50] Chondrocytes have a typically spheroid form and, unexceptionally, are surrounded by a pericellular matrix compartment. However, the large interterritorial region is exceptional in that it contains an abundance of collagen fibrils and thick fibers[44] which are mainly of type I, but types III and V may also be present[45] (this contrasts to the situation existing in all hyaline cartilages, where types II, IX, and XI only are universally encountered).

4. Rodent Respiratory Tract Cartilage

Hyaline cartilage within the walls of the respiratory tract also fulfills a primarily skeletal function (for patency of airways).[80] Here, chondrocytes exhibit an unusual richness in lipids, particularly in rodents. These molecules are contained within fat vacuoles, which sometimes constitute practically the whole cytoplasmic space and are easily identified (Figure 24). The basis for these inclusions is, however, not understood.

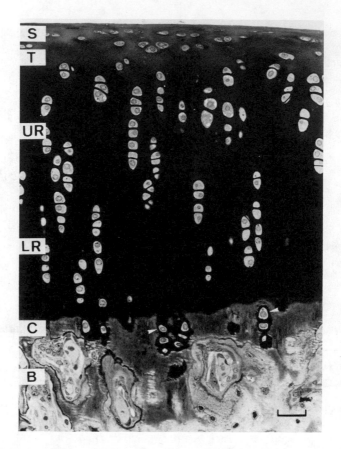

FIGURE 17. Light micrograph of mature mammalian articular cartilage (8-month-old rabbit; femoral condyle). Adult articular cartilage exhibits a well-defined structural organization into horizontal layers: superficial zone (S), tangential zone (T), upper (UR) and lower (LR) radial zones, zone of calcified cartilage (C) containing cells which are structurally (and functionally) integral (⇨), and subchondral bone plate (B). Bar = 40 μm. (From Hunziker, E. B., in *Articular Cartilage and Osteoarthritis*, Kuettner, K. E., Schleyerbach, R., Peyron, J. G., and Hascall, V. C., Eds., Raven Press, New York, 1992, 183. With permission.)

V. CHONDROCYTE CULTURE SYSTEMS

Chondrocytes maintained *in vitro* readily undergo changes in both their morphological and functional characteristics. Young chondrocytes may also lose their capacity to undergo further differentiation. Culture conditions need to be individually optimized according to the chondrocyte type used.[81,82]

Examples of a mature and an immature chondrocyte type which have been successfully cultured and maintain approximately their physiologic phenotype are presented in Figures 25 and 26, respectively. Both types have been shown to be useful for probing various questions suited to experimental analysis in

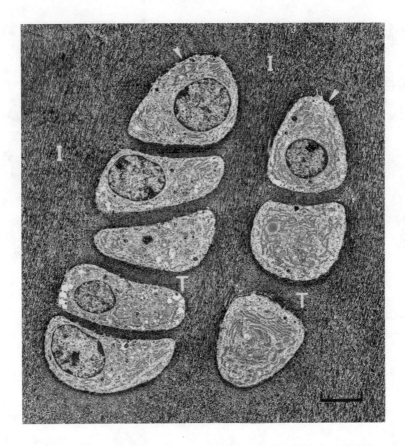

FIGURE 18. Electron micrograph of chondrocyte groups (chondrones) within the radial zone of mature rabbit articular cartilage. Each chondrocyte is surrounded by a pericellular rim ($\Rightarrow$) and a territorial matrix compartment (T), and it is the fibrillar network of the latter which delimits and gives coherence to the chondrone as a unit. Between chondrones run the parallel-orientated fibrils of the interterritorial matrix (I). Bar = 4 µm. (From Hunziker, E. B., in *Articular Cartilage and Osteoarthritis,* Kuettner, K. E., Schleyerbach, R., Peyron, J. G., and Hascall, V. C., Eds., Raven Press, New York, 1992, 183. With permission.)

culture systems.[83,84] They are, however, influenced considerably by a number of factors which have as yet not been well defined; until these hazy areas are elucidated,[85] the usefulness of such systems as models for the physiological situation is severely limited.

ACKNOWLEDGMENTS

This work was supported by the AO-/ASIF-Foundation (Switzerland) and the Swiss National Science Foundation. The author is indebted to Ceri England for her editorial advice, to Eva Kapfinger for her technical expertise, and to Verena Rickli for her secretarial assistance.

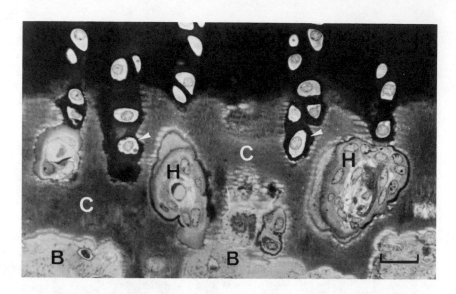

FIGURE 19. Light micrograph of the calcified zone within mature rabbit articular cartilage. The calcified layer (C) is bordered on one side by articular cartilage tissue containing viable chondrocytes (⇨) and on the other by a subchondral bone plate (B). The calcified cartilage tissue is remodeled by Haversian remodeling processes (H) which lead to a continuous narrowing of this zone and its replacement by bone tissue during postnatal adult life. Bar = 20 μm. (From Hunziker, E. B., in *Articular Cartilage and Osteoarthritis,* Kuettner, K. E., Schleyerbach, R., Peyron, J. G., and Hascall, V. C., Eds., Raven Press, New York, 1992, 183. With permission.)

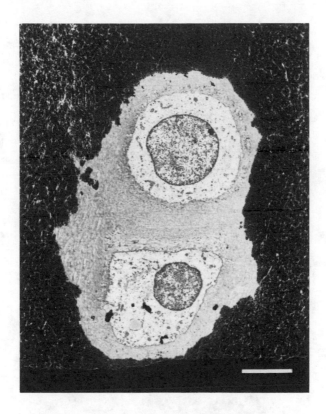

FIGURE 20. Electron micrograph of a pair of chondrocytes in the calcified zone of mature rabbit articular cartilage. These cells induce and control the degree of mineralization in the outer territorial and interterritorial matrix compartments and maintain their pericellular and inner territorial compartments mineral-free. In this activity phase, chondrocytes exhibit a relative paucity in organelles. Bar = 5 μm. (From Hunziker, E. B., in *Articular Cartilage and Osteoarthritis,* Kuettner, K. E., Schleyerbach, R., Peyron, J. G., and Hascall, V. C., Eds., Raven Press, New York, 1992, 183. With permission.)

FIGURE 21. Light micrograph of adult human articular cartilage (28-year-old; medial femoral condyle). The structural organization of chondrocytes into axial columns and horizontal layers is basically similar to that in other mammals, although the relative height of each zone is species specific. Chondrocytes are also somewhat more ellipsoidal than spindle-like in shape. Bar = 100 μm. (From Hunziker, E. B., in *Articular Cartilage and Osteoarthritis,* Kuettner, K. E., Schleyerbach, R., Peyron, J. G., and Hascall, V. C., Eds., Raven Press, New York, 1992, 183. With permission.)

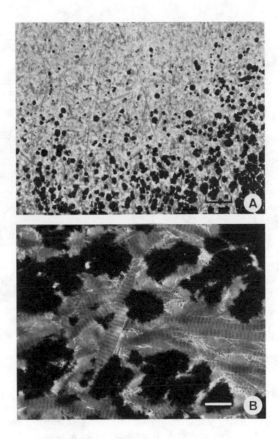

FIGURE 22. Low (A) and high (B) power electron micrographs of the mineralization front ("tide mark") in a human adult joint (28-year-old; medial femoral condyle). Calcium hydroxyapatite crystals form, grow (A), and progressively fill the interfibrillar space (B). Bars = 2 μm (A) and 0.3 μm (B). (From Hunziker, E. B., in *Articular Cartilage and Osteoarthritis,* Kuettner, K. E., Schleyerbach, R., Peyron, J. G., and Hascall, V. C., Eds., Raven Press, New York, 1992, 183. With permission.)

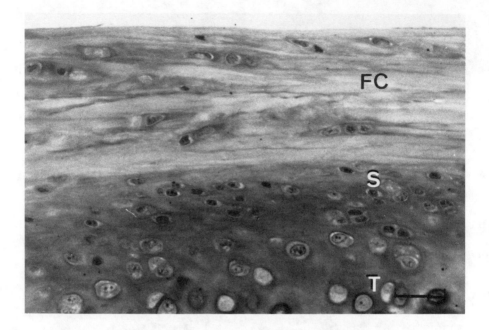

FIGURE 23. Light micrograph of mature primate mandibular articular cartilage. In primates, the superficial zone of articular cartilage in this joint is composed of fibrous connective tissue and a layer of fibrocartilage (FC); the latter consists of ellipsoidal- and spherically-shaped chondrocytes, which produce a fibrous intercellular matrix. It is continuous with a lower layer of hyaline cartilage, the superficial (S) and transitional zones (T) of which are illustrated. Bar = 20 μm. (From Hunziker, E. B. and Herrmann, W., in *Ultrastructure of Skeletal Tissues,* Bonucci, E. and Motta, P. M., Eds., Kluwer Academic, New York, 1990, 79. With permission.)

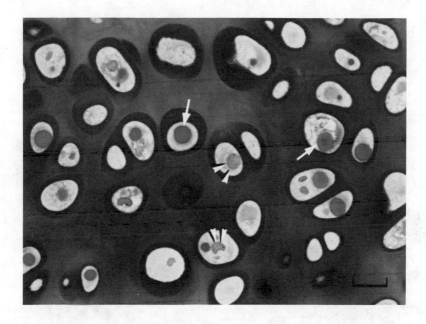

FIGURE 24. Light micrograph of mature rabbit hypertrophic laryngeal cartilage. The chondrocytes of respiratory tract cartilage in rodents are frequently rich in lipid vacuoles (↑). Nuclei are indicated in some instances (⇨⇨). Thick (1 μm) section, stained with Toluidine Blue 0. Bar = 20 μm. (From Hunziker, E. B. and Kuettner, K. E., in *The Lung: Scientific Foundations,* Crystal, R. G. and West, J. B., Eds., Raven Press, New York, 1991, 451. With permission.)

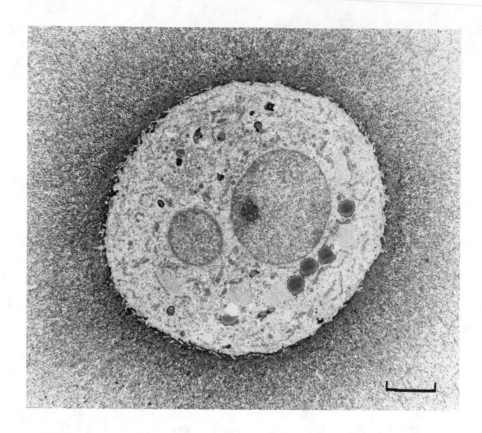

FIGURE 25. Electron micrograph of a chondrocyte in agarose culture, 3 weeks after isolation from bovine nasal cartilage. The chondrocyte is surrounded by a structured matrix consisting of pericellular and territorial compartments. (Material kindly provided by Dr. M. Aydelotte, Chicago, IL.) Bar = 3 μm.

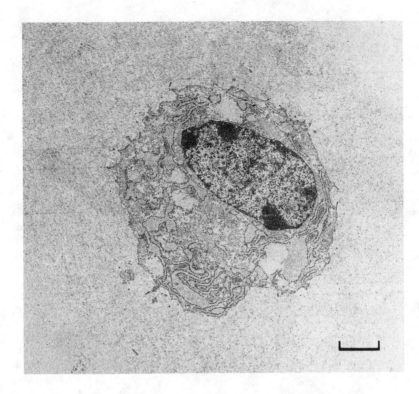

FIGURE 26. Electron micrograph of a chondrocyte in agarose culture, 4 weeks after isolation from fetal calf articular cartilage. The immature structural characteristics of cell and matrix are also mimicked *in vitro*. Bar = 2 μm. (Material kindly provided by Dr. A. Grodzinsky, Cambridge, MA.)

REFERENCES

1. **Mow, V. C., Ratcliffe, A., and Woo,** S. L. Y., *Biomechanics of Diarthrodial Joints,* I and II, Springer-Verlag, New York, 1990.
2. **Maroudas, A. and Kuettner, K.,** *Methods in Cartilage Research,* Academic Press, London, 1990.
3. **Woo, S. L. Y. and Buckwalter, J. A.,** *Injury and Repair of the Musculoskeletal Soft Tissues,* American Academy of Orthopaedic Surgeons, Park Ridge, IL, 1988.
4. **Cormack, D. H.,** *Ham's Histology,* J. B. Lippincott, Philadelphia, 1987.
5. **Von der Mark, K. and Conrad, G.,** Cartilage cell differentiation: review, *Clin. Orthop.,* 139, 185, 1979.
6. **Elmer, W. A.,** Development cues in limb bud chondrogenesis, *Coll. Relat. Res.,* 2, 257, 1982.
7. **Solursh, M., Jensen, K. L., Zanetti, N. C., Linsenmayer, T. F., and Reiter, R. S.,** Extracellular matrix mediates epithelial effects on chondrogenesis in vitro, *Dev. Biol.,* 105, 451, 1984.
8. **Stockwell, R. A.,** *Biology of Cartilage Cells,* Cambridge University Press, Cambridge, 1979.
9. **Thompson, R. C. and Robinson, H. J.,** Current concepts review. Articular cartilage matrix metabolism, *J. Bone Joint Surg. (Am.),* 63, 327, 1981.

10. **Mow, V. and Rosenwasser, M.,** Articular cartilage: biomechanics, in *Injury and Repair of the Musculoskeletal Soft Tissues,* Woo, S. L. Y. and Buckwalter, J. A., Eds., American Academy of Orthopaedic Surgeons, Park Ridge, IL, 1988, 427.

11. **Urban, J. P., Maroudas, A., Bayliss, M. T., and Dillon, J.,** Swelling pressures of proteoglycans at the concentrations found in cartilaginous tissues, *Biorheology,* 16, 447, 1979.

12. **Broom, N. D.,** Further insights into the structural principles governing the function of articular cartilage, *J. Anat.,* 139, 275, 1984.

13. **Cornwall, M. W.,** Biomechanics of noncontractile tissue. A review, *Phys. Ther.,* 64, 1869, 1984.

14. **Armstrong, C. G. and Mow, V. C.,** The mechanical properties of articular cartilage, *Bull. Hosp. Joint Dis. Orthop. Inst.,* 43, 109, 1983.

15. **Lee, R. C., Frank, E. H., Grodzinsky, A. J., and Roylance, D. K.,** Oscillatory compressional behavior of articular cartilage and its associated electromechanical properties, *J. Biomech. Eng.,* 103, 280, 1981.

16. **Myers, E. R. and Mow, V. C.,** Biomechanics of cartilage and its response to biomechanical stimuli, in *Cartilage: Structure, Function and Biochemistry,* Vol. 1, Hall, B. K., Ed., Academic Press, New York, 1983, 313.

17. **Heinegard, D. and Oldberg, A.,** Structure and biology of cartilage and bone matrix noncollagenous macromolecules, *FASEB J.,* 3, 2042, 1989.

18. **Eyre, D. R. and Muir, H.,** Quantitative analysis of types I and II collagens in human intervertebral discs at various ages, *Biochim. Biophys. Acta,* 492, 29, 1977.

19. **Somer, L. and Somer, T.,** Is the meniscus of the knee joint a fibrocartilage?, *Acta Anat. (Basel),* 116, 234, 1983.

20. **Heeger, P. and Rosenbloom, J.,** Biosynthesis of tropoelastin by elastic cartilage, *Connect. Tissue Res.,* 8, 21, 1980.

21. **Quintarelli, G., Starcher, B. C., Vocaturo, A., Di Gianfilippo, F., Gotte, L., and Mecham, R. P.,** Fibrogenesis and biosynthesis of elastin in cartilage, *Connect. Tissue Res.,* 7, 1, 1979.

22. **Cotta Pereira, G., Del Caro, L. M., and Montes, G. S.,** Distribution of elastic system fibers in hyaline and fibrous cartilages of the rat, *Acta Anat. (Basel),* 119, 80, 1984.

23. **Takagi, M., Parmley, R. T., Denys, F. R., Kageyama, M., and Yagasaki, H.,** Ultrastructural distribution of sulfated complex carbohydrates in elastic cartilage of the young rabbit, *Anat. Rec.,* 207, 547, 1983.

24. **Hunziker, E. B., Herrmann, W., Schenk, R. K., Mueller, M., and Moor, H.,** Cartilage ultrastructure after high pressure freezing, freeze substitution, and low temperature embedding. I. Chondrocyte ultrastructure—implications for the theories of mineralization and vascular invasion, *J. Cell. Biol.,* 98, 267, 1984.

25. **Hunziker, E. B., Herrmann, W., and Schenk, R. K.,** Ruthenium hexammine trichloride (RHT)-mediated interaction between plasmalemmal components and pericellular matrix proteoglycans is responsible for the preservation of chondrocytic plasma membranes in situ during cartilage fixation, *J. Histochem. Cytochem.,* 31, 717, 1983.

26. **Kashiwa, H. K., Luchtel, D. L., and Park, H. Z.,** Chondroitin sulfate and electron lucent bodies in the pericellular rim about unshrunken hypertrophied chondrocytes of chick long bone, *Anat. Rec.,* 183, 359, 1975.

27. **Stockwell, R. A.,** Chondrocytes, *J. Clin. Pathol. (R. Coll. Pathol.),* (Suppl.) 12, 7, 1978.

28. **Hunziker, E. B., Herrmann, W., and Schenk, R. K.,** Improved cartilage fixation by ruthenium hexammine trichloride (RHT). A prerequisite for morphometry in growth cartilage, *J. Ultrastruct. Res.,* 81, 1, 1982.

29. **Hunziker, E. B. and Herrmann, W.,** Ultrastructure of cartilage, in *Ultrastructure of Skeletal Tissues,* Bonucci, E. and Motta, P. M., Eds., Kluwer Academic, New York, 1990, 79.

30. **Hunziker, E. B. and Schenk, R. K.,** Cartilage ultrastructure after high pressure freezing, freeze substitution, and low temperature embedding. II. Intercellular matrix ultrastructure — preservation of proteoglycans in their native state, *J. Cell Biol.,* 98, 277, 1984.

31. **Hunziker, E. B. and Herrmann, W.,** In situ localization of cartilage extracellular matrix components by immunoelectron microscopy after cryotechnical tissue processing, *J. Histochem. Cytochem.,* 35, 647, 1987.

32. **Hunziker, E. B., Ludi, A., and Herrmann, W.,** Preservation of cartilage matrix proteoglycans using cationic dyes chemically related to ruthenium hexammine trichloride, *J. Histochem. Cytochem.,* in press, 1992.

33. **Engfeldt, B. and Hjertquist, S. O.,** Studies on the epiphysial growth zone. I. The preservation of acid glycosaminoglycans in tissues in some histotechnical procedures for electron microscopy, *Virchows Arch. (B),* 1, 222, 1968.

34. **Hunziker, E. B., Schenk, R. K., and Cruz Orive, L. M.,** Quantitation of chondrocyte performance in growth-plate cartilage during longitudinal bone growth, *J. Bone Joint Surg. (Am.),* 69, 162, 1987.

35. **Hunziker, E. B. and Schenk, R. K.,** Structural organization of proteoglycans in cartilage, in *Biology of Proteoglycans,* Wight, T. N. and Mecham, R. P., Eds., Academic Press, Orlando, FL, 1987, 155.

36. **Eggli, P. S., Herrmann, W., Hunziker, E. B., and Schenk, R. K.,** Matrix compartments in the growth plate of the proximal tibia of rats, *Anat. Rec.,* 211, 246, 1985.

37. **Arsenault, A. L. and Hunziker, E. B.,** Electron microscopic analysis of mineral deposits in the calcifying epiphyseal growth plate, *Calcif. Tissue Int.,* 42, 119, 1988.

38. **Hunziker, E. B., Herrmann, W., Cruz Orive, L. M., and Arsenault, A. L.,** Image analysis of electron micrographs relating to mineralization in calcifying cartilage: theoretical considerations, *J. Electron Microsc. Tech.,* 11, 9, 1989.

39. **Bonucci, E.,** Fine structure of early cartilage calcification, *J. Ultrastruct. Res.,* 20, 33, 1967.

40. **Anderson, H. C.,** Vesicles associated with calcification in the matrix of epiphyseal cartilage, *J. Cell Biol.,* 41, 59, 1969.

41. **Buckwalter, J. A., Mower, D., and Schaeffer, J.,** Differences in matrix vesicle concentration among growth plate zones, *J. Orthop. Res.,* 5, 157, 1987.

42. **Engfeldt, B., Hjerpe, A., Reinholt, F. P., and Wikstrom, B.,** Distribution of matrix vesicles in the lower part of the epiphyseal growth plate, in *The Chemistry and Biology of Mineralized Tissues,* Butler, W. T., Ed., EBSCO Media, Birmingham, AL, 1985, 356.

43. **Reinholt, F. P. and Wernerson, A.,** Septal distribution and the relationship of matrix vesicle size to cartilage mineralization, *Bone Miner.,* 4, 63, 1988.

44. **Hunziker, E. B.,** Articular cartilage structure in humans and experimental animals, in *Articular Cartilage and Osteoarthritis,* Kuettner, K. E., Schleyerbach, R., Peyron, J. G. and Hascall, V. C., Eds., Raven Press, New York, 1992, 183.

45. **Eyre, D. R., Wu, J. J., and Apone, S.,** A growing family of collagens in articular cartilage: identification of 5 genetically distinct types, *J. Rheumatol.,* 14 Spec. No., 25, 1987.

46. **van der Rest, M. and Mayne, R.,** Type IX collagen proteoglycan from cartilage is covalently cross-linked to type II collagen, *J. Biol. Chem.,* 263, 1615, 1988.

47. **Burgeson, R. E., Hebda, P. A., Morris, N. P., and Hollister, D. W.,** Human cartilage collagens. Comparison of cartilage collagens with human type V collagen, *J. Biol. Chem.,* 257, 7852, 1982.

48. **Niyibizi, C. and Eyre, D. R.,** Identification of the cartilage alpha 1(XI) chain in type V collagen from bovine bone, *FEBS Lett.,* 242, 314, 1989.

49. **Mendler, M., Eich Bender, S. G., Vaughan, L., Winterhalter, K. H., and Bruckner, P.,** Cartilage contains mixed fibrils of collagen types II, IX, and XI, *J. Cell Biol.,* 108, 191, 1989.

50. **Mizoguchi, I., Nakamura, M., Takahashi, I., Kagayama, M., and Mitani, H.,** An immunohistochemical study of localization of type I and type II collagens in mandibular condylar cartilage compared with tibial growth plate, *Histochemistry,* 93, 593, 1990.

51. **Mayne, R. and Von der Mark, K.,** Collagens of cartilage, in *Cartilage, Structure, Function and Biochemistry,* Vol. 1, Hall, B. K., Ed., Academic Press, New York, 1983, 181.

52. **Doege, K. J., Sasaki, M., Kimura, T., and Yamada, Y.,** Complete coding sequence and deduced primary structure of the human cartilage large aggregating proteoglycan, aggrecan. Human-specific repeats, and additional alternatively spliced forms, *J. Biol. Chem.,* 266, 894, 1991.

53. **Lutfi, A. M.,** Mode of growth, fate and functions of cartilage canals, *J. Anat.,* 106, 135, 1970.
54. **Ranz, F. B., Aceitero, J., and Gaytan, F.,** Morphometric study of cartilage dynamics in the chick embryo tibia. II. Dexamethasone-treated embryos, *J. Anat.,* 154, 73, 1987.
55. **Howlett, C. R.,** The fine structure of the proximal growth plate and metaphysis of the avian tibia: endochondral osteogenesis, *J. Anat.,* 130, 745, 1980.
56. **Lutfi, A. M.,** ^{35}S-sulphate uptake by the growing tibia in the domestic fowl, *J. Anat.,* 107, 567, 1970.
57. **Massague, J.,** The transforming growth factor-ß family, *Ann. Rev. Cell. Biol.,* 6, 597, 1990.
58. **Iwamoto, M., Shimazu, A., Nakashima, K., Suzuki, F., and Kato, Y.,** Reduction in basic fibroblast growth factor receptor is coupled with terminal differentiation of chondrocytes, *J. Biol. Chem.,* 266, 461, 1991.
59. **Rosier, R. N., O'Keefe, R. J., Crabb, I. D., and Puzas, J. E.,** Transforming growth factor ß: an autocrine regulator of chondrocytes, *Connect. Tissue Res.,* 20, 295, 1989.
60. **Visco, D. M., Hill, M. A., Van Sickle, D. C., and Kincaid, S. A.,** Cartilage canals and lesions typical of osteochondrosis in growth cartilages from the distal part of the humerus of newborn pigs, *Vet. Rec.,* 128, 221, 1991.
61. **Hunziker, E. B. and Schenk, R. K.,** Physiological mechanisms adopted by chondrocytes in regulating longitudinal bone growth in rats, *J. Physiol. (London),* 414, 55, 1989.
62. **Engfeldt, B.,** Studies on the epiphysial growth zone. III. Electronmicroscopic studies on the normal epiphysial growth zone, *Acta Pathol. Microbiol. Scand.,* 75, 201, 1969.
63. **Holtrop, M. E.,** The ultrastructure of the epiphyseal plate. I. The flattened chondrocyte, *Calcif. Tissue Res.,* 9, 131, 1972.
64. **Brighton, C. T.,** Structure and function of the growth plate, *Clin. Orthop.,* 136, 22, 1978.
65. **Stanescu, R., Stanescu, V., and Maroteaux, P.,** Homozygous achondroplasia: morphologic and biochemical study of cartilage, *Am. J. Med. Genet.,* 37, 412, 1990.
66. **Arsenault, A. L., Ottensmeyer, F. P., and Heath, I. B.,** An electron microscopic and spectroscopic study of murine epiphyseal cartilage: analysis of fine structure and matrix vesicles preserved by slam freezing and freeze substitution, *J. Ultrastruct. Mol. Struct. Res.,* 98, 32, 1988.
67. **Glimcher, M. J.,** Mechanism of calcification: role of collagen fibrils and collagen-phospho-protein complexes in vitro and in vivo, *Anat. Rec.,* 224, 139, 1989.
68. **Boyan, B. D., Schwartz, Z., Swain, L. D., and Khare, A.,** Role of lipids in calcification of cartilage, *Anat. Rec.,* 224, 211, 1989.
69. **Poole, A. R., Matsui, Y., Hinek, A., and Lee, E. R.,** Cartilage macromolecules and the calcification of cartilage matrix, *Anat. Rec.,* 224, 167, 1989.
70. **Hunter, W. L. and Arsenault, A. L.,** Vascular invasion of the epiphyseal growth plate: analysis of metaphyseal capillary ultrastructure and growth dynamics, *Anat. Rec.,* 227, 223, 1990.
71. **Agrawal, P., Kulkarni, D. S., and Atre, P. R.,** The participation of cartilage canals in the ossification of the human fetal calcaneum, *J. Anat.,* 147, 135, 1986.
72. **Cole, A. A. and Wezeman, F. H.,** Perivascular cells in cartilage canals of the developing mouse epiphysis, *Am. J. Anat.,* 174, 119, 1985.
73. **Gray, M. L., Pizzanelli, A. M., Grodzinsky, A. J., and Lee, R. C.,** Mechanical and physiochemical determinants of the chondrocyte biosynthetic response, *J. Orthop. Res.,* 6, 777, 1988.
74. **Sah, R. L., Kim, Y. J., Doong, J. Y., Grodzinsky, A. J., Plaas, A. H., and Sandy, J. D.,** Biosynthetic response of cartilage explants to dynamic compression, *J. Orthop. Res.,* 7, 619, 1989.
75. **Eggli, P. S., Hunziker, E. B., and Schenk, R. K.,** Quantitation of structural features characterizing weight- and less-weight-bearing regions in articular cartilage: a stereological analysis of medial femoral condyles in young adult rabbits, *Anat. Rec.,* 222, 217, 1988.
76. **Szirmai, J. A.,** Structure of cartilage, in *Thule International Symposium on Ageing of Connective and Skeletal Tissue,* Engel, A. and Larsson, T., Eds., Nordiska Boekhandelns, Stockholm, 1969, 163.

77. **Poole, C. A., Matsuoka, A., and Schofield, J. R.,** Chondrons from articular cartilage. III. Morphologic changes in the cellular microenvironment of chondrons isolated from osteoarthritic cartilage, *Arthritis Rheum.,* 34, 22, 1991.
78. **Benninghoff, A.,** Uber den funktionellen Bau des Knorpels, *Anat. Anz. Erg. Heft.,* 55, 250, 1922.
79. **Clark, J. M.,** The organisation of collagen fibrils in the superficial zones of articular cartilage, *J. Anat.,* 171, 117, 1990.
80. **Hunziker, E. B. and Kuettner, K. E.,** Cartilage, in *The Lung: Scientific Foundations,* Crystal, R. G. and West, J. B., Eds., Raven Press, New York, 1991, 451.
81. **Kuettner, K. E., Memoli, V. A., Pauli, B. U., Wrobel, N. C., and Thonar, E. J.,** Synthesis of cartilage matrix by mammalian chondrocytes in vitro. II. Maintenance of collagen and proteoglycan phenotype, *J. Cell Biol.,* 93, 751, 1982.
82. **Nakahara, H., Dennis, J. E., Bruder, S. P., Haynesworth, S. E., Lennon, D. P., and Caplan, A. I.,** In vitro differentiation of bone and hypertrophic cartilage from periosteal-derived cells, *Exp. Cell Res.,* 195, 492, 1991.
83. **Aydelotte, M. B., Greenhill, R. R., and Kuettner, K. E.,** Differences between sub-populations of cultured bovine articular chondrocytes. II. Proteoglycan metabolism, *Connect. Tissue Res.,* 18, 223, 1988.
84. **Buschmann, M. D., Gluzband, Y. A., Grodzinsky, A. J., Kimura, J. H., and Hunziker, E. B.,** Chondrocytes in agarose gel synthesize a mechanically functional extracellular matrix, submitted, 1992.
85. **Bruckner, P., Horler, I., Mendler, M., Houze, Y., and Winterhalter, K. H.,** Induction and prevention of chondrocyte hypertrophy in culture, *J. Cell Biol.,* 109, 2537, 1989.
86. **Hunziker, E. B.,** Struktur und funktion des gelenkknorpels, in *Arthroskopie bei Knorpelschaden und bei Arthrose,* Glinz, W., Kieser, C., and Munzinger, U., Eds., Ferdinand Enke Verlag, Stuttgart, 1990, 1.
87. **Hunziker, E. B.,** Growth plate structure and function, in *Cartilage Degradation: Basic and Clinical Aspects,* Woessner, J. F., Jr. and Howell, D. S., Eds., Marcel Dekker, New York, 1992, in press.

Chapter 2

CARTILAGE COLLAGENS

**Beatrice Petit, Anne-Marie Freyria, Michel van der Rest,
and Daniel Herbage**

TABLE OF CONTENTS

I. INTRODUCTION

The name *collagen* designates a family of structural proteins of the extracellular matrix-forming supramolecular aggregates that involve lateral interactions between characteristic triple-helical (COL) domains.[1] These triple helices are formed by the intertwining of three polypeptide chains, each forming a left-handed helix, into a right-handed superhelix (Figure 1).[2] This particular conformation is only possible if the primary structure of the constituent polypeptide chains have a repetitive primary structure $(Gly-Xaa-Yaa)_n$. Every third amino acid of each chain occupies the center of the helix, a position that, for steric reasons, only glycine (Gly) can occupy without disruption of the triple-helical conformation.

The stability of this structure, as demonstrated by its denaturation temperature, is a function of the percentage of the Xaa and Yaa positions occupied by imino acid residues (proline and hydroxyproline).[2] In most collagens, this proportion is approximately 30% for each position. Hydroxyproline residues are formed by a posttranslational hydroxylation of proline residues in the Yaa position. This enzymatic hydroxylation has been shown to increase the denaturation temperature of the helix above physiological conditions.[3] Since prolyl hydroxylase is only active on non-triple-helical polypeptides, the level of hydroxylation of prolines is limited by the rate of helix formation.[4,5]

Some 14 different collagen types, constituted by 27 distinct polypeptide chains, have been described thus far (Table 1)[1] and others are currently being studied in several laboratories. In addition, molecules that are not constituents of the extracellular matrix have been shown to contain triple-helical domains of the collagen type. It is thus becoming evident that "collagen-like" triple-helical domains are more common in proteins than usually thought.[6] These domains have two main functions: (a) they act as relatively rigid rods that can physically separate other domains in a protein; and (b) since the side chains of all Xaa and Yaa amino acid residues point outside of the helix and are thus at the protein surface, they have a very large potential for lateral interactions, in particular with other triple helices. The interactions between triple helices can be parallel or antiparallel, staggered or nonstaggered, between identical or different molecules (Figure 2).

All collagen molecules include non-triple-helical (NC) domains. These domains are extremely diverse in size and structure.[6] Depending on the collagen type, the NC domains may represent between 3 and 92% of the mature collagen molecule. While the function of some of these domains, such as the NC1 domain of type IV collagen, has been well established,[7] in most cases it is still a matter of speculation. It is clear, however, that NC domains play essential roles in the organization of the supramolecular aggregates formed by collagen molecules.

The diversity of macromolecular aggregates involving collagen molecules is demonstrated in Figure 2. The most abundant and "classical" supramolecular structure formed by collagens is that of the striated fibrils found in most connective tissues. These fibrils are made of staggered molecules and are responsible for the tensile strength of extracellular matrices.[8] A major recent advance in our understanding of extracellular matrices, and of cartilage in particular (Table 2), stems from the finding that collagen fibrils are formed by mixtures of different collagen types and are thus heterotypic.[9-11] This heterogeneity is thought to be responsible for many properties of the fibrils, such as size and diameter.

The fibrils are made up of, in large part, a subgroup of closely related molecules (collagen types I, II, III, V, and XI) that are all able, at least *in vitro*, to form striated fibrils by themselves.[1] These molecules are often referred to as the fibrillar or interstitial collagens. Other molecules, unable by themselves to form striated fibrils and structurally very different from the fibrillar collagens, have been shown to be associated with the fibrils and form the subgroup of fibril-associated collagens with interrupted triple-helices (FACITs) (collagen types IX, XII, and XIV).[12] The other collagen types participate in distinct histological structures, such as basement membranes, anchoring fibrils, Descemet's membrane, or beaded filaments (Figure 2).[1]

The biomechanical properties of cartilage and, in particular, its ability to act as a shock absorbing tissue have been described as resulting from the opposing forces of the swelling pressure of the highly charged proteoglycans and of the tensile strength of the network of collagen fibrils in which the proteoglycans are entrapped.[13] The cartilage collagen fibrils are made up of, in large part, type II collagen molecules. The regular staggering of these molecules in a quasi-crystalline array gives rise to the 67-nm axial periodicity characteristic of collagen fibrils. It has recently been demonstrated that cartilage collagen fibrils are actually mixtures of several collagen types, namely types II, IX, and XI.[11] These three types are characteristic of cartilaginous tissues, although they have been observed at low levels in some other extracellular matrices. Their levels of expression are frequently used as markers of the chondrocyte phenotype in opposition to the biosynthesis of type I collagen, which is the predominant collagen type in most other tissues. Type X collagen is also cartilage specific, but its expression is restricted to the hypertrophic chondrocyte.

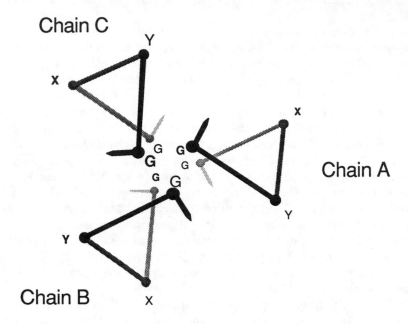

FIGURE 1. Schematic representation of the conformation of the collagen triple helix. The α carbons of the amino-acyl residues are represented by solid balls and the peptide bonds by straight sticks. The distance from one glycine to the next one of the same chain is ~0.9 nm and the diameter of the triple helix is ~1.5 nm. Every X and Y residue is at the surface of the molecule and will have its side chain (not represented) pointing outward of the helix. Each polypeptide forms a left-handed helix. The three chains assemble into a coiled structure to form a right-handed superhelix.

While these four collagen types are usually considered as cartilage-specific, it has been clearly demonstrated that other more ubiquitous collagen types are also present in cartilage and may play an important role in cartilage differentiation and pathology. These collagens have thus been included in our presentation. Very recently, Keene et al.[14] have published a detailed immunolocalization study of two FACITs which they called type XII-like collagens (TL-A and TL-B). TL-A, which is actually type XII collagen (R. E. Burgeson, personal communication), was observed in cartilage. It is quite possible that cartilage contains small amounts of yet undescribed collagen molecules.

II. STRUCTURE AND ORGANIZATION OF CARTILAGE COLLAGENS

A. TYPE II COLLAGEN
1. Introduction

Type II collagen was first isolated by Miller and co-workers[15] who extracted the collagenous material from cartilage by limited pepsin digestion. Type II

TABLE 1
The Different Types of Collagens

Collagen type	Molecular form	Tissue distribution	Characteristics
I	$\alpha1(I)_2\alpha2(I)$	Skin, tendon, bone, cornea, and most of the connective tissues	Thick structured fibers, low level of Hyl and carbohydrates
I trimer	$\alpha1(I)_3$	Embryonnic and pathological tissues. Low level in normal tissues	Hyl in higher level than in collagen type I
II	$\alpha1(II)_3$	Hyaline cartilage, intervertebral disc, vitreous humor	Thin fibers, high level of Hyl, Glc-Gal-Hyl and Gal-Hyl
III	$\alpha1(III)_3$	"Coexists" with type I in skin, wall vessels, liver, placenta, embryonnic tissues	Presence of Cys and intramolecular disulfide bridges
IV	$\alpha1(IV)_2\alpha2(IV)$ $\alpha3(IV)$? $\alpha4(IV)$? $\alpha5(IV)$?	Basal membranes	Regular meshwork; presence of Cys; high level of Hyp-Hyl, low level of Ala, Arg; some 3-Hyp residues; high level of glycosylation
V	$\alpha1(V)_2\alpha2(V)$ or $\alpha1(V)\alpha2(V)\alpha3(V)$ $[\alpha1(V)]_3$	Almost ubiquitous	Associated with collagen type I fibers; low level of Ala, high level of basic amino acids
VI	$\alpha1(VI)\alpha2(VI)\alpha3(VI)$	Aorta intima, placenta, uterus, skin, fœtal cartilage, nucleus pulposus, lung, liver	High molecular weight aggregates; beaded filaments and microfibrils; important noncollagenous domains; rich in Cys-Tyr; high level of Hyl and glycosylated-Hyl, low level of Ala and Hyp
VII	$\alpha1(VII)_3$	Dermo-epidermal junction	Long triple-helical domain ($\times1.5$ that of type I collagen); rich in Cys located in interchain disulfide bridges; anchoring fibrils
VIII	$\alpha1(VIII)_3$	Endothelial cell culture, Descemet membrane	Rich in 3-Hyp, in glycosylated-Hyl; no disulfide bridge although a high level of Cys; strong proteases; susceptibility short-chain collagen"; strong homology "with collagen type X
IX	$\alpha1(IX)\alpha2(IX)\alpha3(IX)$	Cartilaginous tissues, intervertebral disc, vitreous humor	FACIT collagen; rich in Cys; presence of GAG; located on the surface of collagen type II fibers

TABLE 1
The Different Types of Collagens (continued)

Collagen type	Molecular form	Tissue distribution	Characteristics
X	α1(X)$_3$	Synthesized by hypertrophic chondrocytes	Rich in Tyr, Met; 25% of iminoacids; low level of Arg; high melting temperature
XI	α1(XI)α2(XI)α3(XI)	Cartilaginous tissues, chondrosarcoma	α3(XI) similar to α1(II) but more glycosylated; isoform of type V?; associated with type II fibers
XII	α1(XII)$_3$	Tendon, ligament, skin, cartilage?	FACIT collagen; associated with type I fibers; large noncollagenous domain
XIII	?	Skin fibroblast culture	Described from cDNA; frequent alternative splicing; existence of 4 mRNAs
XIV	α1(XIV)$_3$	Tendon, skin	FACIT collagen; similar structure as type XII collagen

collagen represents about 90 to 95% of the total collagen content of the extracellular matrix of articular cartilage[16] and is localized throughout this matrix by indirect immunofluorescence; type I and type III collagens are only localized on the surface of the cartilage (Figure 3 a, b, and c). It belongs to the subgroup of fibrillar collagens which are the constituents of banded fibrils and are characterized by molecules with long continuous triple helices with short N and C telopeptides.

The type II collagen molecule consists of three identical chains known as α1(II); each chain has a molecular weight of about 95 kDa. The entire molecule looks like a thin rod, 300 nm long and 1.5 nm wide, and has a molecular weight of 295 kDa.

Type II collagen, like the other collagens, is synthesized in a precursor form known as a procollagen. Procollagen molecules consist of three major domains: (1) a long triple-helical region with the repetition of (Gly–X–Y) triplets characteristic of collagens. The X and Y amino acids are 33% proline and hydroxyproline. In nonfibrillar collagens, this triple helix is interrupted by globular domains; (2) a N-propeptide which contains triple-helical segments and globular regions; and (3) a C-propeptide which has no triple-helical region but is rich in cysteines.

A number of posttranslational modifications take place in the cell before the triple helix is formed. The proline and lysine residues become hydroxylated: hydroxylation of proline is important for the stability of the triple helix, while that of lysine is important for the formation of intermolecular cross-links and for hydroxylysyl glycosylations. These glycosylations result in the covalent binding of D-galactose and D-glucose on collagen chains; the role of these disaccharide units remains unclear, although they may be involved in regulation of fibril diameter (for review, see Reference 17).

After these modifications, three pro-α chains are linked through their C-propeptides and fold into a triple-helical procollagen molecule. When this molecule is excreted into the extracellular matrix, the propeptides are cleaved by specific N and C proteinases and the molecule is then formed by the triple-helical domain with N and C telopeptides. In some collagen types (type XI), however, the N-propeptide is not cleaved and may be involved in fibril formation.[17]

All of the fibrillar collagens (types I, II, III, V, and XI) form organized fibrils and fibers. The collagen molecules are arranged in a quarter-stagger array, the fibers presenting a cross-striation period of 67 nm. All fibers appear to be heterotypic and are formed of types I and III collagens,[9] types I and V collagens,[10] or types II and XI collagens.[11] The role of these heterotypic fibrils is not yet clear. Another posttranslational modification important for the stability of the fibrils is formation of reducible intermolecular cross-links. These cross-links occur between the lysine and hydroxylysine residues of the different molecules of collagen that form the fibrils, stabilizing the supramolecular structure. Reducible cross-links disappear from human cartilage by 10 to 15 years of age and are replaced by hydroxypyridinium residues which are their maturation products. Mature cross-links are formed rapidly and they are stable and dominant.[18] During aging, the cross-links are gradually converted into nonreducible compounds.[17]

Type II collagen is a genetically distinct member of the fibril-forming collagens and the gene is present in a single copy in the genome.[19] The structure of this gene is very similar to that of other fibrillar collagen genes and will be described in detail below.

2. Structure of Type II Collagen
a. Triple-Helical Domain

Gene structure — In all fibrillar collagens, the triple-helical coding region contains about 44 exons which present a common organization which is highly conserved. All the exons are 54 bp in length or related to 54 bp unit (54bp, 99bp, 108bp, or 162bp); they start with a complete codon for glycine and end with a complete codon for the amino acid in the Y position, resulting in an in-frame "cassette" type of organization.[20] This suggests that the triple-helical coding domain evolved by processes of duplication, fusion, and deletion of an ancestral 54-bp unit.[21] The beginning and the end of the triple-helical portion are coded by junction exons, which also code for telopeptide sequences.[20] In comparison with other genes that have been characterized for fibrillar collagens, the exon-intron boundaries in chicken α1(II) procollagen gene are conserved, except for the presence of an additional intron between the two exons that code for amino acids 568–585 and 586–603.[22] The same organization was shown in the human α1(II) gene.[23] Only the pro-α1(I) gene presents a single 108-bp exon in this region; in

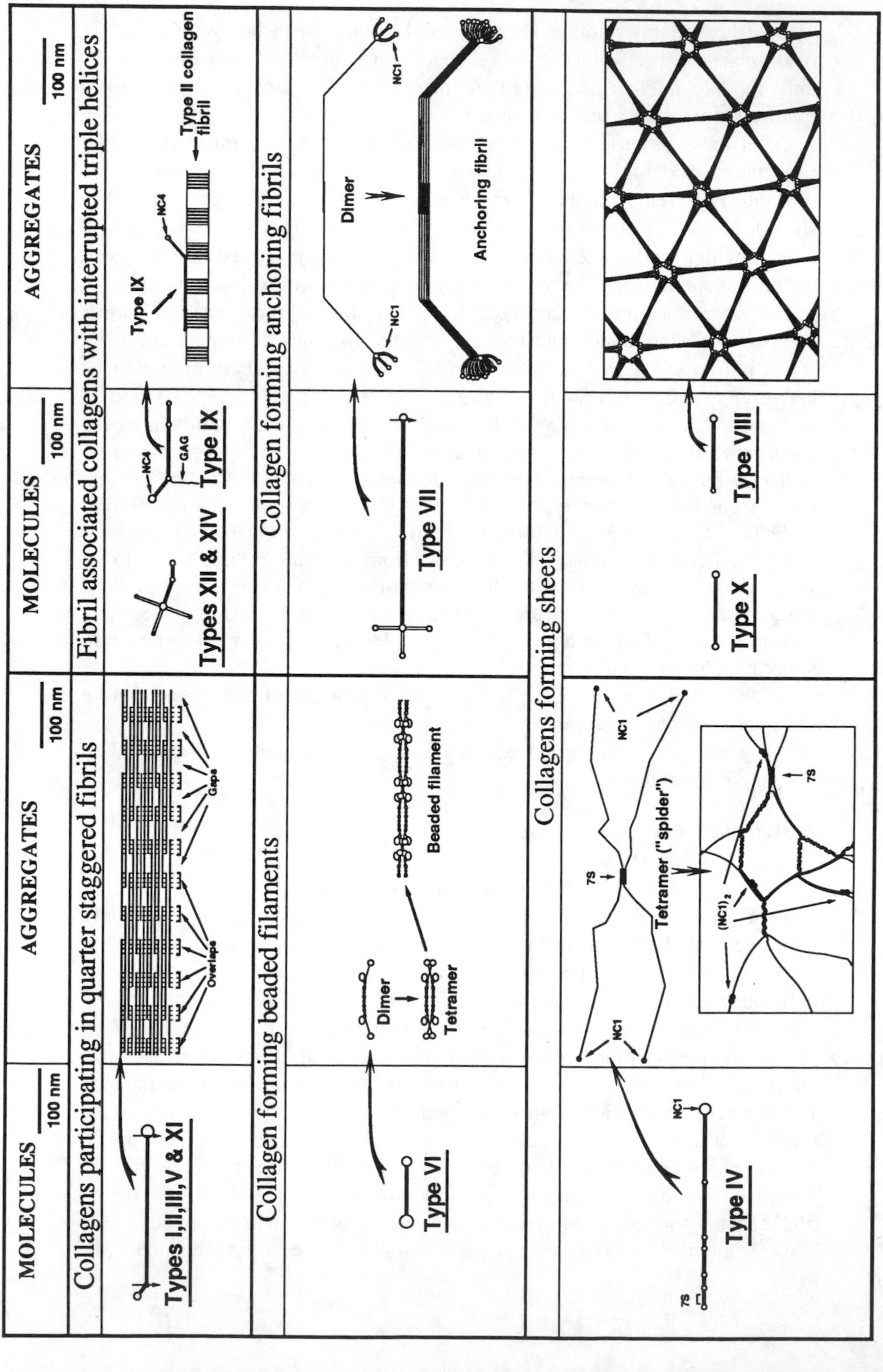
MOLECULES
100 nm
AGGREGATES
100 nm
MOLECULES
100 nm
AGGREGATES
100 nm
Collagens participating in quarter staggered fibrils
Types I,II,III,V & XI
Overlaps
Gaps
Fibril associated collagens with interrupted triple helices
Types XII & XIV
NC4
GAG
Type IX
Type IX
NC4
Type II collagen fibril
Collagen forming beaded filaments
Type VI
Dimer
Tetramer
Beaded filament
Collagen forming anchoring fibrils
Type VII
NC1
Dimer
NC1
Anchoring fibril
Collagens forming sheets
Type IV
7S
NC1
NC1
7S
NC1
Tetramer ("spider")
(NC1)2
7S
Type X
Type VIII

FIGURE 2. Supramolecular aggregates formed by the various collagen types. In this figure, the various collagen molecules and their aggregates are represented on the same scale. The thickness of the triple helical domains and the sizes of the globular domains are, however, exaggerated for easier visualization. The collagen molecules are oriented with their N-termini to the left. The triple helical (COL) domains are represented as solid bars, while non-triple-helical (NC) domains are shown as double lines or empty circles. The globules that are visible by rotary shadowing electron microscopy are represented by the larger circles. In the known collagens, they are always located at one or both ends of the molecule. Small open circles represent interruptions of the triple helices (over 6 nonrepetitive residues). Imperfections of the triple helix (insertion or deletion of one or two residues in the repetitive sequence) are not indicated. A vertical arrow on the drawing of a molecule indicates a site of cleavage by one of the processing proteinases. GAG stands for the glycosaminoglycan side chain which attached to the α2 chain of the type IX collagen molecule. 7S designates the N-terminal triple-helical domain of the type IV collagen by which four molecules are assembled to form the tetrameric aggregate known as the "spider".

Type I, II, III, V, and XI collagen molecules have the same overall molecular structure. The exact processing sites are known only for collagen types I, II, and III. Partial processing has been demonstrated for collagen types V and XI, but the cleavage sites are not precisely known. All these collagens participate in the formation of quarter-staggered fibrils, which are probably always heterotypic.

Type IX, XII, and XIV collagen molecules contain several triple helical domains interspersed by short non-triple-helical (NC) domains. Types XII and XIV collagens have very large cross-shaped N-terminal domains. Type IX collagen has been shown to bind covalently to the surface of type II collagen fibrils. The globular NC4 domain and the triple helical COL3 domain project out of the fibril. In some tissues, the NC4 domain is almost completely absent due to the use of an alternate promoter.

Type IV collagen molecules assemble into tetramers ("spiders") via their 7S region. Tetramers further interact by their C-terminal NC1 domains. The triple-helical domains from the "legs" of the spider laterally aggregate in a staggered fashion to form the complex, branching network shown on the figure inset.

Types VIII and X collagen molecules are dumbbell-shaped and probably interact laterally and by their extremities to form hexagonal lattices. For type VIII collagen, such structures have been observed in Descemet's membrane. For type X collagen, they have only been observed *in vitro*.

Type VI collagen molecules have a short triple-helical domain, with very large globular extremities. They assemble into tetramers, formed by lateral association of dimers which are themselves composed of two antiparallel molecules. Beaded filaments are formed by the association of tetramers by their extremities.

Type VII collagen molecules have a N-terminal domain with three fingers similar to those of collagen types XII and XIV. (Early reports suggested an opposite orientation for this molecule.) The C-terminal globular domain is cleaved during maturation to permit the antiparallel formation of dimers. Several dimers aggregate laterally in register to form anchoring fibrils observed at the dermo-epidermal junction and are responsible for stabilizing the interaction between basement membranes and the underlying stroma.

pro-α2(I) and pro-α1(III) genes, as in the pro-α1(II) gene, amino acids 568–603 (exon 19) are encoded by two exons. Thus, the exon fusion observed in the pro-α1(I) gene occurred some time after the duplication and divergence of the four fibrillar collagen genes.[23]

The chicken type II procollagen gene differs from other fibrillar procollagen genes in that its introns are smaller (average size, about 80 bp).[22] The human gene is longer than the chicken gene, with a size of 30 kb;[24] the human procollagen α1(II) gene spans six large Eco RI fragments with the distribution shown in Figure 4. This gene was completely sequenced by Baldwin,[25] who found that the triple-helical coding region is more like the α1(I) coding region than the α2(I)

TABLE 2
Collagens Present in Cartilage

Type	Molecular organization	Molecular weight (kDa)	Length of helical domain	*in situ* organization
Major fibril constituents				
II	$\alpha1(II)_3$	290	300 nm	Striated fibrils D = 67 nm
IX	$\alpha1(IX)\ \alpha2(IX)\ \alpha3(IX)$	250	170 nm (interrupted)	Associated with type II on the surface of the fibrils
XI	$\alpha1(XI)\ \alpha2(XI)\ \alpha3(XI)$	300	320 nm	Within fibrils
Collagens present in specialized structures				
VI	$\alpha1(VI)\ \alpha2(VI)\ \alpha3(VI)$	500–550	105 nm	Filaments D = 100 nm
X	$\alpha1(X)_3$	170	140 nm	Associated with hypertrophic chondrocytes. Hexagonal lattice?
Collagens occasionally reported in cartilage				
I	$\alpha1(I)_2\ \alpha2(I)$	290	300 nm	Striated fibrils D = 67 nm
III	$\alpha1(III)_3$	290	300 nm	Striated fibrils D = 67 nm
V	$\alpha1(V)_2\ \alpha2(V)$ $\alpha1(V)\ \alpha2(V)\ \alpha3(V)$ other forms	300	305	Associated with type-I fibrils
XII	$\alpha1(XII)_3$	660	75 nm	?
XIV	$\alpha1(XIV)_3$	660	75 nm	?

coding region or the $\alpha1(III)$ coding region. The exons that code for the triple helix are identical in size to the exons of the chicken gene, with a high G+C content. Comparison of clones from three alleles of the human gene defined several neutral variations in coding and non-coding sequences.[26] An important mutation was found in the triple-helical region of the human gene, which converted the Arg at position 519 to a Cys. It was recently shown that this conversion concerns about one quarter of the $\alpha1(II)$ chains present in the cartilage of patients who express an inherited form of osteoarthritis associated with mild chondrodysplasia. One hypothesis is that the Arg–Cys substitution allows formation of disulfide-bonded $\alpha1(II)$ dimers; this implies that osteoarthritis is a consequence of this molecular abnormality of collagen.[26,27]

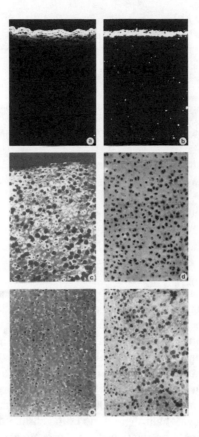

FIGURE 3. Indirect immunofluorescence of sections of fetal bovine articular cartilage with polyclonal antibodies to: type I collagen (a), type III collagen (b), type II collagen (c), type IX collagen (d), and type XI collagen (e and f). Staining for types I and III collagens is restrictied to the articular surface (a and b). Staining for types II and IX collagens is observed throughout the matrix (c and d). For type XI collagen, only pericellular staining is observed when the sections are not treated with pepsin (e), while after pepsinization this staining is distributed throughout the matrix (f).

Protein structure — The triple-helical domain of the $\alpha1(II)$ chain contains 1014 amino acids. In the NH_2-terminal helical portion of the bovine $\alpha1(II)$ chain, the level of homology between the $\alpha1(I)$ and $\alpha1(II)$ chains is about 80%; there are three glycosylgalactosyl hydroxylysines (positions 87, 99, and 108) in $\alpha1(II)$, while there is only one in the $\alpha1(I)$ chain at position 87. In the amino acid sequence of residues 363–551 of bovine $\alpha1(II)$ chains, about 73% of the residues are identical to those of $\alpha1(I)$ chains. If the invariant glycines are not considered, the level of homology of the X and Y positions of $\alpha1(I)$ and $\alpha1(II)$ chains in this region is about 60%.[28]

Further studies have shown that the triple-helical domains of type II and type I collagens are very similar. Cross-linking sites are localized on regions with low

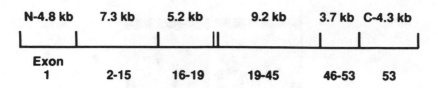

FIGURE 4. Distribution of exons in six major EcoR1 fragments of the human procollagen α1(II) gene. Vertical lines represent EcoR1 sites. (From Huang M. C., Seyer J. M., Thompson J. P., Spinella D., Cheah K., and Kang A., *Eur. J. Biochem.*, 195, 3, 1991. With permission.)

contents of proline and hydroxyproline and are characterized by the presence of the Hyl–Gly–His–Arg sequence, which is important as an attachment site for lysyl oxidase.[17]

The intact triple helix is resistant to proteolytic attack. In the fibril, the triple-helical molecule can be cleaved only by mammalian collagenase, and the cleavage site is located in an area with a relatively low content of imino acids.

b. C-Propeptide

All C-propeptides in fibrillar collagens are characterized by three highly conserved elements: the C-protease cleavage site, the cysteinyl residues, and the N-linked carbohydrate attachment site.

Intracellular folding of procollagens into the triple helix begins with the formation of inter- and intrachain disulfide bonds in this region.[21]

Curran and Prockop[29] were among the first to characterize the C-propeptide of type II procollagen in the sternal cartilage of chick embryos. The C-propeptide is a disulfide-linked trimer of about 100 kDa. After reduction, monomers of 34 kDa were obtained, in which there is no collagen-like domain. Although the domain is similar to the C-propeptide of type I procollagen, some differences can be seen: in the 246 amino acid residues of the α1(II) C-propeptide, an inserted residue is found at position 7 from the C-protease cleavage site compared with pro-α1(I) chains, and a deleted residue at position 101. While the pro-α1(I) chain contains only one site for N-linked oligosaccharide side-chains, the pro-α1(II) chain contains two.[30]

Like the C-propeptide of type I collagen, the C-propeptide of type II collagen contains eight cysteinyl residues in the same position as in the pro-α1(I) chain. Six of these cysteines are involved in intrachain disulfide bridges, which stabilize the globular structure. The two other cysteinyl residues form interchain bridges that are important for stabilization of the three-chain complex.[17]

The C-propeptide domain of the pro-α1(II) collagen chain is encoded by four exons (numbers 51–54).[31] Sandell et al.[32,33] were among the first to characterize this region in the chicken gene by isolating a clone that contains four exons which code for the last 15 amino acids of the triple-helical domain and 273 amino acids of the C-telopeptide and the C-propeptide. This region shows close similarities to that of the chicken pro-α1(I) collagen gene and differs from those of the α2(I) and $\alpha 1$(III) procollagen genes. One of the four exons contains 54 bp that code for Gly–X–Y repeats plus an additional sequence for the telopeptide and C-propeptide. This exon is known as the "junction-exon" since it encodes the end of the triple helix, the C-telopeptide, and the beginning of the propeptide. The three remaining exons code for the rest of the C-propeptide and for the 3' non-coding sequence. In exon 53 (which codes for amino acids 171^c–186^c), there is high conservation of the nucleotide sequence compared with other procollagen genes. Within this sequence there is a site for carbohydrate attachment.

The 3' untranslated region of type II procollagen mRNA (about 525 bp) is longer than corresponding regions in $\alpha 2$(I) mRNA or $\alpha 1$(I) mRNA. This region contains unusually long repetitions, consisting of one or two bases. Although there is no sequence that corresponds to the canonical site for polyadenylation, there are two closely related sequences 32 and 41 bases before the presumptive poly(A) site.[31,33]

The bovine pro-$\alpha 1$(II) collagen gene exhibits 85% homology with the avian gene. The potential site for glycosylation found in avian C-propeptide is absent in the bovine C-propeptide, suggesting that this site is not used in the chicken. The 3' untranslated region is 436 nucleotides long in the bovine gene, and the polyadenylation signal is found 21 bases upstream from the poly-A tail. The nucleotide sequence in this region is highly divergent in both the avian and the bovine pro-$\alpha 1$(II) mRNAs.[23]

The human pro-$\alpha 1$(II) collagen gene shows, respectively, 85 and 96% homology in the C-propeptide domain with the corresponding sequence in avian and bovine genes. There is complete identity with the homologous region of the bovine gene in the 3' untranslated region,[23] and the polyadenylation site is found 23 bases upstream from the poly-A tail.[34]

The C-propeptide of type II collagen is chemically stable in the extracellular matrix after its cleavage by C-protease. "Chondrocalcin" was first identified as a distinct protein involved in the calcification of cartilage[35] and was later demonstrated to be the C-propeptide of type II procollagen.[36,37] This propeptide was found in equal amounts as a prominent component of immature cartilage and intervertebral disc tissue.[38] The finding that it is retained in newly formed fibrils in chick embryo chondrocyte cultures confirms that it has a role during the initial steps of collagen fibril formation.[39]

c. N-Propeptide

Two functions have been proposed for this region of the a chain: regulation of fibril diameter and/or feedback inhibition of collagen biosynthesis (for review, see Reference 17).

In contrast to the two other domains, which are very similar in fibrillar collagen, the N-propeptides differ in size and composition within the group. N-propeptide consists of three domains: a cysteine-rich globular segment, a collagenous sequence, and a short, nonhelical domain. At the amino terminus

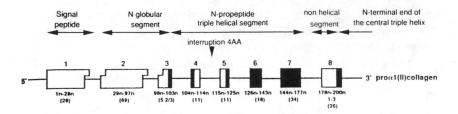

FIGURE 5. Region of the human type II collagen gene coding for the N-propeptide domain. Exons, boxes, introns, lines, and n numbers correspond to the amino acids encoded within the N-propeptide sequence. Other numbers refer to the beginning of the major triple-helical domain. Numbers in parentheses refer to the number of residues encoded within each exon. (From Ninomya, Y., *Extracellular Matrix Genes,* Sandell L. J. and Boyd C. D., Eds., Academic Press, New York, 1990. With permission.)

of all N-propeptides is a 23-amino acid-long signal peptide which is cleaved when the chain has passed the membrane.[17] A feature common to the pro-α1(II) chain and the pro-α1(V) and pro-α1(XI) chains is interruption of their collagenous sequences in the N-propeptide by a short, noncollagenous sequence.[40]

Collagen type II N-propeptide is encoded by seven exons, like the majority of N-propeptide of fibrillar collagens (Figure 5). The main difference between type II N-propeptide and the N-propeptides of other fibrillar collagens is that in procollagen II this region undergoes alternative splicing. Analysis of human clones reveals that exon 2 encodes the 69-amino acid cysteine-rich domain. Use of oligonucleotide probes and polymerase chain reaction amplification provides direct evidence for two distinct pro-α1(II) collagen mRNAs, resulting from the alternative splicing of exon 2. Thus, a single copy gene, COL2A1, produces two pro-α1(II) collagen mRNA transcripts: one in which exon 1 is spliced to exon 2 and one in which exon 1 is spliced to exon 3.[41] *In situ* hybridization reveals that some populations of cells express one of the two mRNAs, while some cartilaginous tissues express both (human fetal chondrocytes, costal chondrocytes, or bovine articular chondrocytes).[31] Ryan and Sandell concluded that the expression of exon 2 may be a marker for a distinct population of chondrocytes and that the negative correlation between expression of the N-propeptide and matrix deposition is consistent with the regulatory role proposed for this domain. The coding sequence contained within exon 2 was not previously identified because the coding region was established by comparison with rat chondrosarcoma cDNA and not with genomic DNA from normal chondrocytes.[31]

Type II procollagen is now well characterized, and the gene has been located by *in situ* hybridization in band q,13-1-q,13-2 of chromosome 12.[42] The promoter of this gene is controlled by both negative and positive elements. While the triple-helical region and the C-propeptide are better conserved than in other fibrillar collagens, the N-propeptide shows a higher degree of divergence in its organization and is the first example of alternative splicing of mRNAs in fibrillar collagens. It would be interesting to establish why and when this splicing occurs, as another collagen chain, $\alpha3(XI)$, is identical to $\alpha1(II)$. The mechanism of alternative splicing described here may also be important in the procollagen XI.

B. TYPES XI AND V COLLAGENS
1. Structure

Type XI collagen was first discovered by Burgeson and Hollister[43] during differential salt precipitation of pepsin digests of human and bovine hyaline cartilage. This collagen contains three different chains, first known as 1α, 2α, and 3α — and now called $\alpha1(XI)$, $\alpha2(XI)$, and $\alpha3(XI)$. The $\alpha3(XI)$ chain is an overglycosylated form of the $\alpha1(II)$ chain of type II collagen. These three collagenous chains have been found in cartilaginous tissue from other species (like chicken and pig), but they present the same general features in all species.[44,45] Type XI collagen has a molecular weight of 300 kDa and is 300 nm long.

Type V collagen is distributed in various connective tissues (bone, tendon, placenta, and skin) and is associated with type I collagen in heterotypic fibrils. It has a molecular weight of 300 kDa, with a triple-helical domain 305 nm long (for review, see Reference 46). Four chains have been described in this collagen: $\alpha1(V)$, $\alpha1'(V)$, $\alpha2(V)$, and $\alpha3(V)$. In some tissues, the most prevalent form is $[\alpha1(V)]_2\alpha2(V)$; but in others, the $\alpha1(V)$ $\alpha2(V)$ $\alpha3(V)$ form is found. Different compositions of the chain impart different stabilities and flexibilities to the molecule that may be better suited to their particular functions.[46] Pro-$\alpha1'(V)$ and pro-$\alpha1(V)$ chains may be the products of two similar genes which differ in the noncollagenous peptide region, or they may arise from one gene by different splicing of RNA. The Pro-$\alpha1'(V)$ chain can be associated in a homotrimer and in heterotrimers [pro-$\alpha1(V)$, pro-$\alpha1'(V)$, pro-$\alpha2(V)$]. $\alpha1'(V)$ is obtained from chick embryo tendon and from human epitheloid lung cells (for review, see Reference 46).

Type XI collagen is often compared with type V collagen because both show common biochemical characteristics, and also because $\alpha1(XI)$ and $\alpha1(V)$ chains can be co-expressed in bone and cartilage (see below for details). It was found recently, by immunolocalization, that these two collagens are present at distinct locations in rat cartilage.[47]

Type XI collagen resembles type V collagen in the following ways:

1. Both collagens precipitate with 1.2 M NaCl from acid solutions.
2. They have the same electrophoretic mobility on SDS-PAGE electrophoresis.
3. The alanine content and the high content of hydroxylysine suggest a closer relationship between type XI and type V chains than with other fibrillar collagens.
4. $\alpha1(V)$ has 82% homology in its triple-helical domain with the $\alpha1(XI)$

chain,[48] (This high percentage must be kept in mind for producing specific antibodies and for immunolocalization).
5.   The two types bind to heparin.[49,50]

Comparison of cyanogen bromide peptides and V8 protease peptides, however, indicates differences in the primary structure and shows that they are products of different collagen genes.[51,52] Strong homologies were found between the chains of type XI and type V collagens (see References 48, 53, 55, and van der Rest, unpublished results):

•α1(XI)/α1(V): 85%,
•α1(XI)/α2(XI): 78%,
•α3(V)/α1(XI): 70%, and
•α2(V)/α2(XI): less than 50%

Comparison of these percentages shows first that the often cited resemblance between α2(V) and α2(XI) does not exist. Kimura et al.[53] showed that the pro-α2(XI) sequence is related more closely to that of pro-α1(XI) than to any other known fibrillar collagen chain. Second, it shows that the α1′(V) chain could be the α1(XI) chain. Third, type XI and type V collagens may be a single type of collagen formed by differential association of several chains of either type XI or type V collagen. This hypothesis is reinforced by the observation of a heterotrimer $[\alpha1(XI)]_2\,\alpha2(V)$ in cultured A204 rhabdomyosarcoma cells (J. P. Kleman, manuscript in preparation).

a. The Three Chains of Type XI Collagen
    Cloning and sequencing of α1(XI) and α2(XI) show that type XI collagen belongs to the family of fibrillar collagens, each chain presenting features characteristic of this class of collagen.
    Pro-α1(XI) is now well characterized; its genomic organization has been studied, and the gene is found to be located on the short arm of chromosome 1 in the 1p.21 region.[54] The α1(XI) chain presents a triple-helical domain 1017 amino acid residues long. This domain is coded by exons that are multiples of 54 bp, like other fibrillar collagens. In addition, sequencing of this chain shows the presence of cross-linking sites. The lysines are present at positions 87 and 927 (in other fibrillar collagens, the lysines are at positions 87 and 933); the only other chain that preserves the lysyl residues in the same place as α1(XI) is the α1(V) chain.[48] A unique characteristic of the α1(XI) chain is the presence of a cysteinyl residue located in approximately the middle of the triple-helical domain. Reduction of the pepsin-treated molecule does not affect migration of the type XI chain on SDS-PAGE, indicating that this cysteine is not used for formation of an interchain disulfide bond. The reason for the presence of this amino acid remains unclear.
    The C-propeptide of the α1(XI) chain is similar to those of other fibrillar collagen chains and presents 77% homology with the C-propeptide of the α1(V)

chain.[48] It contains two glycosylation sites, one of which has the sequence (Asn–Phe–Thr) and is a potential acceptor for N-asparaginyl-linked carbohydrate attachment, like other C-propeptides of the fibrillar group. Differences have been detected, however, in the cysteine contents of $\alpha1(I)$, $\alpha1(II)$, $\alpha1(III)$, and $\alpha1(V)$ chains: the C-propeptide of $\alpha1(XI)$ contains only seven cysteinyl residues, while the other C-propeptides contain eight. In pro-$\alpha1(XI)$, the number 2 cysteine is missing; this is due, like in $\alpha2(I)$ and $\alpha2(V)$, to a single change in the nucleotide sequence, whereby the cysteine codon is changed to a serine codon. The number 2 cysteine in pro-α chains of $\alpha1(I)$, $\alpha1(II)$, $\alpha1(III)$, and $\alpha1(V)$ is involved in interchain disulfide bonds, and this residue is lacking in all chains involved exclusively in heterotrimeric molecules.[55]

The $\alpha1(XI)$ N-propeptide is highly divergent from those of other fibrillar molecules except for the $\alpha1(U)$ chain. It appears to be divided into three consecutive elements after the signal peptide: a long globular domain, an interrupted collagenous region, and a short nonhelical segment. The globular region is 383 amino acids long, and the presence of three cysteines in the middle of this zone demarcates a transition from a basic into a highly acidic region. It represents, with that of the pro-$\alpha1(V)$ chain, the longest globular region found in a fibrillar collagen N-propeptide and a unique arrangement in the sequence. Pro-$\alpha1(V)$ N-propeptide contains also a large globular domain with 4 cysteinyl residues and its collagenous sequence contains 25 interrupted Gly–X–Y triplets.[48] The collagenous segment of the pro-$\alpha1(XI)$ N-propeptide is constituted by 24 (Gly–X–Y) repeats with two noncollagenous interruptions. The cross-linking lysine residue is absent in the short nonhelical segment, but a potential N-proteinase cleavage site is found. Another particular feature of pro-$\alpha1(XI)$ is that its N-telopeptide, consisting of 17 amino acids, lacks a cross-linking site.[56] Recently, a protein rich in proline and arginine (called PARP), weighing 24 kDa, was isolated from bovine nasal and articular cartilage. This protein is 49% identical to the NH_2 noncollagenous end of the pro-$\alpha1(XI)$ chain and contains four cysteines involved in two disulfide bridges and conserved with $\alpha1(XI)$. One hypothesis to explain the presence of this PARP is that it is a collagen fragment which is removed during processing in the same manner as chondrocalcin from type II collagen.[57]

The expression of the $\alpha1(XI)$ chain is not restricted to cartilaginous tissue. Yoshioka and Ramirez[56] have shown that pro-$\alpha1(XI)$ mRNAs can occur in human rhabdomyosarcoma cell lines with pro-$\alpha2(V)$ mRNAs. Another example of coexpression of $\alpha1(XI)$ and $\alpha1(V)$ chains is in bovine bone.[58] A preparation of type V collagen from bovine bone was shown to contain a third chain; peptide mapping of the three chains established that this third chain is $\alpha1(XI)$. In adult bone, the type V collagen fraction is richer in $\alpha1(XI)$ than fetal tissue. The ratio between the three chains suggests the formation of cross-type heterotrimers between types V and XI chains. The opposite situation is also observed, in that type XI collagen from articular cartilage contains an $\alpha1(V)$ chain.[52] How these polypeptides are organized into the native molecule is not clear.

The second chain of type XI collagen, $\alpha2(XI)$, also has the characteristics of a fibrillar collagen chain. The gene has been located on the p212 region of chromosome 6.[53] By cloning cDNA and genomic DNA from a human chondrocyte cDNA library, Kimura et al.[53] found clones that encode for 257 uninterrupted Gly–X–Y triplets and about 200 amino acids of the C-telopeptide and C-propeptide. They showed that the Gly–X–Y triplets represent about 80% of the triple-helical domain of the chain. In this region, the pro-α2(XI) sequence is related more closely to that of the pro-α1(XI) chain than to any other known fibrillar collagen chain. One important difference is found in the carboxy-terminal triplets of the triple-helical domain, since in the α2(XI) gene, the last 108-bp exon (exon 48 in other fibrillar collagens) has been replaced by exons of 54, 36, and 54 bp with only 18 bp of the triple-helical domain in the C-terminal joining exon. There is no cysteinyl residue in the middle of the triple-helical domain as is found in the α1(XI) chain.

As in the triple-helical region, the C-propeptide domain of the pro-α2(XI) chain has a sequence similarity that is higher for the pro-α1(XI) chain than for other fibrillar collagens. Some differences that can be noted in the exon structure are that exon 3 (from the 3′ end) is only 113 bp in length, while in other fibrillar collagen genes it has a size range of 186 to 191 bp, and exon 2 is similarly smaller, with 207 bp compared with an invariant 243 bp in other fibrillar collagen genes. It has been suggested that chains containing four "variable" cysteinyl residues in the C-propeptide may be able to form hetero-and/or homotrimeric molecules, whereas pro-α chains, with three "variable" cysteines, can only form heterotrimeric molecules. The known sequence of pro-α2(XI) C-propeptide contains four variable and three invariant residues. The sequence comprising the position of the last cysteinyl residue has not been determined yet. On this basis, Kimura et al.[107] speculated that the pro-α2(XI) chain may form homotrimeric molecules, in addition to the heterotrimeric type XI collagen.

Only one site of glycosylation has the sequence Asn–X–Thr in this C-propeptide and is located after the fourth cysteinyl residue. The other potential site of glycosylation, observed in the pro-α1(XI) chain just before the sixth cysteinyl residue, does not exist. This can be explained by the fact that the highly conserved sequence (Asn–X–Thr) surrounding the N-linked glycosylation sites in other fibrillar collagen chains is replaced in pro-α2(XI) by Asp–Val–Ser. Thus, the pro-α2(XI) chain cannot be glycosylated at this location.[53]

The sequence and gene organization of the N-propeptide of the pro-α2(XI) chain have not yet been described.

The third chain of type XI collagen is an overglycosylated form of the α1(II) chain. For its primary structure and genomic organization, see the section on type II collagen.

b. Structure of the Type XI Heterotrimer

The structural properties of the three chains of type XI collagen indicate that the α1(XI) chain may be associated with other collagen chains and that the α2(XI) and α3(XI) chains can form homotrimeric molecules. The presence of the heterotrimeric form α1(XI) α2(XI) α3(XI) in cartilaginous tissue was confirmed by Morris et al.[59] By introducing formaldehyde-derived cross-links into the native molecules, those authors showed that the three chains form a predominantly heterotrimeric molecule. This molecule is susceptible to trypsin at all temperatures, suggesting the presence of only one kind of triple helix. Extraction without pepsin digestion from chick sternal cartilage revealed that the matrix form is larger than the corresponding pepsin products, thus indicating that all three chains contain non-triple-helical pepsin-sensitive domains, which could be the unprocessed N-telopeptides.[52]

One property of the type XI collagen heterotrimer that is common to type V collagen (both forms: $[\alpha1(V)]_2\alpha2(V)$ and $\alpha1(V)\,\alpha2(V)\,\alpha3(V)$), is the presence of unfolding intermediates during denaturation of the molecules. The thermal denaturation of type II collagen yields a single sharp transition midpoint of $41°C$ which reflects the cooperative nature of the unfolding process; whereas the thermal denaturation of type XI collagen yields two main transitions, at 38.5 and $41.5°C$, and a smaller transition at $40.1°C$. Similarly, type V_x collagen (form $[\alpha1(V)]_2\alpha2(V)$) has two main transitions, at 38.2 and $42.9°C$, and two smaller ones at 40.1 and $41.3°C$; type V_y (form $\alpha1(V)\,\alpha2(V)\,\alpha3(V)$) shows two main transitions at 36.4 and $38.1°C$ and two minor ones at 40.5 and $42.9°C$.[60] The presence of unfolding intermediates could be explained by the existence in the molecule of triple-helical domains of different stabilities.

The intermediates of type XI collagen are composed of two fragments of about 135 nm in length, flanked by regions with lower stability at $38°C$ and start at positions 495 and 519, respectively; at $40°C$, the length is reduced to 113 nm with starting positions at residues 585 and 618. Thus, the more stable domain starts near the middle of the triple helix and extends close to the carboxy-terminal end.

The intermediates of type V_x collagen are composed of a stable, 160-nm long stretch of triple helix starting at position 430 at $38°C$, which gives a fragment 126 nm long at $40°C$.

These different thermal stabilities might be due to differences in the amino acid sequences of type XI and type V collagens compared to other fibrillar collagens. The only appreciable difference is the small number of tripeptide units containing alanine; it has been shown that units like Gly–Ala–Hyp and Gly–Pro–Ala can form stable triple helices and probably contribute to the stability of collagens. The role of these differently stable domains remains unclear: they may be important for the flexibility of the molecule.[60]

Even if the type XI molecule belongs to the fibrillar group of collagens, it forms heterotrimeric molecules with specific structural properties similar to those of type V collagen. The entire structure of type XI is not yet known; the sequence of the $\alpha2(XI)$ N-propeptide has not been elucidated, the presence of the N-propeptides in the matrix form is not proven. The formation of heterotrimeric molecules introduces two important questions. First, with regard to the $\alpha3(XI)$ chain, if type II collagen represents 95% of the collagen content in cartilaginous tissues, while type XI represents only 2 to 3%,[16] how are some pro-$\alpha1(II)$ chains selected preferentially to form type XI collagen? Second, as type XI and V collagen chains form a number of trimeric combinations between themselves, could this high flexibility at the molecular level play a significant role in the diversity of fibers observed *in vivo*?

2. Synthesis and Degradation of Type XI Collagen

a. Synthesis of Type XI Collagen

The synthesis of type XI collagen follows the same process as that of other collagens. Type XI procollagen chains are the products of cultures of chondrocytes and cartilaginous tissues. In contrast to type II collagen, type XI procollagen is preferentially associated with the cell layer rather than with the culture medium in chondrocyte cultures.[61,62] It is retained at the chondrocyte surface and may be involved in the organization of the pericellular matrix.[63] An interesting finding is that the switch in synthesis from type XI to type V collagen occurs in chondrocyte cultures under the same experimental conditions in which type II collagen synthesis is replaced by type I synthesis.[52,61,64,65]

b. Protease Susceptibility

None of the type XI collagen chains is cleaved by human collagenase at 25 or 37°C under conditions in which types I, II, and III collagens are cleaved.[66] In order to explain the finding that $\alpha3(XI)$ is resistant to this collagenase while $\alpha1(II)$ is not, Eyre and Wu[52] proposed that the specific cleavage site in this chain is inaccessible when it is folded in heterotrimeric molecules together with $\alpha1(XI)$ and $\alpha2(XI)$ chains.

Gadher and co-workers[67,68] showed that type XI collagen is degraded by neutrophil elastase and more generally by conditioned culture medium derived from interleukin-1 activated human articular chondrocytes.

Gelatinase (matrix metalloproteinase 2) can cleave type XI collagen of different forms: native type XI in solution, denatured type XI or reconstituted fibers. The activity of this enzyme is much stronger in cartilage from osteoarthritic joints than in that from normal joints. In each $\alpha1(XI)$ and $\alpha2(XI)$ chain from humans and rats, there are about 40 favored cleavage sites, the enzyme prefering Leu–Gly and Ileu–Gly bonds. One possible function of the gelatinase in tissues is to digest denatured fragments of type II collagen. Its ability to digest type XI collagen could be very important for the turnover of the extracellular matrix of cartilage.[69]

The cysteine proteinases cathepin B and L can also degrade type XI collagen. Cathepsin B cleaves a large fragment from $\alpha1(XI)$, while cathepsin L cleaves large fragments from $\alpha1(XI)$ and $\alpha2(XI)$. The action of the cathepsins is the same at 20 and 30°C at a pH of about 5.5. The cleavage may occur preferentially in telopeptide regions.[70]

3. Fibril Formation and Interactions with Other Molecules

Type XI collagen can form fibrils *in vitro:* Smith et al.[50,71] showed that under certain conditions (0.15 M NaCl, 0.008 M phosphate buffer, 4°C), this collagen can form fibrils 20 nm in diameter or thinner. The larger fibrils had a periodicity of 60 to 70 nm, characteristic of fibrillar collagens.

Such fibrillogenesis can be modified by the action of polyanions because native type XI collagen and fibers can interact with these polyanions. These interactions are unique among the fibrillar collagens. Type XI binds to the high-density proteoglycans of cartilage *in vitro,* apparently in a nonspecific manner. This binding is stable in increasing concentrations of NaCl until the collagen fibrils begin to dissolve. The effect of these interactions may be to terminate growth of type XI collagen fibers.

The interactions between proteoglycans and type XI collagen are strongly inhibited by heparin. The fact that the interaction between this collagen and heparin is stronger than that between heparin and other collagens provides the basis for a good method for preparing type XI collagen without type II. Each molecule of type XI contains at least one binding site for polyanions. The binding sites for proteoglycans and for heparin are probably identical since the two molecules compete for binding to fibers.[50,71,72]

4. Conclusion

Type XI collagen, one of the minor cartilaginous collagens, is a heterotrimeric, fibrillar molecule. The structure of each chain is not completely known, although much progress has been made in the last few years. Some points remain unclear, such as the reasons for the presence of the $\alpha 1(XI)$ chain in other tissues. Figure 3 (e and f) shows that type XI is colocated with type II collagen when the collagen network is submitted to pepsin digestion. The major function of type XI collagen, illustrated by its immunolocalization in the fibers of type II could be to participate in the formation of heterotypic fibrils and thus to regulate fibrillogenesis *in vivo.*[11] (B. Petit, manuscript in preparation).

C. TYPE IX COLLAGEN
1. Introduction

Burgeson and Hollister[43] described for the first time the new collagenous chains (1α, 2α, 3α) corresponding to type XI collagen. In the same paper, they reported that early fetal and (to a lesser extent) neonatal cartilage contained small amounts of two additional collagenous materials. Subsequently, several laboratories attempted to isolate these chains from pepsin extracts of hyaline cartilage of different origins. The first report of what was later called type IX collagen was that of Shimokomaki et al.[45,73] who extracted a new disulfide-bonded collagen (designated type M) from pig cartilage. Similar chains were isolated from chicken cartilage (designated HMW or LMW,[44,74] or M1 and M2[75]), from bovine nasal cartilage[76,77] (designated CPS_1 and CPS_2), and from bovine epiphyseal cartilage in our laboratory[78,79] (chains designated X_1 to X_7).

A model for the arrangement of these pepsin-resistant fragments emerged from further biochemical analyses and observations by rotary shadowing of these fragments[79] and of intact molecules isolated from chondrocyte cell culture or cartilage organ cultures;[80-83] in addition, the cDNA clones that encode portions of two of the constituent chains from chicken cartilage were sequenced.[84-86]

In parallel, Noro et al.[87] isolated a proteoglycan (PG-Lt) from chick embryo cartilage which contained disulfide-bonded collagenous polypeptides. The similarity between this proteoglycan and type IX collagen was demonstrated later,[88] as well as the attachment site for the glycosaminoglycan side-chain.[89-91]

Preparation of monoclonal and polyclonal antibodies against pepsin-resistant fragments of type IX collagen made it possible to immunolocalize this collagen at both the light and electron microscopic level in cartilage and corneal tissues (Figure 3).[78,83,92-97]

The association between type IX and type II collagen molecules was demonstrated both by biochemical analysis of the cross-linking region[98-100] and by direct visualization by rotary shadowing of type IX collagen molecules attached to type II collagen fibrils.[97]

The partial intron-exon structure of chicken, rat, and human $\alpha 1(IX)$ and $\alpha 2(IX)$ collagen genes was described,[90,101-106] and tissue-dependent variation in the utilization of transcription start sites in the $\alpha 1(IX)$ chain gene was demonstrated.[43,102,106-109] Thus, over the last 10 years, the molecular structure of the type IX collagen molecule and its organization within cartilage matrices have been elucidated. The precise biological role and the mechanisms that regulate its synthesis and degradation, however, remain virtually unknown. We describe here in more detail the structure and macromolecular organization of the type IX collagen molecule, the structure of the corresponding genes, and available data on its synthesis and degradation.

2. Structure and Macromolecular Organization

Type IX collagen (Figure 6) contains three genetically distinct chains — designated $\alpha 1(IX)$, $\alpha 2(IX)$, and $\alpha 3(IX)$ — which are associated in a 1/1/1 ratio in a molecule with seven distinct domains: three triple-helical domains (COL1, COL2, and COL3) and four noncollagenous domains (NC1, NC2, NC3, and NC4). The biosynthetic form of the type IX collagen molecule (average length, 190 nm) is shorter than the type II procollagen molecule (average length, 335 nm) and appears on sodium dodecylsulfate-polyacrylamide gel electrophoresis (SDS-PAGE) as a single, broad band of apparent M_r 250,000. This apparent heterogeneity of the molecule arises from the chondroitin sulfate chain which is attached to the NC3 domain of the $\alpha 2(IX)$ chain at a serine residue.[90,91] After reduction of the intact molecules, this chain appears as a fuzzy band at M_r 115,000 on gel electrophoresis and is converted into a M_r 68,000 band after digestion by chondroitinase AC and ABC.[80-83] Under the same electrophoretic conditions, the $\alpha 1(IX)$ chain appears as a M_r 84,000 band and the $\alpha 3(IX)$ chain as a M_r 68,000 band. This difference arises from the presence of a large NC4 domain in the $\alpha 1(IX)$ chain (243 amino acid residues) which is virtually absent in the $\alpha 2(IX)$ and $\alpha 3(IX)$ chains.[104] This NC4 domain corresponds to the distinctive knob observed at one end of the molecule with rotary shadowing, and it is used as a marker of the amino-terminal end.[83] A distinct form of the $\alpha 1(IX)$ chain lacking the NC4 domain is dominant in noncartilage tissue (primary

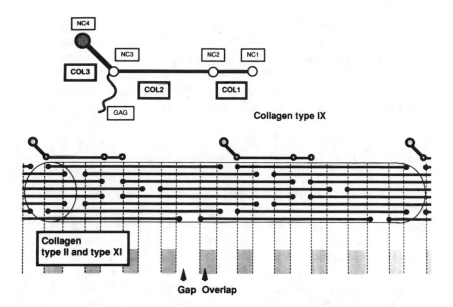

FIGURE 6. Type IX collagen molecule and its association with type II and XI collagen molecules.

corneal stroma and vitreous humor) and is probably also present at low concentration in cartilage. The single chondroitin sulfate chain, present at the NC3 domain of the $\alpha2(IX)$ chain, is approximately 10 times longer in chicken vitreous humor than in cartilage[108] (M_r ~350,000).

The complete amino acid sequence of chicken $\alpha1(IX)$ and $\alpha2(IX)$ chains was derived from the nucleotide sequence of several cDNA clones. Partial sequences of the rat, bovine, and human $\alpha1(IX)$ chain and of the chicken $\alpha3(IX)$ chain are now available. The presence of several cysteine residues was thus demonstrated in the NC1, NC2, and NC3 domains, forming disulfide interchain bridges, and in the NC4 domain of the $\alpha1(IX)$ chain, forming intrachain disulfide bonds. Several imperfections in the Gly–X–Y triplet structure were observed in the COL1 and COL3 domains. It is worth noting[79] that the triple helix in the COL3 domain, which is rich in Gly–Pro–Hypro sequences, shows greater thermal stability (Td = 42.8°C) than the COL2 domain (Td = 35.5°C).

Maturation of type IX collagen involves the formation of pyridinoline cross-links.[98] By isolating and sequencing the pyridinoline-containing peptides from bovine and chicken cartilage type IX collagen, it was shown that type IX and type II collagens are covalently cross-linked.[99,100] For example, one major site of cross-linking is located in COL2 close to NC3 and involves the $\alpha2(IX)$ and the amino-terminal of type II collagen. The existence of this covalent cross-linking explains why only a small fraction of type IX collagen can be solubilized without pepsin digestion of the cartilage. This treatment results in partial cleavage of the noncollagenous portion of the molecule and permits solubilization of several triple-helical fragments of type IX collagen. These fragments are isolated from type II and type XI collagens by precipitation at high concentration of NaCl (2 and 3 M) under acidic conditions (0.5 M acetic acid); the COL2, COL2+NC2+COL1, and COL3 domains precipitate at 2 M NaCl and the COL1 domain at 3 M NaCl (Figure 7). These pepsin-resistant fragments of type IX collagen were shown to be immunogenic (by preparation of polyclonal and monoclonal antibodies in mice and rabbits)[92-97] and arthritogenic in mice.[110]

3. Gene Structure

As suggested previously by its protein structure, the recent characterization of a partial sequence of the $\alpha 1$(IX) and $\alpha 2$(IX) chain genes in different species (human, mouse, and chicken) confirms that type IX collagen is a member of a distinct class within the collagen superfamily (review in Reference 111). The sizes of the exons in these genes are strikingly dependent on their location within the gene: the exons that encode the triple-helical domains are 54 bp or related to 54 bp, but those in the 5′ and 3′ regions differ in size from the exons that code for the propeptides of the fibrillar collagen. The $\alpha 1$(XI) chain gene, in humans located in chromosome 6. In chicken, it is almost 100 Kb and probably contains 38 exons, while the $\alpha 2$(IX) chain gene is only about 10 Kb and contains 32 exons. Except for the exon that encodes the glycosaminoglycan attachment site in $\alpha 2$(IX) and the exons that encodes the NC4 domain of $\alpha 1$(IX), these two genes have homologous exon structures and probably arise from duplication of a common precursor gene. Regulation of the formation of two alternative transcripts from the $\alpha 1$(IX) chain gene will be described later.

4. Synthesis and Degradation

Newly synthesized type IX collagen[80-83] was isolated from organ cultures or cell extracts and media of cultured chondrocytes of different origins, under different culture conditions (monolayer, on or within collagen or agarose gels) in the presence of radioactive amino acids (^{3}H or ^{14}C proline, ^{3}H glycine) or Na_2 $^{35}SO_4$ (for labeling the chondroitin sulfate chain). Radiolabeled type IX collagen was isolated from other collagenous proteins by immunoprecipitation or after pepsin treatment by differential salt precipitation. The results of these studies indicate that the molecule is not processed extracellularly. Furthermore, synthesis of type IX collagen is not affected by treatment of chondrocytes with ß-xyloside, demonstrating that the mechanism by which sugar residues are added to the $\alpha 2$(IX) chain is different from that in other cartilage proteoglycans.[112]

Parallel studies of collagen gene expression and chondrocyte differentiation-dedifferentiation *in vitro* and cartilage development *in vivo* show that the synthesis of type II and type IX collagens is coordinated and closely related to the differentiated cartilage phenotype.[113]

The catabolic processing of type IX collagen *in vivo* is poorly understood. Experiments *in vitro* with native and pepsin-treated type IX collagens have shown that synovial collagenase (MMP1) and gelatinase (MMP2) do not degrade type IX collagen, but that human neutrophil elastase, matrix metalloproteinase B (MMP3 or stromelysin), cathepsin B and L readily degrade it into smaller fragments.[70,114,115] Conditioned culture media derived from

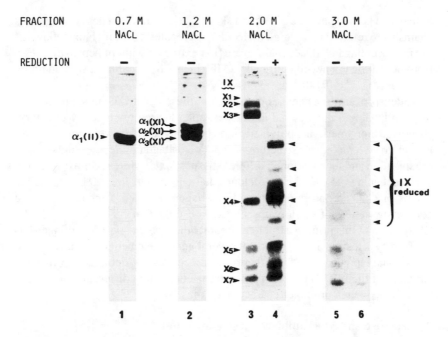

FIGURE 7. SDS-PAGE on 7.5 % (lanes 1 to 4) and 6.75 % (lanes 5 and 6) polyacrylamide gels of the precipitates obtained by salt fractionation of the pepsin digest from fetal calf cartilage. Lane 1; type II collagen; lane 2; type XI collagen; lanes 3 to 6; type IX collagen. (From Ronziere, M. C. et al., *Biochim. Biophys. Acta,* 1038, 222, 1990. With permission.)

interleukin 1α-activated human articular chondrocytes contained cartilage collagen-degrading activities (type II, IX, X, and XI collagens). These chondrocytes have the potential for catabolizing each cartilage collagen species;[116] however, the exact protease susceptibility of the different collagens (II, IX, and XI) that are associated (cross-linked) in the same fibrils *in vivo,* remains unknown.

Furthermore, type IX collagen and its fragments are potent inducers of prostaglandin E_2 and interleukin 1 production by human monocyte and macrophages.[117] This may be important in the destructive process in articular joints during inflammatory diseases. Antibodies to type IX collagen were found in a high proportion of serum samples from rheumatoid arthritis patients.[118]

D. TYPE VI COLLAGEN
1. Introduction

Type VI collagen is a unique component within the family of collagenous proteins, forming small microfilaments that occur ubiquitously in soft connective tissue and cartilage. Composed of three polypeptide chains α1(VI), α2(VI), and α3(VI), its molecule contains a short triple-helical-domain (~105 nm in length) and a large globular domain at each end (see review in References 119 and 120).

Type VI collagen fragments were first isolated by Chung et al.[121] from a pepsin digest of the intimal layer of human blood vessels. Similar pepsin-resistant chains were prepared from other tissues, such as cirrhotic liver,[122] human and bovine placenta,[123,124] calf skin,[125] and uterus.[126] These collagenous chains were referred to as "high molecular aggregates", "short chain collagen," or "intima collagen". Intact chains were first extracted without pepsin digestion from nuchal ligament cell cultures[127] (MFP1) and from cultured skin and lung fibroblasts[128,129] (GP140).

Analysis of the biochemical properties of these isolated chains, preparation of polyclonal and monoclonal antibodies, and electron microscopy of rotary-shadowed monomeric and polymeric molecules together allowed rapid comprehension of the molecular structure of type VI collagen molecules and of their organization and localization in different tissues.[130-141] More recently, isolation and sequencing of several cDNA clones that encode the entire $\alpha 1$, $\alpha 2$, and $\alpha 3$ chains of chicken and human type VI collagen revealed a mosaic structure containing at least five domains already described in other proteins.[142-148] Elucidation of this primary structure supports the idea that type VI collagen has evolved by selective incorporation of several genetic elements from a variety of evolutionary precursors.[146] Furthermore, recent results demonstrate that the structure of the $\alpha 2(VI)$ collagen promoter is completely different from that of any other collagen promoter characterized thus far.[149]

2. Structure and Macromolecular Organization

The $\alpha 1(VI)$ and $\alpha 2(VI)$ collagen genes are both located on chromosome 21, in band q 223; the $\alpha 3(VI)$ collagen gene is located on chromosome 2 in the 2q 37 region. Three other extracellular matrix genes [$\alpha 1(III)$ and $\alpha 2(V)$ collagen chains and fibronectin] have been mapped to the same region (2q) of chromosome 2.[150] The gene domain that encodes for the triple-helical region subunit of chicken type VI collagen is assembled from a variety of short exons that are integral multiples of 9 bp, with a predominant size of 63 bp.[145] This organization is quite different from that of fibrillar collagen genes (primordial building block of 54 bp) and that of basement membrane collagen genes.

The cDNA sequences published recently[142,146,148,149] demonstrate that the $\alpha 1(VI)$, $\alpha 2(VI)$, and $\alpha 3(VI)$ chains of chicken type VI collagen contain 1019, 1015, and 2914 amino acid residues, respectively, with a collagenous domain of 336 amino acid residues. These chains have apparent M_r in SDS-PAGE of about 150,000 ($\alpha 1$), 140,000 ($\alpha 2$), and 260,000 ($\alpha 3$). After digestion of the native molecule with pepsin, the intact subunits of the triple-helical domain give rise to three small fragments of apparent M_r 70,000 ($\alpha 1$), 55,000 ($\alpha 2$), and 40,000 ($\alpha 3$). Multiple Arg–Gly–Asp (RGD) sequences have been demonstrated in these fragments, and their presence explains the cell attachment properties of the triple helix of type VI collagen.[151]

As shown in Figure 8, the N- and C-terminal domains of the type VI collagen chains contain a mosaic structure, with von Willebrand factor, glycoprotein Ib-like, fibronectin type III, and Kunitz modules.[146] For example, the $\alpha 3(VI)$ chain contains 11 repeats similar to the type A repeats of von Willebrand factor, with collagen binding properties. Thus, the presence of specific domains along type

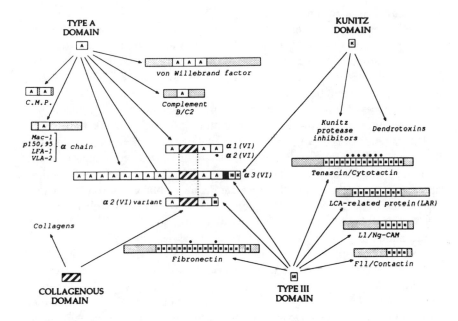

FIGURE 8. Relationship of type VI collagen modules with homologous modules of other proteins. (From Bonaldo, P. and Colombati, A., *J. Biol. Chem.,* 264, 20235, 1989. With permission.)

VI collagen molecules with cell attachment and type-I collagen binding properties indicates a bridging role for this collagen in connective tissues.[148]

Recently, multiple forms of chicken $\alpha3(VI)$ chain were described,[152] generated by alternative splicing in these type A-repeats. This finding probably explains why different forms of the $\alpha3(VI)$ chain (M_r between 230,000 and 180,000) were extracted from bovine,[132] avian and human[153] tissues. Furthermore, alternative splicing of the $\alpha2(VI)$ collagen gene generates multiple mRNA transcripts, which predict three protein variants with distinct carboxyl termini.[154] In accordance with these data, isoforms of type VI collagen were detected using specific monoclonal antibodies in different chicken embryo tissues.[152] Little is known at present about the biological significance of the tissue distribution of the specific type VI collagen isoforms.

Although heterotrimeric molecules undoubtedly predominate in most adult connective tissues, recent data show that alternative type VI collagen assemblies, composed of $\alpha1(VI)$ and $\alpha2(VI)$ chains, occur in some developing tissues, such as skin.[155]

The chain composition of type VI collagen is thus subject to modulation at the level of transcription, as demonstrated by these results and by the observation of uncoordinated regulation of gene expression of $\alpha3(VI)$ chain and of $\alpha1(VI)$ and $\alpha2(VI)$ chains in human skin fibroblasts grown in collagen gels[156] or treated with γ-interferon[157] and in skin and corneal fibroblasts and several tumor cell lines.[158]

3. Tissular Organization

Results obtained using methods that avoid aggregation and insolubilization of the tissue form of type VI collagen show that this type represents a large fraction of connective tissue collagen.[135,153] Its molecules can assemble intracellularly into dimers and tetramers which can form microfilaments by end-to-end association after secretion.[119,120] Lacking the classical lysine-derived cross-linking,[159] type VI collagen was extracted from intact tissue by 6 M guanidine-HCl or by 6 M urea in a disulfide-bonded tetrameric form (M_r ~2,000,000).[153,159]

The presence of type VI collagen in cartilage was first detected by Ayad et al.[160] using indirect immunofluorescence techniques, in a pericellular position around adult bovine nasal cartilage chondrocytes; the location was confirmed by labeling the pericellular capsule of isolated canine tibial chondrons.[161] A low concentration (1 to 2 %) of type VI collagen was recovered and was characterized biochemically from fetal and adult articular cartilage.[162-164] Electron microscopic immunolocalization of type VI collagen was described in human femoral head and costal cartilage: it is distributed as randomly arranged, periodic fibrils in the cartilage matrix, with a higher concentration immediately adjacent to chondrocytes.[165] We observed a greater concentration of type VI collagen in guanidinium chloride extracts in osteoarthritic cartilage than in normal tissue. This finding was corroborated by electron microscopic observations of the same samples[164] (Figure 9); abundant periodic fibrils (100 nm) were observed in the disorganized pericellular capsule of cloned cells in osteoarthritic cartilage. These broad-banded aggregates are formed in tissues with high activity of neutral proteases,[166] which is the case in osteoarthritic cartilage.[167,168] In normal tissues, the pericellular zone was more compact and contained only a few such banded fibrils. These data confirm the results obtained in dogs with surgically induced osteoarthritis.[169]

The molecular structure of type VI collagen composed of multifunctional domains and the ultrastructure and distribution of its fibrillar network indicate that it probably forms a flexible network that anchors connective tissue cells into the surrounding extracellular matrix. We do not yet know, however, the exact chain composition of type VI collagen in cartilage or its temporal or spatial variation during development, aging, and pathological degradation of this tissue.

E. TYPE X COLLAGEN

Type X collagen was described in 1982 by Schmid and Conrad[170,171] as a unique collagenous molecule of 59 kDa per chain in cultures of chondrocytes from developing bones. Gibson et al.[172] and Capasso et al.[173] reported in the same year parallel observations on a molecule they called G collagen and 64K collagen, respectively. A mammalian homologue isolated from primary cultures of rabbit growth plate was reported in 1983 by Remington et al.[174]

It was soon recognized that this molecule has a very limited and transient pattern of expression *in vivo*. It has been immunolocalized exclusively in

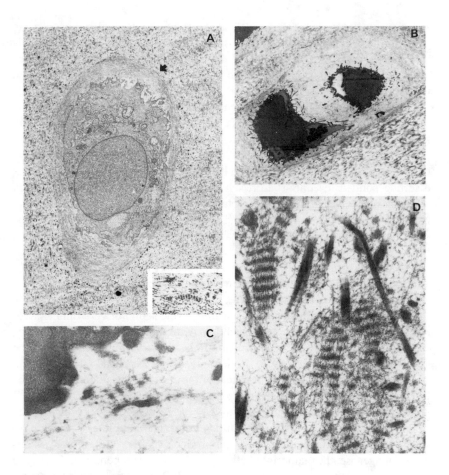

FIGURE 9. (A) Normal cartilage, middle and deep layers. Chondrocyte surrounded by a fibrous pericellular capsule. Banded aggregates are sometimes visible (arrow). (× 5400) The insert is an enlargement of a banded aggregate. (× 12,000) (B) A chondrocyte clone of the upper layers of osteoarthitic cartilage. The pericellular capsule appears less structured than in normal cartilage . (× 3300) (C) Fibrous aggregate near a chondrocyte plasma membrane of osteoarthritic cartilage. (× 45,000) (D) Large, cross-striated fibrous aggregates in osteoarthritic cartilage. (× 45,000). (From Ronziere, M. C. et al., *Biochim. Biophys. Acta,* 1038, 222, 1990. With permission.)

cartilaginous tissues undergoing endochondral ossification, in the vicinity of hypertrophic chondrocytes of the growth plate, in the cephalic region of the sternum of the chick embryo, and in fracture callus.[175-182] It is also a component of the matrix of the notochord, but it appears late compared to type II collagen and only at sites where bone will form.[183] Type X collagen is synthesized only by hypertrophic chondrocytes.[184-189] These observations have raised a lot of interest in this molecule and led to suggestions that type X collagen might be involved in the calcification process. While type X collagen is expressed simultaneously with other markers of calcification such as alkaline phosphatase in several systems,[190] studies aimed at evaluating the potential role of this molecule in initial phases of mineral deposition did not support this hypothesis.[191] In fact, the exact function and the mode of aggregation of type X collagen in tissues are not yet fully understood.

At the protein level, type X collagen is detected as a homotrimeric molecule of three 59-kDa polypeptide chains. Partial degradation products of 49 and 45 kDa were observed in tissue extracts.[192] There is, however, no evidence of a proteolytic processing of the molecule.[180,187] Upon pepsin treatment, the size of the polypeptide chain is reduced to 45 kDa. The bovine pepsinized molecule is disulfide bonded in contrast to its chicken counterpart.[187] Type X collagen molecules have been visualized by rotary shadowing electron microscopy and have a length of 132 to 148 nm.[186,193] A globular domain is observed at one extremity, which is now known to be the C-terminal end. The purified molecule has a very high denaturation temperature (47°C), in contrast with fibrillar collagens that denature just above physiological temperature.[192] The denaturation temperatures inside the cell and in the matrix appear to be still much higher — up to 55°C and 65 to 67.5°C, respectively.[194]

The molecule contains two sites sensitive to the action of vertebrate collagenase.[195] These two sites, located near the two extremities of the triple helix, have recently been identified as corresponding to two imperfections of the Gly–Xaa–Yaa–Xaa–Yaa type.[196] Both cleavages occur at the Gly–Xaa bond and, in a situation quite analogous to that of the cleavage sites of collagen types I, II, and III, the Xaa residues are hydrophobic, being Leu[93] at one site and Ile[421] at the other. The molecule is also sensitive to the 72-kDa type IV collagenase at or near the same two sites only. It is, however, completely resistant to stromelysin.[196] Contrary to the cleavage products of other collagens by vertebrate collagenase, the main triple-helical 32-kDa fragment generated from type X collagen has a denaturation temperature of 43°C, indicating that it would not spontaneously unfold at physiological temperature.[195] The consequences of this observation in terms of degradation of type X collagen remain to be established.

Type X collagen cDNAs have been isolated and characterized from chicken[197,198] and bovine libraries.[199] The deduced amino acid sequence of the chicken type X collagen reveals a primary translation product of 674 residues. The signal peptide is 18 amino acids long. The central triple-helical domain is 460 residues long and is bordered by N- and C-terminal non-triple-helical domains of 34 and 162 residues, respectively. The triple helix is characterized by the presence of eight imperfections in the Gly–Xaa–Yaa repetitive sequence. Three imperfections are of the Gly–Xaa–Yaa–Xaa–Yaa type and five of the Gly–Xaa–Gly–Xaa–Yaa type. The bovine sequence is highly homologous to its chicken counterpart (73% amino acid sequence identity). It shows the presence of two cysteines in the triple helix which are absent from the chicken molecule and its triple helix is shorter by three residues.

The type X collagen gene strikingly differs from other collagen genes by the fact that most of the protein (the C-terminal domain, the triple-helical domain and the extremity of the N-terminal domain) is encoded by a single exon, 2137 nucleotides long.[197] Two other exons, coding for the 5′ untranslated portion of the RNA, the signal peptide and most of the N-terminal domain, complete the structure of the coding region of the gene.[200] This structure is remarkable in view of the highly interrupted exon structure characteristic of other collagen genes.

Very recently, however, the cloning and the characterization of cDNAs encoding the α1 and the α2 chains of type VIII collagen has revealed that this collagen is highly homologous to type X collagen, thereby forming a distinct subgroup in the collagen family of proteins.[201,202] The homology is also noted in the exon structures of these genes.[203] This subgroup has been referred to as the short chain collagens. The sequence similarity between rabbit α1(VIII) and chicken α1(X) is about 60%, both at the nucleotide and at the amino acid level.

Since type VIII collagen has been shown to participate in the formation of the hexagonal lattice of Descemet's membrane, it can be hypothesized that type X collagen could participate in similar aggregates. However, the reported immunoelectron micrographs do not show association of type X collagen labeling with clear hexagonal lattices, but rather with thin fibrils and with some poorly organized mat structures.[191,204] Very recently, the ability of type X collagen to organize itself *in vitro* into an hexagonal lattice closely homologous to the one observed in Descemet's membrane was reported by Kwan et al.[205] One can therefore hypothesize that type X collagen initially aggregates to form a sheet with an hexagonal lattice and that the mats and fibrils observed by electron microscopy do actually represent partially degraded structures. One attractive possibility would be that type X collagen would form a temporary scaffold around hypertrophic chondrocytes and stabilize this highly metabolically active extracellular matrix. It has been suggested by Schmid et al.[192] that type X collagen could serve also to target matrices that will undergo profound remodeling. The intriguing ability of type X collagen molecules to make long-range moves through the cartilage matrix and to become fibril associated, as recently demonstrated by Chen et al.,[206] suggests that this molecule may also play a role by modifying the preexisting fibrils.

The expression of type X collagen, which is regulated at the transcriptionnal level,[207] is now considered a marker of the differentiation of chondrocytes into hypertrophic chondrocytes (sometimes called terminally differentiated chondrocytes). This further chondrocyte differentiation can actually be achieved *in vitro* even for cells collected from tissues in which chondrocytes would not normally hypertrophy, such as the caudal portion of the chick sternum.[208] This process of terminal chondrocyte differentiation, as detected by the expression of type X collagen, is exquisitely sensitive to the culture conditions, such as cell density, the presence of a collagen gel, the addition of ascorbate or of calcium ß-glycerophosphate or of levamisole, the presence or absence of serum.[172,173,190,209-211] Recently, Adams et al.[212,213] have demonstrated that the switch of chick vertebral chondrocytes from monolayer culture where they have a flattened shape to suspension culture where they take a round shape is accompanied by a rapid 10-fold increase in type X collagen protein synthesis. This increase correlates with an increased transcription rate. This suggests that part of the transcriptional control of type X collagen synthesis may actually depend on the organization of the cytoskeleton.

III. CONTROL OF COLLAGEN GENE EXPRESSION

The 14 currently recognized collagen types[20,111] have a characteristic tissue distribution and play different functional roles in embryogenesis and morphogenesis. Their expression is also dependent on a variety of exogenous factors, including inflammatory response mediators, growth factors, and oncogene products. Most collagen genes thus appear to be specifically regulated, and a given tissue or cell type expresses only a specific subset of collagen genes. Control of the expression of the different collagen genes is thus complex, varying with the degree of development and with the tissue being analyzed.

Changes in collagen gene transcription represent a principal component in changes in gene expression; the stability of collagen mRNAs and their level of translation into proteins may also affect net collagen production. Many mechanisms are probably involved in the transcriptional and posttranscriptional control of collagen genes. They contain sequence elements in common with other genes and specific elements for controlled gene expression. Generally, the region responsible for the rate of transcription is situated at the 5' end of the gene upstream from the transcription start site. Promoters recognized by RNA polymerase II require DNA regulatory elements for modulation of transcription by different processes. Sequences in collagen DNA contain the necessary *cis*-acting elements, which, coupled with the defined *trans*-acting protein factors, determine efficient transcription of collagen genes.

In order to study regulatory elements in DNA, chimeric genes consisting of 5' end sequences of the gene and a portion of its first intron fused to a bacterial marker gene, such as chloramphenicol acetyl transferase (CAT) gene,[214] were introduced into cultured cells by DNA transfection or into animals at the stage of the fertilization of oocytes. The functions of these DNA elements can be further examined by generating site-specific mutations. Furthermore, studies based on nucleic acid hybridization techniques using the corresponding DNA probes give the levels of different collagen mRNAs. Nuclear run-on experiments can be performed to distinguish between an increase in mRNA stability and an increase in transcription rate in specific collagen genes.

A review of the factors involved in regulation of collagen gene expression was published recently.[215] Here, we review briefly the various controls of collagen gene expression, emphasizing type II and type I collagens as markers of the differentiated and dedifferentiated states of chondrocytes. We also report on the effects of exogenous factors in the regulation of collagen genes during development and pathological metabolism of the cartilage.

A. REGULATORY SEQUENCES IN COLLAGEN GENES

Reports on the molecular level of control of collagen gene expression have been limited mainly to the first four collagen types. Control and regulatory elements of human and mouse type I collagens have been reviewed recently.[111,214,216]

1. Promoter Regulatory Elements

Promoter sequences have been reported for six collagen genes: $\alpha1(I)$, $\alpha2(I)$, $\alpha1(II)$, $\alpha1(III)$, and $\alpha1(IV)$, $\alpha2(IV)$, with substantial conservation of regulatory sequences between mammalian species. The first four genes contain a TATA-like sequence located 20 to 30 bases 5' from the start of transcription.

A CCAAT motif (binding site 1:BS1) is present at approximately −120 on the $\alpha1(I)$ gene and at −80 on the $\alpha2(I)$ gene in mice and at −77 on the $\alpha2(I)$ gene in chicken.[111] Single mutations introduced at this site decrease the binding of CCAAT binding factor and the transcriptional activity of the collagen promoter, suggesting an activating role of transcription for the binding factor. Two other regulatory elements lie further upstream and act as binding sites for transcriptional inhibitors in the promoter, as indicated by point mutations in these sites.[214] Three additional binding sites have been identified at −250 (BS2), −300 (BS3), and −400 (BS4) for the $\alpha2(I)$ gene. BS3 is a binding site for a protein called nuclear factor 1.[214] Point mutations at BS2 and BS4 suggest that the cognate factors act as transcriptional activators.[214] Since $\alpha2(I)$ and $\alpha1(I)$ collagen genes contain common regulatory elements, their coordinated expression may be regulated by these factors.

In $\alpha1(II)$ gene,[217] the 690 bp of the 5' flanking sequence of the human promoter shows four highly conserved regions. One region, extending 400 bp upstream from the start site, includes a viral core enhancer, four copies of the hexanucleotide 5'GGCGG3', and one copy of its reverse complement. A sequence similar to the CAAT box, but reversed, is located at −100 in both human and rat $\alpha1(II)$ genes. In human and rat type II collagen promoters, the TATA box and three other sequences from nucleotides −501 to −649 are conserved. These three regions may correspond to negative regulatory regions since, in transfection assays[218] with a construct containing 1800 bp of the 5' flanking sequence of the rat type II promoter, no promoter activity was measured. In addition, this gene region contains two silencers which regulate the expression of this gene[219] in various cell types, but not in chondrocytes.

2. Intronic Elements

The substantial degree of identity that exists between the intron of the $\alpha1(I)$ collagen gene in humans and rats indicates the presence of regulatory sequences. In fact, both positively and negatively acting elements have been identified in this gene region and in other collagen genes.

Insertion of the murine Moloney leukemia virus (MOV 13) into $\alpha1(I)$ collagen modifies gene expression; and insertion of the retrovirus into the first intron of COL1A1 inactivates transcription of the mouse gene.[220] Interestingly, no transcription of COL1A1 was observed following insertion of MOV 13 in any cell except odontoblasts,[221] suggesting tissue-specific regulation for this collagen type.

In humans, inhibition of transcription was described in one 274-bp intron in α1(I) collagen gene with an orientation-dependent effect; in addition, enhancer activity was identified in 782 bp.[222]

Different enhancers are present in the first intron of collagen genes. One transcriptional enhancer found in mouse COL1A2 exerts its effect in conjunction with the promoter and directs tissue-specific expression of COL1A1; another enhancer was identified in rat COL2A1[218] which acts as a cartilage-specific enhancer. This enhancer contains[217] three sites that can bind Sp1 (specific factor), a serum-responsive element (SRE), a recognition sequence for nuclear factor 1, and a viral core enhancer sequence, as illustrated in Figure 10. The presence of an alternating purine/pyrimidine sequence at +1133 to +1162 only in the human α1(II) gene is proposed to play a role in the expression of the type II collagen gene.

3. DNA Methylation

One of the mechanisms responsible for the inhibition of collagen gene expression has been reported to be DNA methylation in COL1A1 and COL1A2.[223] Bacterial methylation of the pro-α1(I) promoter/enhancer construct significantly reduced its transcriptional activity. A role for DNA methylation in the alteration of gene expression is suggested[224] by the finding that, in MOV 13-infected mice, the absence of COL1A1 expression is associated with *de novo* methylation within 1 kb 5′ from the insertion site. The same pattern of hypomethylation of COL2A1[225] in embryonic chondrocytes was present, however, whether or not the cells expressed type II collagen (differentiated vs. dedifferentiated cells), indicating that collagen gene methylation may be complex.

4. Regulatory Elements in Other Gene Regions

Many 3′ untranslated regions of collagen genes contain multiple polyA signals; their differential use leads to considerable length variation in the size of mRNAs. Studies of the 3′ noncoding regions of COL1A1 show the presence of regulatory elements within this region of the gene.[226] The two highly conserved sequences (pA1 and pA2) bind cell-specific nuclear proteins, suggesting that these DNA regions are necessary for the cell-specific gene expression.

B. DEVELOPMENTAL AND TISSUE-SPECIFIC CONTROL OF COLLAGEN GENES

Developmental expression of collagen genes is controlled in time-, tissue-, and cell type-specific manners. Examples of this mode of ontological regulation are provided by collagen type I during chondrogenesis and chondrocyte dedifferentiation and redifferentiation,[21,113,227-230] by the presence of collagen type II mainly in chondrocytes, by the presence of collagen type IX in both cartilage and in cornea, and by the presence of collagen type X only in hypertrophic chondrocytes.

The α2(I) collagen chain, although present in prechondrogenic mesenchymal cells, is not found in most cartilage. The mechanisms responsible for this

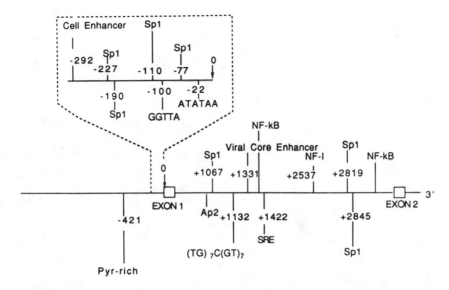

FIGURE 10. Relative locations of the putative transcriptional control sequence motifs in the promoter region and the first intron of the human αl(II) collagen gene. Consensus sequences for binding motifs are defined as follows: Sp1, specific factor 1; SRE, serum responsive element; NF-1, nuclear factor 1; NF-kB, nuclear factor kB. Distances are calculated from the start site of transcription. (From Ninomya, Y., *Extracellular Matrix Genes,* Sandell, L. J. and Boyd, C. D., Eds. Academic Press, New York, 1990. With permission.)

phenomenon may include incomplete processing of the α1(I) chain and a severe reduction in the translation elongation rate of the α2(I) chain.[231] A shorter form of α2(I) collagen mRNA is found in chondrocytes when compared with fibroblasts.[227] It was demonstrated recently that the nonexpression of COL1A2 in chondrocytes corresponded to the use of (1) another start site of transcription of the gene within intron 2 and (2) splicing of exon A directly to exon 3,[229] as illustrated in Figure 11. The resulting transcripts no longer encode for α2(I) collagen because they contain at least four open reading frames out of frame with the collagen coding sequence. Among the peptides subsequently coded for, a 71-amino acid peptide exhibiting the characteristics of a nucleic acid-binding protein is probably synthesized by the chondrocytes. It is interesting to note that chondrocytes transcribe a poorly translatable form of mRNA for α1(I) collagen when cultured in suspension and a translatable form with a different 5′ end when cultured as a monolayer,[232] indicating that the composition of the extracellular matrix is involved in gene expression. The cell shape is not, however, linked to the expression of type I and II collagen during cultures of chicken chondrocytes, and this appears to be regulated primarily by mechanisms that control mRNA levels.[233]

The tissue-specific expression of type II collagen follows alternative exon usage, as reported earlier in this chapter. Developmental expression of type II collagen during chondrogenesis shows a significant accumulation of mRNA during the cell condensation stage as a result of an increase in the rate of gene

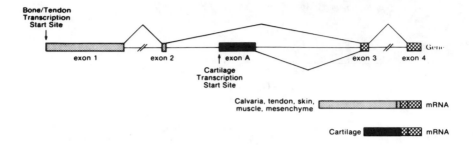

FIGURE 11. Structure of a portion of the chick α2(I) collagen gene and of the two resulting transcripts. The bone/tendon and cartilage transcription start sites are indicated. *Lightly shaded boxes* represent the exons specific for the mRNAs that initiate transcription at the bone/tendon start site; the *solid box* represents the exon specific for the mRNAs that initiate transcription at the cartilage start site; and the *hatched boxes* represent exons common to all mRNAs. (From Bennett, V. D., *J. Biol. Chem.*, 265, 1990. With permission.)

transcription.[234] While the cells are synthesizing the cartilage-specific matrix, the mRNA level increases steadily corresponding to a stabilization of this mRNA.[234] It is interesting that these regulatory mechanisms appear at the cytoplasmic level and are related only to type II collagen.

Whereas α2(IX) collagen mRNA appears to be of the same size in cornea and cartilage, α1(IX) collagen expression depends on an alternate transcription start site. Tissue-specific promoters direct the synthesis of the two forms of the protein;[235] one promoter directs transcription in cartilage and the second directs transcription in primary stroma of corneal epithelium,[101] as illustrated in Figure 12.[111] The two size classes of the mRNAs in the two tissues are due, in part, to the utilization of two clusters of polyadenylated signals at the 3′ end of the gene.[235]

The final stage of chondrocyte differentiation (obtained from cartilage zone undergoing mineralization) corresponds to a hypertrophic cell. Expression of type X collagen is a phenotypic characteristic of these chondrocytes. Cell-specific expression of type X collagen in chick embryo chondrocytes was recently reported to be determined at the translational level.[207]

C. EXOGENOUS FACTORS THAT REGULATE COLLAGEN GENE EXPRESSION

Many exogenous factors[222,236] participate in the regulation of the expression of collagen genes during chondrogenesis and during pathological events. Their effects on the collagens found in cartilage are summarized in Table 3.

Transforming growth factor ß (TGF-ß), which has been studied extensively, increases type I collagen synthesis and steady-state mRNA levels in mouse fibroblasts.[214] For type I collagen, this increase is mainly due to stimulation of transcription of the corresponding genes. A small DNA segment in the promoter of COL1A2 was found to mediate the effects of the growth factor[237] and to overlap with the nuclear factor 1 binding site. However, since TGF-ß does not alter the

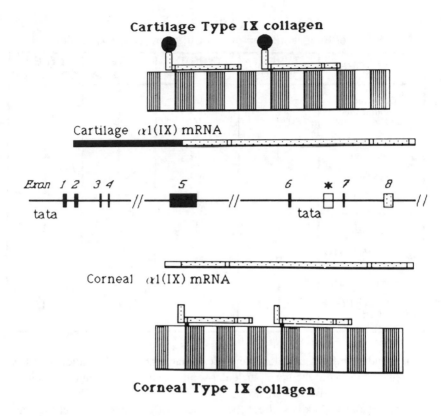

FIGURE 12. Exons of the 5′ portion of the chicken α1(IX) collagen gene (center). Exons are numbered starting at the 5′ end. Exons 1 to 7 (solid black bars) encode the N-terminal non-triple-helical domain NC 4 of the cartilage α1(IX) chain (at the top of the figure), while exon 8 (stippled) contains the coding information of the last amino acid residue of NC4 and the N-terminal portion of the triple-helical domain COL3. The intron between exons 6 and 7 contains a TATA box and an alternative exon (open box, labeled with star), which encodes the 5′ portion of the corneal form of α1(IX) mRNA (at the bottom of the figure). This alternative exon is spliced to the sequence of exon 8 in the corneal mRNA. The assembly of type IX with type II collagen is represented in the cartilage (top of the figure) and in the cornea (bottom of the figure). (From Ninomya, Y., *Extracellular Matrix Genes,* Sandell, L. J. and Boyd, C. D., Eds., Academic Press, New York, 1990. With permission.)

binding of nuclear factor 1,[214] it has been postulated that interaction of the growth factor with its cellular receptor triggers a signaling pathway that leads to activation of nuclear factor 1 and subsequently to a higher rate of COL1A2 transcription. Moreover, TGF-ß may regulate the expression of collagen genes differentially, depending on the growth state of the cells studied. An increase in the stability of mRNA was seen in confluent fibroblasts but not in subconfluent cells.[222] TGF-ß was also found to cause a fivefold increase in promoter activity of COL1A1 in DNA transfection experiments,[214] but no responsive regulatory sequence has been yet identified. In chicken and rat chondrocytes, however, TGF-ß decreased type II collagen synthesis and the steady-state level of the

TABLE 3
Effect of Exogenous Factors on Collagen Gene Expression

FACTORS	TYPE II COLLAGEN	TYPE IX COLLAGEN	TYPE X COLLAGEN	TYPE I COLLAGEN
Chondrocyte Differentiated Dedifferentiated	+ 0	+ 0	+* ND	0 +
Vitamin A	↓	ND	↑	↑
Vitamin C	↑	ND	↑	↑
Vitamin D3	↓	↑	↑	↑
Interleukin 1	↓	↓	ND	↑
Tumor Necrosis Factor	↑↓	ND	ND	↑
Gamma-Interferon	↓	ND	ND	↓
Transforming Growth Factor	↓	ND	ND	↑

* Present only in hypetrophic cells

corresponding mRNA in a dose-dependent manner.[238] Further studies will be needed to determine whether a particular sequence in the promoter or enhancer of COL2 is responsive to TGF-ß.

Inflammatory cytokines affect the synthesis of collagen genes at the transcriptional level. The inflammatory factor interleukin 1 (IL-1ß) stimulates the synthesis of collagen type I and III in synovial fibroblasts and human chondrocytes, only in the presence of indomethacin which blocks production of prostaglandin E2 (an inhibitor of collagen synthesis).[239] In contrast, in chondrocytes, IL-1ß suppresses synthesis of type II collagen and reduces the levels of types II, IX, and XI collagen steady-state mRNAs.[240-243] This reduction in collagen synthesis involves DNA regulatory sequences which control COL2A1 expression,[241] but it remains to be elucidated whether the promoter or enhancer or both are affected by IL-1.

Unlike IL-1, tumor necrosis factor alpha (TNF-α) suppresses COL1 expression in fibroblasts and osteoblasts;[244] however, transient increases in both protein and mRNA level indicate that the inhibitory effect may be mediated at the posttranslational level. Like IL-1, TNF-α suppresses expression of type II collagen mRNA in human chondrocytes.[239]

γ-Interferon, an immunoregulatory agent, has a more drastic effect on collagens, decreasing their synthesis in fibroblasts, myofibroblasts, and articular and costal chondrocytes.[245] This suppression is associated with inhibition of transcription for types I, III, and II, with a greater decrease of $\alpha1(I)$ and $\alpha1(II)$ mRNAs than of $\alpha2(I)$ mRNA in human articular chondrocytes, whereas the three mRNAs decreased to the same extent in costal chondrocytes, suggesting a specific tissue regulation.[246]

Vitamin A analogs such as retinoic acid are known to alter the state of differentiation of many cell types, including chondrocytes.[247,248] Several mechanisms have been postulated to account for these modulations of cellular phenotype, and alteration of gene expression has been recently proposed.[249] The expression of a retinoic acid nuclear receptor specific of cartilage during limb development indicates the precise mechanism of action of this molecule. After exposure to retinoic acid, the chondrocytes stopped synthesizing type II collagen as a consequence of inhibition of enhancer activity;[250] the synthesis and the mRNA level of the pro-α1 chain of collagen III were, however, increased. The presence of retinoic acid[251] also induces increased expression of type X collagen by chick vertebral chondrocytes after 48 h of treatment, indicating that this molecule has an important role in chondrocyte phenotypic expression.

1,25-Dihydroxyvitamin D_3, another agent that affects cellular differentiation, has been recently reported to be active in cartilage development by modifying collagen gene expression.[252] During hormone treatment, chicken chondrocytes show decreased synthesis of type II collagen and of its mRNA, whereas both α1 and α2 type I mRNA levels are increased with no detectable amount of type I collagen, indicating translational control of protein synthesis. Over the same time course, an early, transient increase in type IX collagen mRNA and a later increase in type X collagen mRNA levels were also recorded, the former being the result of a direct hormonal effect and the latter more of a secondary hormonal effect. The changes observed in fact are related to a secondary hormonal response, since the nonadherent 1,25-dihydroxyvitamin D_3-treated cells displayed different responses.

Ascorbate (ascorbic acid), which is known to affect the posttranslational processing of procollagen as a cofactor of prolyl and lysyl hydroxylases, also affects types I, II, and X collagen mRNA levels along chondrocyte cultures. In chick tendon fibroblasts, it increases the levels of α2(I) procollagen mRNA[253] as a consequence of an increased transcriptional rate associated with a decrease in the degradation rate of mRNAs. When ascorbate was added to chick embryo chondrocytes, no modification of the level of type II procollagen mRNA and no increase in α1(I) or α2(I) mRNA levels were observed in a 5-d culture. Conversely, in adult bovine joint chondrocytes cultured under the same conditions, it increased the levels of α1(I) and α1(II) collagen mRNAs,[254] pointing to a species difference in the effect of ascorbate on chondrocytes. This compound also appears to be a potent enhancer of type X collagen expression in chick chondrocyte cultures,[190,255] confirming the role of this vitamin during chondrocyte phenotypic changes.

D. CONCLUSIONS

Although the synthesis of type II, IX, and X collagens in differentiating chondrocytes appears to be coordinated temporally, a few common sequences have been identified in the promoters of these collagen genes. Common transcriptional factors or structural elements within the genes must now be sought which might be utilized during differentiation.

Both transcriptional and posttranscriptional controls are operative during chondrocyte differentiation/dedifferentiation, which are associated with interactions between the cell cytoskeleton and the extracellular matrix proteins. Therefore, study of these interactions would be of additional help in understanding the expression of chondrocyte collagen genes.

ACKNOWLEDGMENTS

The personal results presented were supported by grants from the CNAMTS (83-86), MRT (85C0684), and INSERM (880004).

REFERENCES

1. **van der Rest, M. and Garrone, R.,** The collagen family of proteins, *FASEB J.,* 5, 2814, 1991.
2. **Ramachandran, G. N. and Ramakrishnan, C.,** Molecular structure, in *Biochemistry of Collagen,* Ramachandran, G. N. and Reddi, A. H., Eds., Plenum Press, New York, 1976, 45.
3. **Berg, R. A. and Prockop, D. J.,** The thermal transition of a nonhydroxylated form of collagen. Evidence for a role for hydroxyproline in stabilizing the triple helix of collagen, *Biochem. Biophys. Res. Commun.,* 52, 115, 1973.
4. **Prockop, D. J., Kivirikko, K. I., Tuderman, L., and Guzman, N. A.,** The biosynthesis of collagen and its disorders. (First of two parts), *N. Engl. J. Med.,* 301, 13, 1979.
5. **Prockop, D. J., Kivirikko, K. I., Tuderman, L., and Guzman, N. A.,** The biosynthesis of collagen and its disorders. (Second of two parts), *N. Engl. J. Med.,* 301, 77, 1979.
6. **van der Rest, M. and Garrone, R.,** Collagens as multidomain proteins, *Biochimie,* 72, 473, 1990.
7. **Timpl, R., Wiedemann, H., van Delden, V., Furthmayr, H., and Kühn, K.,** A network model for the organisation of type IV collagen molecules in basement membranes, *Eur. J. Biochem.,* 120, 203, 1981.
8. **Hodge, A. J. and Petruska, J. A.,** Recent studies with the electron microscope on the ordered aggregates of the tropocollagen molecule, in *Aspects of Protein Structure,* Ramachandran, G. N., Ed., Academic Press, New York, 1963, 289.
9. **Keene, D. R., Sakai, L. Y., Bächinger, H. P., and Burgeson, R. E.,** Type III collagen can be present on banded collagen fibrils regardless of fibril diameter, *J. Cell Biol.,* 105, 2393, 1987.
10. **Birk, D. E., Fitch, J. M., Barbiaz, J. P., and Linsenmayer, T. F.,** Collagen type I and type V are present in the same fibril in the avian corneal stroma, *J. Cell Biol.,* 106, 999, 1988.
11. **Mendler, M., Eich-Bender, S. G., Vaughan, L., Winterhalter, K. H., and Bruckner, P.,** Cartilage contains mixed fibrils of collagen types II, IX and XI, *J. Cell Biol.,* 108, 191, 1989.
12. **Gordon, M. K. and Olsen, B. R.,** The contribution of collagenous proteins to tissue-specific matrix assemblies, *Curr. Opinion Cell Biol.,* 2, 833, 1990.
13. **Muir, H.,** Macromolecular interactions and connective tissue metabolism, *Biochem. Soc. Trans.,* 5, 397, 1977.
14. **Keene, D. R., Lunstrum, G. P., Morris, N. P., Stoddard, D. W., and Burgeson, R. E.,** Two type XII-like collagens localize to the surface of banded collagen fibrils, *J. Cell Biol.,* 113, 971, 1991.

15. **Miller, E. J., Epstein, E. H., and Piez, K. A.,** Identification of three genetically distinct collagens by cyanogen bromide cleavage of insoluble human skin and cartilage collagen, *Biochem. Biophys. Res. Commun.,* 42, 1024, 1971.
16. **Eyre, D. R., Wu, J.-J., and Apone, S.,** A growing family of collagens in articular cartilage: identification of five genetically distinct types, *J. Rheum.,* 14, 25, 1987.
17. **Kuhn, K.,** The classical collagens: types I, II and III, in *Structure and Function of Collagen Types,* Mayne R. and Burgeson R. E., Eds., Academic Press, New York, 1987, 1.
18. **Eyre, D. R., Dickson, I. R., and Van Ness, K.,** Collagen cross-linking in human bone and articular cartilage, *Biochem. J.,* 252, 495, 1988.
19. **Sangiorgi, F. O., Benson-Chanda, V., de Wet, W. J., Sobel, M. E., and Ramirez, F.,** Analysis of cDNA and genomic clones coding for the pro α(I) chain of calf type II collagen, *Nucleic Acid Res.,* 13, 2815, 1985.
20. **Vuorio, E. and de Crombrugghe, B.,** The family of collagen genes, *Ann. Rev. Biochem.,* 59, 837, 1990.
21. **Ramirez, F., Boast, S., D'Alessio, M., Lee, B., Prince, J., Su, M.-W., Vissing, H., and Yoshioka, H.,** Fibrillar collagen genes: structure and expression in normal and diseased states, *Ann. N.Y. Acad. Sci.,* 580, 74, 1990.
22. **Upholt, W. B. and Sandell, L. J.,** Exon/intron organization of the chicken type II procollagen gene: intron size distribution suggests a minimal intron size, *Proc. Natl. Acad. Sci. U.S.A.,* 2325, 1986.
23. **Sangiorgi, F. O., Benson-Chanda, V., de Wet, W. J., Sobel, M. E., Tsipouras, P., and Ramirez, F.,** Isolation and partial characterization of the entire human proα(II) collagen gene, *Nucleic Acid Res.,* 13, 2207, 1985.
24. **Cheah, K. S. E., Stoker, N. G., Griffin, J. R., Grosveld, F. G., and Solomon, E.,** Identification and characterization of the human type II collagen gene (COL2A1), *Proc. Natl. Acad. Sci. U.S.A.,* 82, 2555, 1985.
25. **Baldwin, C. T., Reginato, A. M., Smith, C., Jimenez, S. A., and Prockop, D. J.,** Structure of cDNA clones coding for human type II procollagen. The α1(II) chain is more similar to the α1(I) chain than two other α chains of fibrillar collagen, *Biochem. J.,* 262, 521, 1989.
26. **Ala-Kokko, L. and Prockop, D. J.,** Completion of the intron-exon structure of the gene for human type II procollagen (COL2A1): variations in the nucleotide sequences of the alleles from three chromosomes, *Genomics,* 8, 454, 1990.
27. **Eyre, D. R., Weis, M. A., and Moskowitz, R. W.,** Cartilage expression of a type II collagen mutation in an inherited form of osteoarthritis associated with a mild chondrodysplasia, *J. Clin. Invest.,* 87, 357, 1991.
28. **Butler, W. B., Finch, J. E., and Miller, E. J.,** Covalent structure of cartilage collagen. Amino acid sequence of residues 363-551 of bovine α1(II) chains, *Biochemistry,* 16, 4981, 1977.
29. **Curran S. and Prockop D. J.,** Isolation and partial characterization of the carboxy-terminal propeptide of type II procollagen from chick embryo, *Biochemistry,* 23, 741, 1984.
30. **Ninomiya, Y., Showalter, A. M., van der Rest, M., Seidah, N. G., Chretien, M., and Olsen, B. R.,** Structure of the carboxyl propeptide of chicken type II procollagen determined by DNA and protein sequence analysis, *Biochemistry,* 23, 617, 1984.
31. **Sandell, L. J., Ryan, M. C., Morris, N. P., Robbins, J., and Goldring, M.,** A new look at type II collagen: alternative splicing generates two procollagens with distinct patterns of expression, *Matrix,* 10, 216, 1990.
32. **Sandell, L. J., Yamada, Y., Dorfman, A., and Upholt, W. B.,** Identification of genomic DNA coding for chicken type II procollagen, *J. Biol. Chem.,* 258, 11617, 1983.
33. **Sandell, L. J., Prentice, H. L., Kravis, D., and Upholt, W. B.,** Structure and sequence of the chicken type II procollagen gene, *J. Biol. Chem.,* 259, 7826, 1984.
34. **Elima, K., Vuorio, T., and Vuorio, E.,** Determination of the single polyadenylation site of the human pro-α1(II) collagen gene, *Nucleic Acid Res.,* 15, 9499, 1987.
35. **Poole, A. R., Pidoux, I., Reiner, A., Choi, H., and Rosenberg, L. C.,** Association of an extracellular protein (chondrocalcin) with the calcification of cartilage in endochondral bone formation, *J. Cell Biol.,* 98, 54, 1984.

36. **van der Rest, M., Rosenberg, L., Olsen, B. R., and Poole, A. R.,** Chondrocalcin is identical with the C-propeptide of type II procollagen, *Biochem. J.,* 237, 923, 1986.
37. **Poole, A. R. and Rosenberg, L.,** Chondrocalcin and the calcification of cartilage, *Clin. Orthop.,* 208, 114, 1986.
38. **Niyibizi, C., Wu, J.-J., and Eyre, D. R.,** The carboxy-propeptide trimer of type II collagen is a prominent component of immature cartilages and intervertebral-disc tissue, *Biochem. Biophys. Acta,* 916, 493, 1987.
39. **Ruggiero, F., Pfäffle, M., von der Mark, K., and Garrone, R.,** Retention of carboxypropeptides in type II collagen fibrils in chick embryo chondrocyte cultures, *Cell Tissue Res.,* 252, 619, 1988.
40. **Su, M. W., Benson-Chanda, V., Vissing, H., and Ramirez, F.,** Organization of the exons coding for pro α1(II) collagen N-propeptide confirms a distinct evolutionary history of this domain of the fibrillar collagen genes, *Genomics,* 4, 438, 1989.
41. **Ryan, M. C. and Sandell, L. J.,** Differential expression of a cysteine-rich domain in the amino-terminal propeptide of type II (cartilage) procollagen by alternative splicing on mRNA, *J. Biol. Chem.,* 265, 10334, 1990.
42. **Takahashi, E., Hori, T., Lawrence, J. B., Mac Neil, J., Singer, R. H., O'Connell, P., Leppert, M., and White, R.,** Human type II collagen gene (COL2A1) assigned to chromosome 12q13. 1-q13. 2 by *in situ* hybridization with biotinylated DNA probe, *Jpn. J. Human Genet.,* 34, 307, 1989.
43. **Burgeson, R. E. and Hollister, D. W.,** Collagen heterogeneity in human cartilage: identification of several new collagen chains, *Biochem. Biophys. Res. Commun.,* 87, 1124, 1979.
44. **Reese, C. A. and Mayne, R.,** Minor collagens of chicken hyaline cartilage, *Biochemistry,* 20, 5443, 1981.
45. **Shimokomaki, M., Duance, V. C., and Bailey, A. J.,** Identification of a new disulfide bonded collagen from cartilage, *FEBS Lett.,* 121, 51, 1980.
46. **Fessler, J. H. and Fessler, L. I.,** Type V collagen, in *Structure and Function of Collagen Types,* Mayne R. and Burgeson R. E., Eds., Academic Press, New York, 1987, 81.
47. **Furuto, D. K., Gay, R. E., Stewart, T. E., Miller, E. J., and Gay, S.,** Immunolocalization of types V and XI collagen in cartilage using monoclonal antibodies, *Matrix,* 11, 144, 1991.
48. **Takahara, K., Sato, Y., Okazawa, K., Okamoto, N., Noda, A., Yaoi, Y., and Kato, I.,** Complete primary structure of human collagen α1(V) chain, *J. Biol. Chem.,* 266, 13124, 1991.
49. **Yaoi, Y., Hashimoto, K., Koitabashi, H., Takahara, K., Ito, M., and Kato, I.,** Primary structure of the heparin-binding site of type V collagen, *Biochim. Biophys. Acta,* 1035, 139, 1990.
50. **Smith, G. N., Williams, J. M., and Brandt, K. D.,** Effect of polyanions on fibrillogenesis by type XI collagen, *Collagen Relat. Res.,* 7, 17, 1987.
51. **Burgeson, R. E., Hebda, P. A., Morris, N. P., and Hollister, D. W.,** Human cartilage collagens: comparison of cartilage collagens with human type V collagen, *J. Biol. Chem.,* 257, 7852, 1982.
52. **Eyre, D. R. and Wu, J.-J.,** Type XI or 1α 2α 3α collagen, in *Structure and Function of Collagen Types,* Mayne R. and Burgeson R. E., Eds., Academic Press, New York, 1987, 261.
53. **Kimura, T., Cheah, K. S. E., Chan, S. D. H., Lui, V. C. H., Mattei, M. G., van der Rest, M., Ono, K., Solomon, E., Ninomiya, Y., and Olsen, B. R.,** The human α2(XI) collagen (COL11A2) chain: molecular cloning of cDNA and genomic DNA reveals characteristics of a fibrillar collagen with differences in genomic organization, *J. Biol. Chem.,* 264, 13910, 1989.
54. **Henry, I., Bernheim, A., Bernard, M., van der Rest, M., Kimura, T., Jeanpierre, C., Barichard, F., Berger, R., Olsen, B. R., and Ramirez, F.,** Mapping of the fibrillar collagen gene, pro-α1(XI) (COL11A1), to the p21 region of chromosome 1, *Genomics,* 3, 87, 1988.
55. **Bernard, M., Yoshioka, H., Rodriguez, E., van der Rest, M., Kimura, T., Ninomiya, Y., Olsen, B. R., and Ramirez, F.,** Cloning and sequencing of pro-α1(XI) collagen cDNA demonstrates that type XI collagen belongs to the fibrillar class of collagens and reveals that the expression of the gene is not restricted to cartilagenous tissue, *J. Biol. Chem.,* 263, 17159, 1988.

56. **Yoshioka, H. and Ramirez, F.,** Pro-α1(XI) collagen: structure of the amino-terminal propeptide and expression of the gene in tumor cell lines, *J. Biol. Chem.,* 265, 6423, 1990.

57. **Neame, P. J., Young, C. N., and Treep, J. T.,** Isolation and primary structure of PARP, a 24-kDa proline- and arginine-rich protein from bovine cartilage closely related to the NH$_2$-terminal domain in collagen α1(XI), *J. Biol. Chem.,* 265, 20401, 1990.

58. **Niyibizi, C. and Eyre, D. R.,** Identification of the cartilage α1(XI) chain in type V collagen from bovine bone, *FEBS Lett.,* 242, 314, 1989.

59. **Morris, N. P. and Bächinger, H. P.,** Type XI collagen is a heterotrimer with the composition (1α, 2α, 3α) retaining non-triple-helical domains, *J. Biol. Chem.,* 262, 11345, 1987.

60. **Morris, N. P., Watt, S. L., Davis, J. M., and Bächinger, H. P.,** Unfolding intermediates in the triple helix to coil transition of bovine type XI collagen and human type V collagens (α1)2α2 and α1, α2, α3, *J. Biol. Chem.,* 265, 10081, 1990.

61. **Mayne, R., Reese, C. A., Williams, C. C., and Mayne, P. M.,** New collagens as marker proteins for the cartilage phenotype, in *Limb Development and Regeneration,* Kelley, C. R. O., Goetinck, P. F., and Mac Cabe, J. A., Eds., Alan R. Liss, New York, 1983, 125.

62. **Clark, C. C. and Richards, C. F.,** Isolation and partial characterization of precursors to minor cartilage collagens, *Collagen Relat. Res.,* 5, 205, 1985.

63. **Smith, G. N., Hasty, K. A., and Brandt, K. D.,** Type XI collagen is associated with the chondrocyte surface in suspension culture, *Matrix,* 9, 186, 1989.

64. **Mayne, R., Elrod, B. W., Mayne, P. M., Snaderson, R. D., and Linsenmayer, T. F.,** Changes in the synthesis of minor cartilage collagens after growth of chick chondrocytes in 5-bromo-2′-deoxyuridine or to senescence, *Exp. Cell Res.,* 151, 171, 1984.

65. **Yasui, N., Benya, P. D., and Nimni M. E.,** Coordinate regulation of type IX and type II collagen synthesis during growth of chick chondrocytes in retinoïc acid or 5-bromo-2′deoxyuridine, *J. Biol. Chem.,* 261, 7997, 1986.

66. **Eyre, D. R., Wu, J.-J., and Woolley, D. E.,** All three chains of 1α 2α 3α collagen from hyaline cartilage resist human collagenase, *Biochem. Biophys. Res. Commun.,* 118, 724, 1984.

67. **Gadher, J. J., Eyre, D. R., Duance, V. C., Wotton, S. F., Heck, L. W., Schmid, T. M., and Woolley, D. E.,** Susceptibility of cartilage collagens type II, IX, X and XI to human synovial collagenase and neutrophil elastase, *FEBS Lett.,* 175, 1, 1988.

68. **Gadher, J. J., Eyre, D. R., Wotton, S. F., Schmid, T. M., and Woolley, D. E.,** Degradation of cartilage collagens type II, IX, X and XI by enzymes derived from human articular chondrocytes, *Matrix,* 10, 154, 1990.

69. **Smith, G. N., Jr., Hasty, K. A., Yu, L. P., Jr., Lamberson, K. S., Mickler, E. A., and Brandt, K. D.,** Cleavage of type XI collagen fibers by gelatinase and by extracts of osteoarthritic canine cartilage, *Matrix,* 11, 36, 1991.

70. **Maciewicz, R. A., Wotton, S. F., Etherington, D. J., and Duance, V. C.,** Susceptibility of the cartilage collagens type II, IX, and XI to degradation by the cysteine proteinases, cathepsins B and L, *FEBS Lett.,* 269, 189, 1990.

71. **Smith, G. N., Jr., Williams, J. M., and Brandt, K. D.,** Interaction of proteoglycans with the pericellular (1alpha, 2alpha, 3alpha) collagens of cartilage, *J. Biol. Chem.,* 260, 10761, 1985.

72. **Smith, G. N., Jr. and Brandt, K. D.,** Interaction of cartilage collagens with heparin, *Collagen Relat. Res.,* 7, 315, 1987.

73. **Shimokomaki, M., Duance, V. C., and Bailey, A. J.,** Identification of two further collagenous fractions from articular cartilage, *Biosci. Rep.,* 1, 561, 1981.

74. **Reese, C. A., Wiedemann, H., Kühn, K., and Mayne, R.,** Characterization of a highly soluble collagenous molecule isolated from chicken hyaline cartilage, *Biochemistry,* 21, 826, 1982.

75. **von der Mark, K., van Menxel, M., and Wiedemann, H.,** Isolation and charcterization of new collagens from chick cartilage, *Eur. J. Biochem.,* 124, 57, 1982.

76. **Ayad, S., Abedin, M. Z., Grundy, S. M., and Weiss, J. B.,** Isolation and characterization of an unusual collagen from hyaline cartilage and intervertebral disc, *FEBS Lett.,* 123, 195, 1981.

77. **Ayad, S., Abedin, M. Z., Weiss, J. B., and Grundy, S. M.,** Characterization of another short-chain disulfide-bonded collagen from cartilage, vitreous and intervertebral disc, *FEBS Lett.,* 139, 300, 1982.

78. **Ricard-Blum, S., Hartmann, D. J., Herbage, D., Payen-Meyran, C., and Ville, G.,** Biochemical properties and immunolocalization of minor collagens in foetal calf cartilage, *FEBS Lett.,* 146, 343, 1982.

79. **Ricard-Blum, S., Tiollier, J., Garrone, R., and Herbage, D.,** Further biochemical and physicochemical characterization of minor disulfide-bonded (Type IX) collagen, extracted from foetal calf cartilage, *J. Cell. Biochem.,* 27, 347, 1985.

80. **Bruckner, P., Mayne, R., and Tuderman, L.,** p-HMW-collagen, a minor collagen obtained from chick embryo cartilage without proteolytic treatment of the tissue, *Eur. J. Biochem.,* 136, 333, 1983.

81. **Duance, V. C., Wotton, S. F., Voyle, C. A., and Bailey, A. J.,** Isolation and characterization of the precursor of type M collagen, *Biochem. J.,* 221, 885, 1984.

82. **von der Mark, K., van Menxel, M., and Wiedemann, H.,** Isolation and characterization of a precursor form embryonic chicken cartilage, *Eur. J. Biochem.,* 138, 629, 1984.

83. **Irwin, M. H., Silvers, S. H., and Mayne, R.,** Monoclonal antibody against chicken type IX collagen: preparation, characterization and recognition of the intact form of type IX collagen secreted by chondrocytes, *J. Cell Biol.,* 101, 814, 1985.

84. **Ninomiya, Y. and Olsen, B. R.,** Synthesis and characterization of cDNA encoding a cartilage-specific short collagen, *Proc. Natl. Acad. Sci. U.S.A.,* 81, 3014, 1984.

85. **Ninomiya, Y., van der Rest, M., Mayne, R., Lozano, G., and Olsen, B. R.,** Construction and characterization of cDNA encoding the $\alpha2$ chain of chicken type IX collagen, *Biochemistry,* 24, 4223, 1985.

86. **van der Rest, M., Mayne, R., Ninomiya, Y., Seidah, N. G., Chretien, M., and Olsen, B. R.,** The structure of type IX collagen, *J. Biol. Chem.,* 260, 220, 1985.

87. **Noro, A., Kimata, K., Oike, Y., Shinomura, T., Maeda, N., Yano, S., Takahashi, N., and Susuki, S.,** Isolation and characterization of a third proteoglycan (PG-Lt) from chick embryo cartilage which contains disulfide-bonded collagenous polypeptide, *J. Biol. Chem.,* 258, 9323, 1983.

88. **Vaughan, L., Winterhalter, K. M., and Bruckner, P.,** Proteoglycan Lt from chicken embryo sternum identified as type IX collagen, *J. Biol. Chem.,* 260, 4758, 1985.

89. **Konomi, H., Seyer, J. M., Ninomiya, Y., and Olsen, B. R.,** Peptide-specific antibodies identify the $\alpha2$ chain as the proteoglycan subunit of type IX collagen, *J. Biol. Chem.,* 261, 6742, 1986.

90. **McCormick, D., van der Rest, M., Goodship, J., Lozano, G., Ninomiya, Y., and Olsen, B. R.,** Structure of the glycosaminoglycan domain in the type IX collagen-proteoglycan, *Proc. Natl. Acad. Sci. U.S.A.,* 84, 4044, 1987.

91. **Huber, S., Winterhalter, K. H., and Vaughan, L.,** Isolation and sequence analysis of the glycosaminoglycan attachment site of type IX collagen, *J. Biol. Chem.,* 263, 752, 1988.

92. **Duance, V. C., Shimokomaki, M., and Bailey, A. J.,** Immunofluorescence localization of type-M collagen in articular cartilage, *Biosci. Rep.,* 2, 223, 1982.

93. **Evans, H. B., Ayad, S., Abedin, M. Z., Hopkins, S., Morgan, K., Walton, W., Weiss, J. B., and Holt, P. J. L.,** Localisation of collagen types and fibronectin in cartilage by immunofluorescence, *Ann. Rheum. Dis.,* 42, 575, 1983.

94. **Hartmann, D. J., Magloire, H., Ricard-Blum, S., Joffre, A., Couble, M., Ville, G., and Herbage, D.,** Light and electron immunoperoxidase localization of minor disulfide-bonded collagens in fetal calf epiphyseal cartilage, *Collagen Relat. Res.,* 3, 349, 1983.

95. **Poole, C. A., Wotton, S. F., and Duance, V. C.,** Localization of type IX collagen in chondrons isolated from porcine articular cartilage and rat chondrosarcoma, *Histochem. J.,* 20, 567, 1988.

96. **Vaughan, L., Mendler, H., Huber, S., Bruckner, P., Winterhalter, K. H., Irwin, M. I., and Mayne, R.,** D-periodic distribution of collagen type IX along cartilage fibrils, *J. Cell Biol.,* 106, 991, 1988.

97. **Vilamitjana, J., Barge, A., Julliard, A. K., Herbage, D., Baltz, T., Garrone, R., and Harmand, M. F.,** Problem in the immunolocalization of type IX collagen in fetal calf cartilage using a monoclonal antibody, *Connect. Tiss. Res.,* 18, 277, 1989.

98. **Wu, J. J. and Eyre, D. R.,** Cartilage type IX collagen is cross-linked by hydroxypyridinium residues, *Biochem. Biophys. Res. Commun.,* 123, 1033, 1984.

99. **Eyre, D. R., Apon, S., Wu, J. J., Ericsson, L. H., and Walsh, K. A.,** Collagen type IX: evidence for covalent linkages to type II collagen in cartilage, *FEBS Lett.,* 220, 337, 1987.

100. **van der Rest, M. and Mayne, R.,** Type IX collagen proteoglycan from cartilage is covalently cross-linked to type II collagen, *J. Biol. Chem.,* 263, 1615, 1988.

101. **Nishimura, I., Muragaki, Y., and Olsen, B. R.,** Tissue-specific forms of type IX collagen-proteoglycan arise from the use of widely separated promoters, *J. Biol. Chem.,* 264, 20033, 1989.

102. **Muragaki, Y., Nishimura, I., Henney, A., Ninomiya, Y., and Olsen, B. R.,** The α1(IX) collagen gene gives rise to two different transcripts in both mouse embryonic and human fetal RNA, *Proc. Natl. Acad. Sci. U.S.A.,* 87, 2400, 1990.

103. **Lozano, G., Ninomiya, Y., Thompson, H., and Olsen, B. R.,** A distinct class of vertebrate collagen genes encode chicken type IX collagen polypeptides, *Proc. Natl. Acad. Sci. U.S.A.,* 82, 4050, 1985.

104. **Vasios, G., Nishimura, I., Konomi, H., van der Rest, M., Ninomiya, Y., and Olsen, B. R.,** Cartilage type IX collagen-proteoglycan contains a large amino-terminal globular domain encoded by multiple exons, *J. Biol. Chem.,* 263, 2324, 1988.

105. **Kimura, T., Mattei, M. G., Stevens, J. W., Goldring, M. B., Ninomiya, Y., and Olsen, B. R.,** Molecular cloning of rat and human type IX collagen cDNA and localization of the α1(IX) gene on the human chromosome 6, *Eur. J. Biochem.,* 179, 71, 1989.

106. **Muragaki, Y., Kimura, T., Ninomiya, Y., and Olsen, B. R.,** The complete primary structure of two distinct forms of human α1(IX) collagen chains, *Eur. J. Biochem.,* 192, 703, 1990.

107. **Svoboda, K. K., Nishimura, I., Sugrue, S. P., Ninomiya, Y., and Olsen, B. R.,** Embryonic chicken cornea and cartilage synthesize type IX collagen molecules with different amino-terminal domains, *Proc. Natl. Acad. Sci. U.S.A.,* 85, 7496, 1988.

108. **Yada, T., Suzuki, I., Kobayashi, K., Kobayashi, M., Hoshino, T., Horie, K., and Kimata, K.,** Occurrence in chick embryo vitreous humor of type IX collagen proteoglycan with an extraodinarily large chondroitin sulfate chain and short α1 polypeptide, *J. Biol. Chem.,* 265, 6992, 1990.

109. **Brewton, R., Wright, D. W., and Mayne, R.,** Structural and functional comparaison of type IX collagen-proteoglycan from chicken cartilage and vitreous humor, *J. Biol. Chem.,* 266, 4752, 1991.

110. **Boissier, M. C., Chiocchia, G., Ronziere, M. C, Herbage, D., and Fournier, C.,** Arthritogenicity of a minor cartilage collagens (type IX and XI) in mice, *Arthritis Rheum.,* 33, 1, 1990.

111. **Sandell, L. J. and Boyd, C. D.,** *Extracellular Matrix Genes,* Academic Press, San Diego, 1990, Chap. 1.

112. **Chandrasekhar, S. and Harvey, A. K.,** Synthesis of type IX collagen: effect of ß-xylosides, *Biochem. Biophys. Res. Commun.,* 146, 1040, 1987.

113. **Castagnola, P., Dozin, B., Moro, G., and Cancedda, R.,** Changes in the expression of collagen genes show two stages in chondrocyte differentiation in vitro, *J. Cell Biol.,* 106, 461, 1988.

114. **Gadher, S. J., Eyre, D. R., Duance, V. C., Wotton, S. F., Heck, L. W., Schmid, T. M., and Wooley, D. E.,** Susceptibility of cartilage collagens type II, IX, X, and XI to human synovial collagenase and neutrophil elastase, *Eur. J. Biochem.,* 175, 1, 1988.

115. **Okada, Y., Konomi, H., Yada, T., Kimata, K., and Nagase, H.,** Degradation of type IX collagen by matrix metalloproteinase 3 (stromelysin) from human rheumatoid synovial cells, *FEBS Lett.,* 244, 473, 1989.

116. **Gadher, S. J., Eyre, D., Wotton, S. F., Schmid, T. M., and Woolley, D. E.,** Degradation of cartilage collagens type II, IX, X, and XI by enzymes derived from human articular chondrocytes, *Matrix,* 10, 154, 1990.

117. **Dayer, J. M., Ricard-Blum, S., Kaufmann, M. T., and Herbage, D.,** Type IX collagen is a potent inducer of PGE2 and interleukin 1 production by human monocytes macrophages, *FEBS Lett.,* 198, 208, 1986.

118. **Charriere, G., Hartmann, D. J., Vignon, E., Ronziere, M. C., Herbage, D., and Ville, G.,** Antibodies to types I, II, IX and XI collagen in the serum of patients with rheumatic diseases, *Arthritis Rheum.,* 31, 325, 1988.
119. **Rauterberg, J., Jander, R., and Troyer, D.,** Type VI collagen. A structural glycoprotein with a collagenous domain, *Front. Matrix Biol.,* 11, 90, 1986.
120. **Timpl, R. and Engel, J.,** Type VI collagen, in *Structure and Function of Collagen Types,* Mayne, R. and Burgeson, R. E., Eds., Academic Press, Orlando, FL, 1987, 105.
121. **Chung, E., Rhodes, R. K., and Miller, E. J,** Isolation of three collagenous components of probable basement membrane origin from several tissues, *Biochem. Biophys. Res. Commun.,* 71, 1167, 1976.
122. **Rojkind, M., Giambrone, M. A., and Biempica, L.,** Collagen types in normal and cirrhotic liver, *Gastroenterology,* 76, 710, 1979.
123. **Furuto, D. K. and Miller, E. J.,** Isolation of a unique collagen fraction from limited pepsin digests of human placental tissue, *J. Biol. Chem.,* 255, 290, 1980.
124. **Jander, R., Rauterberg, J., Voss, B., and von Bassewitz, D. B.,** A cysteine-rich collagenous protein from bovine placenta, *Eur. J. Biochem.,* 114, 17, 1981.
125. **Laurain, G., Delvincourt, T., and Szynamowicz, A. G.,** Isolation of a macromolecular collagenous fraction and AB2 collagen from calf skin, *FEBS Lett.,* 120, 44, 1980.
126. **Abedin, M. Z., Ayad, S., and Weiss, J. B.,** Isolation and native characterisation of cysteine-rich collagens from bovine placental tissues and uterus and their relationship to types IV and V collagens, *Biosci. Rep.,* 2, 493, 1982.
127. **Sear, C. H. J., Kewley, M. A., Jones, C. J., Grant, M. E., and Jackson, D. S.,** The identification of glycoprotein associated with elastic-tissue microfibrils, *J. Biol. Chem.,* 170, 715, 1978.
128. **Carter, W. G. and Hakomori, S.,** A new cell surface detergent-insoluble glycoprotein matrix of human and hamster fibroblast, *J. Biol. Chem.,* 256, 6953, 1981.
129. **Carter, W. G.,** The cooperative role for the transformation-sensitive glycoproteins, GP140 and fibronectin in cell attachement and spreading, *J. Biol. Chem.,* 257, 3249, 1982.
130. **Gibson, M. A. and Cleary, E. G.,** Distribution of CL glycoprotein in tissues: an immuno-histochemical study, *Coll. Rel. Res.,* 3, 469, 1983.
131. **Heller-Harrison, R. A. and Carter, W. G.,** Pepsin-generated type VI collagen is a degradation product of GP140, *J. Biol. Chem.,* 259, 6858, 1984.
132. **Jander, R., Troyer, D., and Rauterberg, J.,** A collagen-like glycoprotein of the extracellular matrix is the undegraded form of type VI collagen, *Biochemistry,* 23, 3675, 1984.
133. **Hessle, H. and Engvall, E.,** Type VI collagen studies on its localisation, structure and biosynthetic form with monoclonal antibodies, *J. Biol. Chem.,* 259, 3955, 1984.
134. **Ayad, S., Chambers, C. A., Shuttleworth, C. A., and Grant, M. E.,** Isolation from bovine elastic tissues of collagen type VI and characterization of its form in vivo, *Biochem. J.,* 230, 465, 1985.
135. **Trüeb, B. and Winterhalter, K. H.,** Type VI collagen is composed of a 200Kd subunit and two 140Kd subunits, *EMBO J.,* 5, 2815, 1986.
136. **Engvall, E., Hessle, H., and Klier, G.,** Molecular assembly, secretion, and matrix deposition of type VI collagen, *J. Cell Biol.,* 102, 703, 1986.
137. **Bruns, R. R., Press, W., Engvall, E., Timpl, R., and Gross, J.,** Type VI collagen in extracellular 100-nm periodic filaments and fibrils: identification by immunoelectron microscopy, *J. Cell Biol.,* 103, 393, 1986.
138. **Linsenmayer, F., Mentzer, A., Irwin, M. H., Waldrep, N. K., and Mayne, R.,** Avian type VI collagen monoclonal antibody production and immunohistochemical identification as a major connective tissue component of cornea and skeletal muscle, *Exp. Cell. Res.,* 165, 518, 1986.

139. **Colombatti, A., Bonaldo, P., Ainger, K., Bressan, G., and Volpin, D.,** Biosynthesis of chick type VI collagen I intracellular assembly and molecular structure, *J. Biol. Chem.,* 262, 14454, 1987.

140. **Colombatti, A. and Bonaldo, P.,** Biosynthesis of chick type VI collagen II processing and secretion in fibroplast and smootle muscle cells, *J. Biol. Chem.,* 262, 14461, 1987.

141. **Trüeb, B., Schneier, T., Bruckner, P., and Winterhalter, K. H.,** Type VI collagen represent a major fraction of connective tissue collagens, *Eur. J. Biochem.,* 166, 699, 1987.

142. **Bonaldo, P., Russo, V., Bucciotti, F., Bressan, G. M., and Colombatti, A.,** α1 chains of chick type VI collagen, *J. Biol. Chem.,* 264, 5575, 1989.

143. **Koller, E., Winterhalter, K. H., and Trüeb, B.,** The globular domains of type VI collagen are related to the collagen-binding domains of cartilage matrix protein and von Willebrand factor, *EMBO J.,* 8, 1073, 1989.

144. **Chu, M. L., Pan, T., Conway, D., Kuo, H. J., Glanville, R., Timpl, R., Mann, K., and Deutzmann, R.,** Sequence analysis of α1(VI) and α2(VI) chains of human type VI collagen reveals internal triplication of globular domains similar to the A domains of von Willebrand factor and two α2(VI) chain variants that differ in the carboxy terminus, *EMBO J.,* 8, 1939, 1989.

145. **Hayman, A. R., Köppel, J., and Trüeb, B.,** Complete structure of the chicken α2(VI) collagen gene, *Eur. J. Biochem.,* 197, 177, 1991.

146. **Bonaldo, P. and Colombatti, A.,** The carboxyl terminus of the chicken α3 chain of collagen VI is a unique mosaic structure with glycoprotein Ib-like fibronectin type III, and Kunitz modules, *J. Biol. Chem.,* 264, 20235, 1989.

147. **Chu, M. L., Zhang, R. Z., Pan, T., Stokes, D., Conway, D., Kuo, H.-J., Glanville, R., Mayer, U., Mann, K., Deutzmann, R., and Timpl, R.,** Mosaic structure of globular domains in the human type VI collagen α3 chain: similarity to von Willebrand factor, fibronectin, actin, salivary proteins and aprotinin type protease inhibitors, *EMBO J.,* 9, 385, 1990.

148. **Bonaldo, P., Russo, V., Bucciotti, F., Dolinia, R., and Colombatti, A.,** Structural and functional features of the α3 chain indicate a bridging role for chicken collagen VI in connective tissues, *Biochemistry,* 29, 1245, 1990.

149. **Koller, E., Haymann, A. R., and Trueb, B.,** The promoter of the chicken α2(VI) collagen gene has features characteristic of house-keeping genes and of proto-oncogenes, *Nucleic. Acids Res.,* 19, 485, 1991.

150. **Weil, D., Mattei, M. G., Passage, E., Van Cong, N., Pribula-Conway, D., Mann, K., Deutzmann, R., Timpl, R., and Chu, M. L.,** Cloning and chromosomal localization of human genes encoding the three chains of type VI collagen, *Am. J. Hum. Genet.,* 42, 435, 1988.

151. **Aumailley, M., Mann, K., von der Mark, H., and Timpl, R.,** Cell attachment properties of collagen type VI and Arg-Glys-Asp dependent binding to its α2 (VI) and α3 (VI) chains, *Exp. Cell Res.,* 181, 463, 1989.

152. **Doliana, R., Bonaldo, P., and Colombatti, A.,** Multiple forms of chicken α3(VI) collagen chain generated by alternative splicing in type A repeated domains, *J. Cell Biol.,* 111, 2197, 1990.

153. **Colombatti, A., Ainger, K., and Colizzi, F.,** Type VI collagen: high yields of a molecule with multiple forms of α3 chain from avian and human tissue, *Matrix,* 9, 177, 1989.

154. **Saitta, B., Stokes, D., Vissing, H., Timpl, R., and Chu, M. L.,** Alternative splicing of the human α2(VI) collagen gene generates multiple mRNA transcripts which predict three protein variants with distinct carboxyl termini, *J. Biol. Chem.,* 265, 6473, 1990.

155. **Kielty, C. M., Boot-Handford, R. P., Ayad, S., Shuttleworth, C. A., and Grant, M. E.,** Molecular composition of type VI collagen. Evidence for chain heterogeneity in mammalian tissues and cultured cells, *Biochem. J.,* 272, 787, 1990.

156. **Hatamochi, A., Aumailley, M., Mauch, C., Chu, M. L., Timpl, R., and Krieg, T.,** Regulation of collagen VI expression in fibroblasts, *J. Biol. Chem.,* 264, 3494, 1989.

157. **Heckmann, M., Aumailley, M., Hatamochi, A., Chu, M. L., Timpl, R., and Krieg, T.,** Down-regulation of α3(VI) chain expression by L-interferon decreases synthesis and deposition of collagen type VI, *Eur. J. Biochem.,* 182, 719, 1989.

158. **Chu, M. L., Mann, K., Deutzmann, R., Pribula-Conway, D., Hsu-Chen, C. C., Bernard, M. P., and Timpl, R.,** Characterization of three constituent chains of collagen type VI by peptide sequences and cDNA clones, *Eur. J. Biochem.,* 168, 309, 1987.

159. **Wu, J. J., Eyre, D. R., and Slayter, H. S.,** Type VI collagen of the intervertebral disc. Biochemical and electron-microscopic characterization of the native protein, *Biochem. J.,* 248, 373, 1987.

160. **Ayad, S., Evans, H., Weiss, J. B., and Holt, L.,** Type VI collagen but not type V collagen is present in cartilage, *Coll. Rel. Res.,* 4, 165, 1984.

161. **Poole, C. A., Ayad, S., and Schofield, J. R.,** Chondrons from articular cartilage. I. Immunolocalization of type VI collagen in the pericellular capsule of isolated canine tibial chondrons, *J. Cell Sci.,* 90, 635, 1988.

162. **Eyre, D. R., Wu, J. J., and Apone, S.,** A growing family of collagens in articular cartilage: identification of 5 genetically distinct types, *J. Rheumatol.,* 514, 25, 1987.

163. **Ayad, S., Marriot, A., Morgan, K., and Grant, M. E.,** Bovine cartilage type VI and IX collagens. Characterization of their forms in vivo, *Biochem. J.,* 262, 753, 1989.

164. **Ronziere, M. C., Ricard-Blum, S., Tiollier, J., Hartmann, D. J., Garrone, R., and Herbage, D.,** Comparative analysis of collagens solubilized from human foetal, and normal and osteoarthritic adult articular cartilage, with emphasis on type VI collagen, *Biochim. Biophys. Acta,* 1038, 222, 1990.

165. **Keene, D. R., Engvall, E., and Glanville, R.,** Ultrastructure of type VI collagen in human skin and cartilage suggests an anchoring function for this filamentous network, *J. Cell Biol.,* 107, 1995, 1988.

166. **Kajikawa, K., Nakanishi, I., and Yamamura, T.,** The effects of collagenase on the formation of fibrous long spacing collagen aggregates, *Lab. Invest.,* 43, 410, 1980.

167. **McDevitt, C. A. and Miller, R. R.,** Biochemistry, cell biology and immunology of osteoarthritis, *Curr. Opin. Rheum.,* 1, 303, 1989.

168. **Mitrovic, D. R.,** Joint structure and physiology of tissues and cartilage, *Curr. Opin. Orthop.,* 1, 271, 1990.

169. **McDevitt, C. A., Pahl, J. A., Ayad, S., Miller, R. R., Uratsuji, M., and Andrish, J. T.,** Experimental osteoarthritic articular cartilage is enriched in guanidine-soluble type VI collagen, *Biochem. Biophys. Res. Commun.,* 157, 250, 1988.

170. **Schmid, T. M. and Conrad, H. E.,** A unique low molecular weight collagen secreted by cultured chick embryo chondrocytes, *J. Biol. Chem.,* 257, 12444, 1982.

171. **Schmid, T. M. and Conrad, H. E.,** Metabolism of low molecular weight collagen by chondrocytes obtained from histologically distinct zones of the chick embryo tibiotarsus, *J. Biol. Chem.,* 257, 12451, 1982.

172. **Gibson, G. J., Schor, S. L., and Grant, M. E.,** Effect of matrix macromolecules on chondrocyte gene expression: synthesis of a low molecular weight collagen species by cells cultured within collagen gels, *J. Cell Biol.,* 93, 767, 1982.

173. **Capasso, O., Gionti, E., Pontarelli, G., Ambesi-Impiombato, F. S., Nitsch, L., Tajana, G., and Cancedda, R.,** The culture of chick embryo chondrocytes and the control of their differentiated functions in vitro. I. Characterization of the chondrocyte-specific phenotypes, *Exp. Cell Res.,* 142, 197, 1982.

174. **Remington, M. C., Bashey, R. I., Brighton, C. T., and Jimenez, S.,** Biosynthesis of a low molecular weight collagen by rabbit growth plate cartilage organ cultures, *Collagen Rel. Res.,* 3, 271, 1983.

175. **Capasso, O., Tajana, G., and Cancedda, R.,** Location of 64K collagen producer chondrocytes in developing chicken embryo tibiae, *Mol. Cell. Biol.,* 4, 1163, 1984.

176. **Schmid, T. M. and Linsenmayer, T. F.,** Immunohistochemical localization of short chain cartilage collagen (type X) in avian tissues, *J. Cell Biol.,* 100, 598, 1985.

177. **Schmid, T. M. and Linsenmayer, T. F.,** Developmental acquisition of type X collagen in the embryonic chick tibiotarsus, *Dev. Biol.,* 107, 373, 1985.

178. **Kwan, A. P. L., Freemont, A. J., and Grant, M. E.,** Immunoperoxidase localization of type X collagen in chick tibiae, *Biosci. Rep.,* 6, 155, 1986.

179. **Gibson, G. J., Bearman, C. H., and Flint, M. H.,** The immunoperoxidase localization of type X collagen in chick cartilage and lung, *Collagen Rel. Res.,* 6, 163, 1986.

180. **Summers, T. A., Irwin, M. H., Mayne, R., and Balian, G.,** Monoclonal antibody to type X collagen. Biosynthetic studies using an antibody to the amino-terminal domain, *J. Biol. Chem.,* 263, 581, 1988.

181. **Grant, W. T., Wang, G. J., and Balian, G.,** Type X collagen synthesis during endochondral ossification in fracture repair, *J. Biol. Chem.,* 262, 9844, 1987.

182. **Haynes, J. S.,** Immunohistochemical localization of type X collagen in the proximal tibiotarsi of broiler chickens and turkeys, *Anat. Rec.,* 227, 307, 1990.

183. **Linsenmayer, T. F., Gibney, E., and Schmid, T. M.,** Segmental appearance of type X collagen in the developing avian notochord, *Dev. Biol.,* 113, 467, 1986.

184. **Gibson, G. J., Beaumont, B. W., and Flint, M. H.,** Synthesis of low molecular weight collagen by chondrocytes from presumptive calcification region of the embryonic chick sterna: the influence of culture with collagen gels, *J. Cell Biol.,* 99, 208, 1984.

185. **Remington, M. C., Bashey, R. I., Brighton, C. T., and Jimenez, S.,** Biosynthesis of disulfide-bonded short-chain collagen by calf growth-plate cartilage, *Biochem. J.,* 224, 227, 1984.

186. **Kielty, C. M., Kwan, A. P. L., Holmes, D. F., Schor, S. L., and Grant, M. E.,** Type X collagen, a product of hypertrophic chondrocytes, *Biochem. J.,* 227, 545, 1985.

187. **Grant, W. T., Sussman, M. D., and Balian, G.,** A disulfide-bonded short chain collagen synthesized by degenerative and calcifying zones of bovine growth plate cartilage, *J. Biol. Chem.,* 260, 3798, 1985.

188. **Gibson, G. J. and Flint, M. H.,** Type X collagen synthesis by chick sternal cartilage and its relationship to endochondral development, *J. Cell Biol.,* 101, 277, 1985.

189. **Reginato, A. M., Lash, J. W., and Jimenez, S. A.,** Biosynthetic expression of type X collagen in embryonic chick sternum cartilage during development, *J. Biol. Chem.,* 261, 2897, 1986.

190. **Leboy, P. S., Vaias, L., Uschmann, B., Golub, E., Adams, S. L., and Pacifici, M.,** Ascorbic acid induces alkaline phosphatase, type X collagen, and calcium deposition in cultured chick chondrocytes, *J. Biol. Chem.,* 264, 17281, 1989.

191. **Poole, A. R., and Pidoux, I.,** Immunoelectron microscopic studies of type X collagen in endochondral ossification, *J. Cell Biol.,* 109, 2547, 1989.

192. **Schmid, T. M., and Linsenmayer, T. F.,** Type X collagen, in *Structure and Function of the Collagen Types,* Mayne, R. and Burgeson, R. E., Eds., Academic Press, Orlando, FL, 1987, 223.

193. **Schmid, T. M., Mayne, R., Bruns, R. R., and Linsenmayer, T. F.,** Molecular structure of short-chain (SC) cartilage collagen by electron microscopy, *J. Ultrastruct. Res.,* 86, 186, 1984.

194. **Linsenmayer, T. F., Gibney, E., and Schmid, T. M.,** Intracellular avian type X collagen in situ and determination of its thermal stability using a conformation-dependent monoclonal antibody, *Exp. Cell Res.,* 166, 15, 1986.

195. **Schmid, T. M., Mayne, R., Jeffrey, J. J., and Linsenmayer, T. F.,** Type X collagen contains two cleavage sites for a vertebrate collagenase, *J. Biol. Chem.,* 261, 4184, 1986.

196. **Welgus, H. G., Fliszar, C. J., Seltzer, J. L., Schmid, T. M., and Jeffrey, J. J.,** Differential susceptibility of type X collagen to cleavage by two mammalian interstitial collagenases and 72-kDa type IV collagenase, *J. Biol. Chem.,* 265, 13521, 1990.

197. **Ninomiya, Y., Gordon, M. K., van der Rest, M., Schmid, T. M., Linsenmayer, T. F., and Olsen, B. R.,** The developmentally regulated type X collagen gene contains a long open reading frame without introns, *J. Biol. Chem.,* 261, 5041, 1986.

198. **Ninomiya, Y., Castagnola, P., Gerecke, D. R., Gordon, M. K., Jacenko, P., McCarthy, M., Muragaki, Y., Nishimura, I., Oh, S., Rosenblum, N., Sato, N., Sugrue, S. P., Taylor, R., Vasios, G., Yamaguchi, N., and Olsen, B. R.,** The molecular biology of collagens with short triple-helical domains, in *Collagen Genes: Structure, Regulation and Abnormalities,* Boyd, C., Byers, P. H. and Sandell, L., Eds., Academic Press, Orlando, FL, 1990, 79.

199. **Thomas, J. T., Boot-Handford, R. P., Marriott, A., Kwan, A. P. L., Ayad, S., and Grant, M. E.,** Isolation of cDNAs encoding bovine type X collagen, *Ann. N.Y. Acad. Sci.,* 580, 477, 1990.

200. **LuValle, P., Ninomiya, Y., Rosenblum, N. D., and Olsen, B. R.,** The type X collagen gene. Intron sequences split the 5′-untranslated region and separate the coding regions for the noncollagenous amino-terminal and triple-helical domains, *J. Biol. Chem.,* 263, 18378, 1988.

201. **Yamaguchi, N., Benya, P. D., van der Rest, M., and Ninomiya, Y.,** The cloning and sequencing of α1(VIII) collagen cDNAs demonstrate that type VIII collagen is a short chain collagen and contains triple-helical and carboxyl-terminal non-triple-helical domains similar to those of type X collagen, *J. Biol. Chem.,* 264, 16022, 1989.

202. **Muragaki, Y., Mattei, M.-G., Yamaguchi, N., Olsen, B. R., and Ninomiya, Y.,** The complete primary structure of the human α1(VIII) chain and assignment of its gene (COL8A1) to chromosome 3, *Eur. J. Biochem.,* 197, 615, 1991.

203. **Muragaki, Y., Jacenko, O., Apte, S., Mattei, M.-G., Ninomiya, Y., and Olsen, B. R.,** The α2(VIII) collagen gene — a novel member of the short-chain collagen family located on the human chromosome 1, *J. Biol. Chem.,* 266, 7721, 1991.

204. **Schmid, T. M. and Linsenmayer, T. F.,** Immunoelectron microscopy of type X collagen: supramolecular forms within embryonic chick cartilage, *Dev. Biol.,* 138, 53, 1990.

205. **Kwan, A. P. L., Cummings, C. E., Chapman, J. A., and Grant, M. E.,** Macromolecular organization of chicken type X collagen in vitro, *J. Cell Biol.,* 114, 597, 1991.

206. **Chen, Q., Gibney, E., Fitch, J. M., Linsenmayer, C., Schmid, T. M., and Linsenmayer, T. F.,** Long-range movement and fibril association of type X collagen within embryonic cartilage matrix, *Proc. Natl. Acad. Sci. U.S.A.,* 87, 8046, 1990.

207. **LuValle, P., Hayashi, M., and Olsen, B. R.,** Transcriptional regulation of type X collagen during chondrocyte maturation, *Dev. Biol.,* 133, 613, 1989.

208. **Castagnola, P., Torella, G., and Cancedda, R.,** Type X collagen synthesis by cultured chondrocytes derived from permanent cartilagenous region of chick embryo sternum, *Dev. Biol.,* 123, 332, 1987.

209. **Castagnola, P., Moro, G., Descalzi-Cancedda, F., and Cancedda, R.,** Type X collagen synthesis during in vitro development of chick embryo tibial chondrocytes, *J. Cell Biol.,* 102, 2310, 1986.

210. **Solursh, M., Jensen, K. L., Reiter, R. S., Schmid, T. M., and Linsenmayer, T. F.,** Environmental regulation of type X collagen production by cultures of limb mesenchyme, mesectoderm, and sternal chonsrocytes, *Dev. Biol.,* 117, 90, 1986.

211. **Thomas, J. T., Boot-Handford, R. P., and Grant, M. E.,** Modulation of type X collagen gene expression by calcium ß-glycerophosphate and levamisole: implications for a possible role for type X collagen in endochondral bone formation, *J. Cell Sci.,* 95, 639, 1990.

212. **Adams, S. L., Pallante, K. M., and Pacifici, M.,** Effects of cell shape on type X collagen gene expression in hypertrophic chondrocytes, *Connect. Tissue Res.,* 20, 223, 1989.

213. **Adams, S. L., Pallante, K. M., Niu, Z., Leboy, P. S., Golden, E. B., and Pacifici, M.,** Rapid induction of type X collagen gene expression in cultured chick vertebral chondrocytes, *Exp. Cell Res.,* 193, 190, 1991.

214. **de Crombrugghe, B., Vuorio, T., and Karsenty, G.,** Control of type I collagen genes in scleroderma and normal fibroblasts, *Scleroderma,* 16, 109, 1990.

215. **Bornstein, P. and Sage, H.,** Regulation of collagen gene expression, in *Progress in Nucleic Acid Research and Molecular Biology,* Cohn, W. E. and Moldave, K., Eds., Academic Press, San Diego, 1989, 67.

216. **Ramirez, F. and Di Liberto, M.,** Complex and diversified regulatory programs control the expression of vertebrate collagen genes, *FASEB J.,* 4, 1616, 1990.

217. **Ryan, M. C., Sieraski, M., and Sandell, L. J.,** The human type II procollagen gene: identification of an additional protein-coding domain and location of potential regulatory sequences in the promoter and first intron, *Genomics,* 8, 41, 1990.

218. **Horton, W. E., Miyashita, T., and Kohno, K.,** Identification of a phenotype-specific enhancer in the first intron of the rat collagen II gene, *Proc. Natl. Acad. Sci. U.S.A.,* 84, 8864, 1987.

219. **Savagner, P., Miyashita, T., and Yamada, Y.,** Two silencers regulate the tissue-specific expression of the collagen II gene, *J. Biol. Chem.,* 265, 6669, 1990.

220. **Schnieke, A., Harbers, K., and Jaenisch, R.,** Embryonic lethal mutation in mice induced by retrovirus insertion in the alpha 1(I) collagen gene, *Nature (London),* 304, 315, 1983.

221. **Kratochwill, K., von der Mark, K., Kollar, E. J., Jaenisch, R., Mooslehner, K., Schwarz, M., Hoase, K., Gmachl, I., and Harbers, K,** Retrovirus-induced insertional mutation in Mov 13 mice affects collagen I expression in a tissue-specific manner, *Cell,* 57, 807, 1989.

222. **Cohn, W. E. and Moldave, K.,** *Progress in Nucleic Acid Research and Molecular Biology,* Academic Press, San Diego, 1989.

223. **Thompson, J. P., Simkevitch, C. P., Holness, M. A., Kang, A. H., and Raghow, R.,** In vitro methylation of the promoter and enhancer of pro-α1(I) collagen gene leads to its transcriptional inactivation, *J. Biol. Chem.,* 266, 2549, 1991.

224. **Jähner, D. and Jaenisch, R.,** Retrovirus-induced *de novo* methylation of flanking host sequences correlates with gene inactivity, *Nature (London),* 315, 594, 1985.

225. **Fernandez, M. P., Young, M. F., and Sobel, M. E.,** Methylation of type II and type I collagen genes in differentiated and dedifferentiated chondrocytes, *J. Biol. Chem.,* 260, 2374, 1985.

226. **Määttä, A., Bornstein, P., and Penttinen, R. P. K.,** Highly conserved sequences in the 3′-untranslated region of the COL1A1 gene bind cell-specific nuclear proteins, *FEBS Lett.,* 279, 9, 1991.

227. **Bennett, V. D., Weiss, I. M., and Adams, S. L.,** Cartilage-specific 5′ end of chick alpha 2(1) collagen mRNAs, *J. Biol. Chem.,* 264, 8402, 1989.

228. **Elima, K. and Vuorio, E.,** Expression of mRNAs for collagens and other matrix components in dedifferentiating and redifferentiating human chondrocytes in culture, *FEBS Lett.,* 258, 165, 1989.

229. **Bennett, V. D. and Adams, S. L.,** Identification of a cartilage-specific promoter within intron 2 of the chick alpha 2(I) collagen gene, *J. Biol. Chem.,* 265, 2223, 1990.

230. **Cates, G. A.,** Analysis of the phosphorylation state of a collagen-binding heat-shock glycoprotein from L6 myoblasts by isoelectric focusing, *Biochim. Biophys. Acta,* 1073, 521, 1991.

231. **Bennett, V. D. and Adams, S. L.,** Characterization of the translational control mechanism preventing synthesis of α2(I) collagen in chicken vertebral chondroblasts, *J. Biol. Chem.,* 262, 14806, 1987.

232. **McDonald, J. A.,** Matrix regulation of cell shape and gene expression, *Curr. Opinion Cell Biol.,* 1, 995, 1989.

233. **Mallein-Gerin, F., Ruggiero, F., and Garrone, R.,** Proteoglycan core protein and type II collagen gene expressions are not correlated with cell shape changes during low density chondrocyte cultures, *Differentiation,* 43, 204, 1990.

234. **Dozin, B., Quarto, R., Rossi, F., and Cancedda, R.,** Stabilization of the mRNA follows transcriptional activation of type II collagen gene in differentiating chicken chondrocyte, *J. Biol. Chem.,* 265, 7216, 1990.

235. **Nishimura, I., Muragaki, Y., Hayashi, M., Ninomiya, Y., and Olsen, B. R.,** Tissue-specific expression of type IX collagen, *Ann. N.Y. Acad. Sci.,* 580, 112, 1990.

236. **Adams, S. L.,** Update collagen gene expression, *Am. J. Resp. Cell Mol. Biol.,* 1, 161, 1989.
237. **Rossi, P., Karsenty, G., Roberts, A. B., Roche, N. S., Sporn, M. B., and de Crombrugghe, B.,** A nuclear factor1 binding site mediates the transcriptional activation of a type I collagen promoter by transforming growth factor-ß, *Cell,* 52, 405, 1988.
238. **Horton, W. E., Higginbotham, J. D., and Chandrasekhar, S.,** Transforming growth factor-beta and fibroblasts growth factor act synergistically to inhibit collagen II synthesis through a mechanism involving DNA regulatory sequences, *J. Cell. Physiol.,* 141, 8, 1989.
239. **Goldring, M. B., Birkhead, J., Sandell, L. J., and Krane, S. M.,** Synergistic regulation of collagen gene expression in human chondrocytes by tumor necrosis factor-α and interleukin-1β, *Ann. N.Y. Acad. Sci.,* 580, 536, 1990.
240. **Goldring, M. B., Birkhead, J., Sandell, L. J., Kimura, T., and Krane, S. M.,** Interleukin 1 suppresses expression of cartilage-specific types II and IX collagens and increases types I and III collagens in human chondrocytes, *J. Clin. Invest.,* 82, 2026, 1988.
241. **Chandrasekhar, S., Harvey, A. K., Higginbotham, J. D., and Horton, W. E.,** Interleukin-1-induced suppression of type II collagen gene transcription involves DNA regulatory elements, *Exp. Cell Res.,* 191, 105, 1990.
242. **Tyler, J. A., Bird, J. L. E., and Giller, T.,** Interleukin-1 inhibits the production of types II, IX, and XI procollagen mRNA in cartilage, *Ann. N.Y. Acad. Sci.,* 580, 512, 1990.
243. **Tyler, J. A. and Benton, H. P.,** Synthesis of type II collagen is decreased in cartilage cultured with IL1 while the rate of intracellular degradation remains unchanged, *Collagen Rel. Res.,* 8, 393, 1988.
244. **Mauviel, A., Daireaux, M., Redini, F., Galera, P., Loyau, G., and Pujol, J. P.,** Tumor necrosis factor inhibits collagen and fibronectin synthesis in human dermal fibroblasts, *FEBS Lett.,* 236, 47, 1988.
245. **Granstein, R. D., Flotte, T. J., and Amento, E. P.,** Interferons and collagen production, *J. Invest. Dermatol.,* 95 (Suppl.), 75S, 1990.
246. **Goldring, M. B., Sandell, L. J., Stephenson, M. L., and Krane, S. M.,** Immune interferon suppresses levels of procollagen mRNA and type II collagen synthesis in cultured human articular and costal chondrocytes, *J. Biol. Chem.,* 261, 9049, 1986.
247. **Horton, W. E. and Hassel, J. H.,** Independence of cell shape and loss of cartilage matrix production during retinoic acid treatment of cultured chondrocytes, *Dev. Biol.,* 115, 392, 1986.
248. **Benya, P. D. and Padilla, S. R.,** Modulation of the rabbit chondrocyte phenotype by retinoic acid terminates type II collagen synthesis without inducing type I collagen: the modulated phenotype differs from that produced by subculture, *Dev. Biol.,* 118, 296, 1986.
249. **Horton, W. E., Yamada Y., and Hassell, J. R.,** Retinoic acid rapidly reduces cartilage matrix synthesis by altring gene transcription in chondrocytes, *Dev. Biol.,* 123, 508, 1987.
250. **Yamada, Y., Miyashita, T., Savagner, P., Horton, W. E., Brown, K. S., Abramczuk, J., Hou-Xiang, X., Kohno, K., and Bruggeman, L.,** Regulation of the collagen II gene in vitro and in transgenic mice, *Ann. N.Y. Acad. Sci.,* 590, 81, 1990.
251. **Oettinger, H. F. and Pacifici, M.,** Type X collagen gene expression is transiently up-regulated by retinoic acid treatment in chick chondrocyte cultures, *Exp. Cell Res.,* 191, 292, 1990.
252. **Gerstenfeld, L. C., Kelly, C. M., von Deck, M., and Lian, J. B.,** Effect of 1,25-dihydroxyvitamin D_3 on induction of chondrocyte maturation in culture: extracellular matrix gene expression and morphology, *Endocrinology,* 126, 1599, 1990.
253. **Lyons, B. L. and Schwartz, R. L.,** Ascorbate stimulation of PAT cells causes an increase in transcription rates and a decrease in degradation rates of procollagen mRNA, *Nucleic. Acids Res.,* 12, 2569, 1984.
254. **Sandell, L. J. and Daniel, J. C.,** Effects of ascorbic acid on collagen mRNA levels in short term chondrocyte cultures, *Connect. Tissue Res.,* 17, 11, 1988.
255. **Gerstenfeld, L. C. and Landis, W. J.,** Gene expression and extracellular matrix ultrastructure of a mineralizing chondrocyte cell culture system, *J. Cell Biol.,* 112, 501, 1991.

Chapter 3

PROTEOGLYCAN, HYALURONAN, AND NONCOLLAGENOUS MATRIX PROTEIN METABOLISM BY CHONDROCYTES

Christopher J. Handley and Chee K. Ng

TABLE OF CONTENTS

I. INTRODUCTION

Cartilage is a tissue made up of cells, chondrocytes, which are embedded in an extracellular matrix made up of collagen, proteoglycans, and matrix proteins. This extracellular matrix gives cartilage its unique mechanical properties of being able to withstand and distribute compressive loads over the subchondrial bone, as well as providing a near frictionless surface.[1] The collagen fiber network gives the extracellular matrix of cartilage its tensile properties and the high negative fixed charge of the large-aggregating proteoglycan of cartilage (aggrecan) is responsible for the compressive load-bearing properties of the tissue.[2] Other macromolecular components have been isolated and characterized from the extracellular matrix of cartilage and include small proteoglycans, hyaluronan, and a number of noncollagenous proteins. Some of these macromolecules are components of the proteoglycan aggregate while others are involved in matrix-matrix and matrix-cell interactions.

In adult cartilage, the tissue levels of macromolecules that make up the extracellular matrix are maintained at constant proportions, which is reflected in stable biomechanical properties of cartilage. The macromolecules of cartilage matrix are constantly being turned over; that is, synthesized, incorporated into the matrix, and then selectively lost from the tissue. In order to maintain the tissue levels of macromolecules within the extracellular matrix of the tissue, the rate of synthesis of a particular macromolecule must equal its rate of catabolism.[3,4] If this balance between synthesis and catabolism fails, then the mechanical properties of the tissue will change. This is seen in various pathological conditions such as inflammatory or degenerative joint diseases, where the tissue levels of one or more macromolecular component of the extracellular matrix of the tissue is lowered, thereby resulting in dramatic changes in the mechanical properties of the tissue.[2]

The aim of this chapter is to review the metabolism of proteoglycans, hyaluronan, and noncollagen proteins by chondrocytes with an emphasis on the mechanisms and regulation of their synthesis, incorporation into, and their loss from the extracellular matrix of cartilage.

II. METABOLISM OF PROTEOGLYCANS

A. CARTILAGE PROTEOGLYCANS

There are three major types of proteoglycans that are synthesized by chondrocytes: (1) large cartilage-specific aggregating proteoglycans (aggrecan), (2) small proteoglycans (decorin, biglycan, and fibromodulin), and (3) cell-associated proteoglycans. These proteoglycans mediate different functions in cartilage.

Aggrecan is the major proteoglycan of cartilage, constituting approximately 10% of the wet weight of the tissue; this macromolecule is a component of the proteoglycan aggregate which is responsible for the compressive load-bearing properties of cartilage.[2] Aggrecan is made up of a core protein with a molecular mass of approximately 210,000.[5-7] At the N-terminal end of the core protein, there are two globular domains, G1 and G2. The G1 region is responsible for the interaction of aggrecan with hyaluronan and link protein,[8] and the G2 domain shows homology with the G1 domain but no function has yet been ascribed to this region.[9] The G1 and G2 domains are separated by the interglobular domain, an extended region of about 135 amino acids. This region has been implicated in the catabolism of the macromolecule.[10-12] The core protein also contains another globular domain, G3, which makes up the C-terminal end of the molecule.[5] Keratan sulfate chains are attached to the core protein in a domain adjacent to the C-terminal side of the G2 globular domain. Chondroitin sulfate chains are attached to the core protein in two distinct domains of the core protein. The chondroitin sulfate attachment domains make up over half the core protein and are situated between the keratan sulfate domain and the C-terminal region of the core protein.[5,6] As well as being substituted with glycosaminoglycan chains, the core protein also contains O- and N-linked oligosaccharides.[13] The amino acid sequence of the core protein of aggrecan from human and rat chondrosarcoma have been determined, as has most of the core protein of bovine aggrecan.[5-7,14,15] Although there is homology between core proteins from different species, there are species differences in the domain sizes. For instance, the keratan sulfate attachment domain is not found in the aggrecan of rat chondrosarcoma chondrocytes. However, the size of this domain in bovine aggrecan is larger than that of the human proteoglycan. In the case of the first chondroitin sulfate attachment domain, the rat core protein was found to be smaller than that of human aggrecan.

The small proteoglycans of cartilage are members of the leucine-rich motif proteoglycan family and are ubiquitous to extracellular matrices of all connective tissues.[16] In mature cartilage, these proteoglycans constitute approximately 5% of total proteoglycans in the tissue and are either substituted with chondroitin/dermatan sulfate or keratan sulfate glycosaminoglycan chains. The core proteins of these proteoglycans are small in comparison with aggrecan, as they have a relative molecular mass of less than 45,000. There are two species of dermatan/chondroitin sulfate proteoglycans in cartilage matrix, decorin and biglycan.[17] The core proteins of these two proteoglycans have been cloned and are different despite close homology between the proteins.[18] The core protein of decorin is substituted with one chondroitin/dermatan sulfate chain, whereas biglycan contains two similar glycosaminoglycan chains.[18,19] The site of covalent substitution of these glycosaminoglycan chains along the core protein are different for the respective proteoglycan. These structural differences are reflected in distinct physiochemical properties of the two proteoglycans. Decorin has been shown to be associated with type II collagen fibers and it has been suggested that this proteoglycan may be involved in the formation and maintenance of collagen fibers.[20-23] On the other hand, biglycan does not appear to have similar functions and its biological role remains to be described.

Chondrocytes produce a third type of small proteoglycan, fibromodulin, whose core protein is related to the small dermatan/chondroitin sulfate proteoglycans, decorin and biglycan.[24] The 375-residue core protein of fibromodulin is substituted with up to four N-linked keratan sulfate chains attached to asparagine residues within the leucine-rich region.[25] A noncollagenous matrix protein with a molecular mass of 59,000 extracted from cartilage was shown to correspond to fibromodulin but without the keratan sulfate chains attached.[26] Cloning and sequencing studies showed that the core protein has a molecular weight of 42,000, as compared to 59,000 for the cartilage-derived protein; this difference is due to extensive glycosylation by N-linked oligosaccharides. Fibromodulin binds to fibril-forming collagen and inhibits fibrillogensis *in vitro,* in a similar manner exhibited by decorin but at a different site on the collagen molecule.[27] The physiological role of the N-linked keratan sulfate chains is not clear. The final group of proteoglycans synthesized by chondrocytes are plasma membrane-associated proteoglycans. Two such macromolecules have been characterized in chondrocyte cultures, a heparan sulfate proteoglycan and a chondroitin/dermatan sulfate proteoglycan.[28] These proteoglycans constitute about 1% of the newly synthesized proteoglycans and represent a group of ubiquitous macromolecules that are associated with plasma membranes of eukaryotic cells and are involved with cell-cell and cell-matrix interactions.[29,30] In addition, these cell surface proteoglycans have been implicated in the binding of growth factors. Membrane heparan sulfate proteoglycan has been shown to bind fibroblast growth factor (FGF) and a membrane proteoglycan containing both dermatan sulfate and heparan sulfate glycosaminoglycans has been implicated in the binding of transforming growth factor-ß (TGF-ß) to cells.[30,31]

B. SYNTHESIS OF PROTEOGLYCANS BY CHONDROCYTES

Aggrecan biosynthesis by chondrocytes has been the most studied proteoglycan synthesis system and is used as a model for the synthesis of other proteoglycans. Analysis of the structure of the human and rat aggrecan core protein gene shows it to be made up of 15 exons.[6] The gene for the rat aggrecan core protein has been estimated to be approximately 80 kb in size; however, the human gene is very much smaller, largely because of smaller introns (especially that between exon 1 and 2). It is noteworthy that each exon which makes up the aggrecan gene corresponds exactly with the respective domain structure of the core protein; exceptions to this observation are the G1, G2, and G3 domains which are coded for by more than three exons. In the formation of the mRNA for the core protein of aggrecan, which has been estimated to be 9 kb in size, there is splicing out of the introns during transcription. There is some evidence that alternate splicing of exons occurs in the G3 domain since it has been shown that two of the five exons coding for G3 are alternatively spliced in the expression of human aggrecan. The biological role of alternate splicing of certain exons is not known, but the regulation of exon expression must be considered as part of the overall modulation of the expression of proteoglycans. The biglycan and decorin genes have been localized at chromosomes X and 12, respectively, in the human genome. However, details of the architecture for small proteoglycan genes are not known. Regulation of gene expression is usually achieved via the promoter and enhancer regions associated with the gene for a particular protein and upon binding by a gene regulatory protein to these regulatory regions, an up- or down-regulation of the transcription of the gene is achieved.[32] Promoter and enhancer regions associated with genes for chondrocyte proteoglycans have yet to be described. Since many of the growth factors and regulatory peptides that modulate proteoglycan synthesis in chondrocytes also influence the activity of the various gene regulatory proteins, it is plausible that the regulation of proteoglycan gene expression involves the promoter/enhancer system.

The analysis of cDNA for aggrecan core protein shows that at the N-terminal end of the core protein contains a region of hydrophobic amino acids that acts as a leader sequence directing the translocation of the core protein into the lumen of the endoplasmic reticulum.[33] In common with other glycoproteins that enter the endoplasmic reticulum, this N-terminal leader sequence on the core protein aggrecan is cleaved from the nascent core protein by a specific proteinase soon after entry into the endoplasmic reticulum.[33] Concomitant with the nascent core protein entering the endoplasmic reticulum, oligosaccharides are transferred from dolichol-bound high-mannose oligosaccharides and added to asparagine residues along the core protein. These oligosaccharides form the precursors of N-linked oligosaccharides associated with the proteoglycan.[34] Other posttranslational events that occur in the endoplasmic reticulum include the formation of disulfide bonds in the globular domains (namely, in the G1, G2, and G3 domains of the core protein) and further processing of the N-linked oligosaccharides.[35]

The core protein then enters the Golgi apparatus where O-linked oligosaccharides and the linkage region of glycosaminoglycan chains are added. The N-linked oligosaccharides are also modified by the sequential removal and addition of monosaccharides within this organelle. While in the *trans* region of the Golgi, polymerization reactions resulting in the formation of glycosaminoglycan chains occur and are shortly followed by the sulfation of glycosaminoglycan chains.[35] Kinetic experiments utilizing inhibitors of transcription and translation, as well as the use of radiolabeled precursors in pulse-chase experiments, have allowed an estimation of the temporal relationships in the synthesis of the aggrecan macromolecule. Using the rat chondrosarcoma cells, the half-life of the precursor pool of the aggrecan core protein was estimated to be approximately 60 to 90 min; 10 to 15 min of this time was taken up by the reactions involved in the polymerization and sulfation of glycosaminoglycan chains.[35,36] For adult articular cartilage, the half-life of the core protein precusor pool was found to be 30 min, which is shorter than that measured in rat chondrosarcoma chondrocytes.[37] These observations indicate that the synthesis of aggrecan is a rapid process, especially the reactions which are involved in the polymerization and sulfation of the glycosaminoglycan chains.

C. INCORPORATION INTO AND THE CATABOLISM OF PROTEOGLYCANS WITHIN THE EXTRACELLULAR MATRIX OF CARTILAGE

Little is known about the events associated with the secretion of proteoglycans from chondrocytes except that it is facilitated through secretory vesicles derived from the Golgi apparatus. After secretion into the extracellular matrix, aggrecan interacts with link protein and hyaluronan to form the proteoglycan aggregate. This aggregation takes place within 15 min after secretion and an intermediate complex consisting of link protein and proteoglycan precedes final aggregation with hyaluronan.[38] There is evidence to suggest that newly synthesized aggrecan molecules undergo a maturation process prior to aggregation with link protein and hyaluronan.[39] These reports suggested that the G1 domain of this population of aggrecan molecules may lack the appropriate disulfide bonds which endow this region the ability to interact with link protein and hyaluronan.[40] If this was the case, aggrecan core protein must have intrinsic mechanisms which preclude disulfide bridges from forming in the lumen of the endoplasmic reticulum.

In the absence of information regarding the incorporation of small proteoglycans into the extracellular matrix of cartilage, it can be suggested that these proteoglycans, once secreted from the chondrocyte, will bind to other structural components of the matrix. This will occur through specific binding domains associated with either these proteoglycans or other macromolecular components present in the extracellular matrix of the tissue.

Proteoglycans that are incorporated into the extracellular matrix of cartilage are specifically degraded and lost from the tissue. This degradation is dependent on metabolically functioning chondrocytes since inhibition of protein synthesis or proteinase activity results in the decrease in the rate of aggrecan loss from the tissue.[41] Cleavage of the core protein of aggrecan has been shown to occur as a consequence of the catabolism of aggrecan, and a number of studies have suggested that a preferential site of proteolytic cleavage of the aggrecan core protein is between the G1 and G2 domains, in the interglobular region.[10-12] Other studies using explant cultures of cartilage have shown that the core protein is also cleaved in the chondroitin sulfate attachment regions, resulting in the appearance of discrete core protein fragments in the medium of such cultures.[11,12] Of interest in these studies was the inability to detect degraded products of aggrecan catabolism in the cartilage matrix, thus indicating a very rapid exit of these products from the tissue.[12,41] Indeed, work using cartilage explants treated with proteinase inhibitors showed that intact aggrecan molecules are rapidly lost from the tissue once the proteoglycan aggregate is dissociated.[41] This work suggests that the formation of the proteoglycan aggregate is important for holding aggrecan within the extracellular matrix of cartilage. It also raises the question as to the physical size of aggrecan within the extracellular matrix of the tissue that allows movement of the complete macromolecule or fragments through the matrix of cartilage. It is probable that aggrecan is secreted from the chondrocyte in a compact form and that this small molecular domain is maintained on its incorporation into the extracellular matrix of cartilage. This compact structure of aggrecan may explain the ability of this macromolecule to move freely out of the extracellular matrix of cartilage.

Neutral metalloproteinases present in the extracellular matrix have been implicated in the catabolism of aggrecan. Indeed, this class of proteinases has been isolated from human and rabbit articular cartilage and other cartilage tissue types.[42-44] Two major types of metalloproteinases are found in cartilage matrix: namely, collagenase and stromelysin. Both have a single Zn^{2+}-binding region, require Ca^{2+} for activity, share regions of sequence homology, and they tend to attack hydrophobic residues at the N-terminal of large proteins.[42-45] Stromelysin is synthesized and secreted in a latent zymogen form that can be activated either by self-cleavage or through a plasminogen activation cascade system during inflammation.[46,47] In tissue, the activity of stromelysin is probably tightly regulated by a balance between synthesis of this enzyme and the presence of tissue inhibitors of metalloproteinase. Through this tight regulation, the steady-state metabolism of proteoglycans (in particular, aggrecan) and possibly other matrix proteins are maintained.

Little is known about the mechanism of catabolism of small proteoglycans by chondrocytes. Using explant cultures of articular cartilage, it was shown that the rate of catabolism of the small proteoglycans was slower than that of aggrecan. Exposure of the cultures to retinoic acid did not stimulate the loss of this population of proteoglycans.[48] This suggests that the mechanism and regulation of the catabolism of the small proteoglycans is different from that of aggrecan and may reflect their function and localization in the extracellular matrix of cartilage.

D. REGULATION OF PROTEOGLYCAN METABOLISM BY CHONDROCYTES

Proteoglycan metabolism by chondrocytes is regulated so as to give adult cartilage constant tissue concentrations of these macromolecules and implicit in this is that the rate of synthesis of proteoglycans is equal to their rate of catabolism. If this balance between synthesis and catabolism was lost, then the mechanical integrity of the tissue would change. A number of growth factors and regulatory peptides have been implicated in the regulation of proteoglycan metabolism by chondrocytes; they include insulin-like growth factors (IGFs), transforming growth factor-ß (TGF-ß), interleukin-1 (IL-1), and tumor necrosis factor-α (TNF-α).[49-51] Details of the effect of these molecules on chondrocyte metabolism, including proteoglycan metabolism, is described in detail elsewhere in this volume. Compared with the regulation of collagen metabolism, very little is known about the molecular mechanism by which growth factors and regulatory peptides elicit their effects on proteoglycan metabolism. The addition of fetal calf serum to explant cultures of articular cartilage results in the stimulation of proteoglycan synthesis and this stimulation is reflected in an increase in the amount of mRNA coding for the core protein of aggrecan.[52,53] This suggests that growth factors present in fetal calf serum regulate aggrecan synthesis by altering the rate of transcription of the core protein gene. Of the growth factors present in serum, the IGFs have been shown to be potent in stimulating aggrecan synthesis by cartilage and both kinetic and hybridization experiments have indicated that this group of growth factors stimulate aggrecan synthesis by modulating the rate of translation of the core protein of aggrecan.[54] These observations suggest that modulation of aggrecan core protein synthesis may occur both at the level of translation and/or transcription. Furthermore, IGFs appear to only stimulate the synthesis of aggrecan by cartilage and do not affect the rate of biosynthesis of the small proteoglycans.[55] This is in contrast to TGF-ß which preferentially stimulates the synthesis of the small proteoglycans by chondrocytes.[56] On the basis of these observations, the synthesis of aggrecan and small proteoglycans are not under the same regulatory control.

From the kinetics of inhibition by IL-1 of aggrecan synthesis in cartilage explants, it would appear that this cytokine is acting at the level of translation of the core protein of the macromolecule.[57] This cytokine also enhances the rate of the catabolism of the proteoglycan aggregate and probably elicits its actions by stimulating the synthesis of proteinases and, indeed, it has been shown that IL-1 stimulates the synthesis of latent stromelysin through positive modulation at the gene transcription level.[58]

III. METABOLISM OF HYALURONAN

A. HYALURONAN IN CARTILAGE

Hyaluronan was first isolated from vitreous humor. This polyanionic polysaccharide is an unbranched glycosaminoglycan made up of a repeating disaccharide-linked ß-(1,4) consisting of glucuronic acid and N-acetylglucosamine-linked ß(1,3).[59] As previously indicated, a role of hyaluronan in articular cartilage is in the formation of the proteoglycan aggregate. The physiological interaction between hyaluronan and aggrecan has a K_d in the range of 10^{-7} to $10^{-8} M$.[60,61] Hyaluronan has been reported to be present in high concentration in the pericellular region of chondrocytes in cartilage, and it has been suggested that this pool of hyaluronan may be involved in the maintenance of the chondrocyte phenotype.[62,63]

The role of hyaluronan in chondrogenesis is important in limb development and, together with peptide hormones, hyaluronan controls and regulates this morphogenic process.[64] Embryonic mesodermal cells from limb buds have an extensive hyaluronan coat which is lost at the onset of cell condensation, a prelude to chondrogenesis; this loss of hyaluronan is the result of cell-mediated hyaluronidase degradation and, possibly, a down-regulation of hyaluronan synthase activity.[65] Addition of exogenous hyaluronan, at concentrations as low as 1 ng/ml, can reverse chondrogenesis, thereby suggesting that the release of hyaluronidase is transient and the amount of enzyme released must be limited.

B. SYNTHESIS OF HYALURONAN BY CHONDROCYTES

For years, there has been controversy about the cellular site of hyaluronan synthesis. This was settled by evidence from proteoglycan biosynthetic studies and analysis of precursor nucleotide pools which suggested that proteoglycan and hyaluronan do not share similar synthetic sites and that hyaluronan biosynthetic machinery differs from that of proteoglycans and glycoproteins.[66,67]

Hyaluronan synthesis in eukaryotic cells has been located to the inner side of plasma membrane by an integral membrane-bound enzyme system termed "hyaluronan synthase".[68-71] Hyaluronan is synthesized by the alternate addition of monosaccharide residues from UDP precursors to the reducing end of the nascent hyaluronan molecule. This is in contrast to the synthesis of glycosaminoglycans attached to proteoglycans where the monosaccharides units are added from their UDP derivatives to the nonreducing end of the molecule. This model of hyaluronan synthesis predicts that as the molecule increases in length, it is then extruded from the cell through a channel in the plasma membrane of the cell into the extracellular space. It has further been suggested that the nascent hyaluronan molecule is anchored to the synthase complex through a phosphodiester bridge originating from the UDP group at the reducing end of the hyaluronan molecule.[72]

C. INCORPORATION INTO AND THE CATABOLISM OF HYALURONAN WITHIN THE EXTRACELLULAR MATRIX OF CARTILAGE

Shedding of the hyaluronan chains from the synthase or cell surface appears to be independent of the molecular size of the molecule. Initially, it was suggested that a phosphodiesterase may be responsible for the release of hyaluronan chains from the synthase; however, no correlation has been found between the activity of this enzyme and the loss of hyaluronan from cells.[72] It appears that this process of loss of hyaluronan from the synthase system is the result of the action of free radicals on the growing hyaluronan chain extruding from the cell surface.[71,72]

Studies on the fate of newly synthesized hyaluronan in explant cultures of articular cartilage have shown that approximately half of the hyaluronan is lost to the culture medium and the rest is retained within the extracellular matrix of tissue.[73] In the same experiments, less than 5% of the newly synthesized aggrecan appeared in the medium. It is not clear as to what is the biological significance of the population of hyaluronan lost to the medium, and may simply represent an excess of the macromolecule necessary for the maintenance of the extracellular matrix of the tissue. It is of interest that the population of hyaluronan lost to the medium consists of both high and low molecular mass species, whereas that which is retained by the tissue is of large molecular mass. It is therefore possible to suggest that the loss of hyaluronan from the tissue is, in part, due to the premature termination of synthesis or to degradation of the molecule. Some of the newly synthesized hyaluronan retained in the tissue is incorporated into the proteoglycan aggregate; evidence for this comes from experiments using inhibitors to suppress aggrecan synthesis which resulted in more of the hyaluronan appearing in the medium of cultures.[73]

The newly synthesized hyaluronan exists, initially, in explant cultures of cartilage as a high molecular mass macromolecule that depolymerizes with time to give a population with same molecular mass distribution to that of the endogenous hyaluronan.[74] If these cultures are treated with inhibitors of protein synthesis or with proteinase inhibitors, depolymerization of newly synthesized hyaluronan is not observed, suggesting that cellular activity is required for this process.[74] This difference in molecular mass between newly synthesized and endogenous chemical levels of hyaluronan has also been reported for human articular cartilage.[75] It is not clear what is the significance of the depolymerization of hyaluronan, but it appears that this event is independent of the catabolism of proteoglycan aggregate and probably reflects the action of free radicals on the macromolecule.[74]

The fate of hyaluronan as the result of the catabolism of the proteoglycan has been investigated in cartilage explants and it appears that the hyaluronan has a similar half-life to that of aggrecan.[76] This, together with the failure to observe any products of hyaluronan catabolism in the medium of explant cultures, suggests that hyaluronan may be internalized and degraded by the chondrocytes.

D. REGULATION OF HYALURONAN METABOLISM BY CHONDROCYTES

A number of growth factors, regulatory peptides, and serum are known to modulate hyaluronan synthesis by chondrocytes, but little is known about the molecular mechanisms involved in the regulation of hyaluronan synthesis by chondrocytes. However, certain characteristics of hyaluronan synthase may point to ways in which this enzyme might be regulated. Hyaluronan synthase can be activated and therefore presumably regulated by the phosphorylation of serine, tyrosine, and threonine residues present in the enzyme.[71] Kinetic studies on hyaluronan synthesis indicated that this enzyme has a short cellular half-life in both chondrosarcoma chondrocytes and in cartilage explants, thus suggesting that changes in the rate of expression of this enzyme complex would be expected to have an immediate effect on the rate of hyaluronan synthesis.[66,77]

IV. METABOLISM OF NONCOLLAGENOUS MATRIX PROTEINS

In recent years, there has been growing acknowledgment of another group of macromolecules present in the extracellular matrix of cartilage referred to as "noncollagenous proteins", which have decisive roles in the regulation of structure and function of the matrix of cartilage.[78] At present, there is general consensus that the functions of these proteins are mainly involved with (1) matrix assembly, probably in cooperative interactions with other major macromolecules such as collagen and proteoglycan, (2) interaction with chondrocytes in areas such as cell attachment and matrix maintenance, and (3) regulation of matrix metabolism. These functions need not be mutually exclusive, but are likely to be interrelated in a complex manner from which chondrocytes can respond to environmental changes within the matrix.

This section will survey recent findings on the metabolism of articular cartilage matrix proteins. Apart from link protein, fibronectin, and anchorin CII, this review will maintain the usage of the molecular mass of the protein concerned as the reference nomenclature.

A. LINK PROTEIN

Link proteins are matrix glycoproteins which are components of the proteoglycan aggregate and are specifically responsible for the stabilization of the interaction between hyaluronan and aggrecan.[79] Link protein binds to aggrecan in the proteoglycan aggregate in a 1:1 stoichiometry.[80] Three isoforms of link proteins, termed LP1, LP2, and LP3, can be isolated from cartilage and separated on the basis of differing molecular mass.[81-83] The largest form of this protein, LP1, has a molecular mass of 50,000; LP2 and LP3 have a mass of 44,000 and 40,000, respectively. The relationship of these isoforms of link protein to each other was resolved by structural and kinetic experiments which showed that both LP1 and LP2 were the same gene product but differed in the degree of glycosylation.[82,84] LP1 is N-glycosylated on asparagine residues 6 and 41 while LP2 is glycosylated only on residue 41. No product-precursor relationship was found between the synthesis of these two isoforms of link protein, suggesting that the chondrocytes synthesize both forms of link protein. The biological significance between these two forms of link protein is not clear. The third isoform, LP3, has been shown to be a degradative product of either LP1 or LP2, and appears to be generated by the action of proteinases in the N-terminal region of these molecules.[85] Indeed, with age there appears to be an increase in the abundance of LP3 in cartilage.[86]

The complete cDNA sequence and primary structure for link protein has been reported.[87] From the cDNA sequence, it was calculated that link protein consists of 339 amino acid residues with a deduced molecular mass of 42,500. Secondary structure predictions from the known sequence indicated that link protein has similar structural subdomains to that seen in the G1 domain of aggrecan.[88] The eukaryote genome contains only one copy of the link protein gene.[89] This gene is made up of five exons, interspersed by large introns, and the total gene size is larger than 80 kb.[90] The first exon contains the leader peptide sequence, whereas exons 3, 4, and 5 together code for protein. Exon 2, however, codes for a 53-amino acid residue sequence that is spliced out in about 95% of the synthesized link protein mRNA.[91] Despite secondary structure similarities to G1, link protein gene architecture is very different from that of G1.[92]

The route of link protein biosynthesis is the same as that of other glycoproteins, including proteoglycans; indeed, link proteins have been co-located with proteoglycan monomers in secretory vesicles in chondrocytes.[92] Once secreted into the extracellular matrix of cartilage, it is assumed that this protein is incorporated into the proteoglycan aggregate. It has been reported that newly synthesized link protein may need to go through a maturation process similar to that reported for aggrecan before it will associate with aggrecan and hyaluronan.[93] The fate of link protein as a consequence of proteoglycan aggregate catabolism is not known; it is not clear whether the generation of LP3 isoform from the two larger link protein species by the action of proteinases is a part of this process. Link protein may be degraded in the same manner as the G1 domain of aggrecan.

Little is known about the regulation of link protein metabolism. Addition of serum or IGF-1 to cartilage explants stimulate the synthesis of link proteins 1 and

2 which suggests that, in mature cartilage, aggrecan synthesis may be coordinately regulated with link protein synthesis.[94] However, in the developing limb, the temporal expression of link protein, aggrecan, and type II collagen are different.[95]

B. ANCHORIN CII

Anchorin CII is a specific type II collagen-binding protein with a molecular mass of 34,000 localized on the plasma membrane and in the pericellular region of chondrocytes.[96] Anchorin CII is composed of 329 residues and possibly contains O-linked but not N-linked oligosaccharides.[97] Analysis of the primary sequence of anchorin revealed four repeating regions which show homology with sequences found in intracellular calcium- and phopholipid-binding proteins. In the C-terminal region of this glycoprotein is a hydrophobic region containing three charged residues which indicates that this molecule may be associated with the plasma membrane of the chondrocyte and is not transmembrane in nature.[98] Electron micrographs have shown interaction between type II collagen and anchorin CII at the chondrocyte cell surface.[98] Indeed, it has been suggested that anchorin CII in binding to both chondrocytes and type II collagen may act as a sensor which monitors mechanical stress of the collagen network of cartilage.[78] Little is known about the metabolism of this matrix macromolecule except that the molecule is released from chondrocyte cultures after a 20-h lag period; this delay is in contrast to the release of newly synthesized anchorin by fibroblasts which takes only 30 min.[99]

C. CHONDRONECTIN

Another cartilage protein which mediates the attachment of chondrocytes to type II collagen is known as chondronectin.[100] As well as being found in serum, chondronectin is also found in the pericellular region of chondrocytes as a minor component. It has a relative molecular mass of 175,000 and is made up of three disulfide-bonded subunits. This protein contains high levels of hydrophobic amino acids and has a different amino acid composition to fibronectin and laminin.[101] Chondronectin is believed to be synthesized by chondrocytes in cartilage where it promotes the binding of chondrocytes to the collagenous network.[100] Recent work has shown that chondronectin also interacts with chondroitin sulfate chains of proteoglycans.[102]

D. FIBRONECTIN

The isoforms of fibronectin, ED-A and ED-B, have been reported to be synthesized by chondrocytes at low levels and that these two forms of the molecule are the result of alternative splicing of mRNA for the protein.[103] Fibronectin ED-A is found coexpressed with the ED-B form in chondrocyte cultures; however, in cartilage tissue cultures, chondrocytes synthesized only the ED-B variant. Cartilage fibronectin is immunologically distinct from those found circulating in plasma.[104] Fibronectin is involved in cell adhesion and migration, as well as morphogenesis and differentiation.[105] Cartilage fibronectin binds to type II collagen and to chondroitin sulfate proteoglycans, but not to hyaluronan. Osteoarthritic articular cartilage contains elevated levels of fibronectin (10 to 20 times that of control patients).[106] It is not known if chondrocytes in degenerative cartilage synthesized higher levels of fibronectin or that the elevated fibronectin was preferentially accumulated in the matrix from extraneous sources as the result of osteoarthrosis. Nonetheless, in articular cartilage, fibronectin may be involved in regulating matrix organization as a consequence of cartilage degeneration.

E. CARTILAGE MATRIX PROTEIN

This matrix protein has a molecular mass of 550,000 and consists of possibly 5 disulfide-bonded subunits with apparent molecular weights of 130,000 and 116,000.[107] This protein constitutes approximately 2 to 4% of total detectable protein in adult bovine and canine articular cartilage. Immunolocalization studies showed that this macromolecule is present in both the interterritorial region of the matrix and the pericellular regions of chondrocytes in adult articular cartilage.[108] In fetal cartilage, this protein shows a uniform distribution within the matrix. Since this protein has been shown to have a fast turnover rate in canine cartilage explants, it has been suggested that this molecule may function as a matrix receptor.[108]

F. OTHER MATRIX PROTEINS OF CARTILAGE

A number of other noncollagenous proteins have been isolated from cartilage and partially characterized. There is little information as to the structure and metabolism of these proteins and it is only possible to speculate about their function in cartilage. It has been reported that mature articular cartilage contains a basic matrix glycoprotein with a molecular mass of 58,000 which is extracted by guanidinium chloride. It has been estimated that this molecule makes up between 0.1 and 0.3% of tissue wet weight.[26,109] A protein of molecular size 55,000 has been isolated from human articular cartilage and has been reported to be present in greater abundance in tissue from older animals. Initial biosynthetic studies showed that this protein is synthesized by chondrocytes at low levels and its presence in the extracellular matrix of cartilage may be due to an accumulation from sources outside cartilage.[110] Finally, a protein of molecular mass 36,000 has been shown to be present in most cartilage types, as well as bone.[78,109] Like other matrix proteins, this protein has high levels of leucine and aspartic acid/asparagine. Of interest is that biosynthetic studies have shown that more radiolabeled leucine is incorporated into this protein compared to the core protein of aggrecan, thus suggesting that this macromolecule may exist in high abundance in the extracellular matrix of cartilage.[111]

G. MATRIX PROTEINS OF HYPERTROPHIC CARTILAGE

A number of noncollagenous proteins have been isolated from the extracellular matrix of hypertrophic cartilage; the presence of this group of molecules reinforce the concept of the special role this region of cartilage plays in bone growth. A small protein of molecular mass 21,000 has been found in hypertrophic cartilage and is coexpressed at the same time as type X collagen.[112] cDNA sequence analysis indicated that this protein is made up of 158 amino acid residues and it is a member of the hydrophobic molecule carrier protein superfamily.[113] Apart from being found in hypertrophic cartilage, it has also been immunolocalized in newly-formed bone matrix. These morphological and sequence findings strongly suggest a role for this protein in cartilage mineralization.

A vitamin K-dependent matrix γ-carboxyglutamic acid protein of molecular mass 14,000 was found in growth plate cartilage.[114,115] This protein is synthesized by chondrocytes and is accumulated in cartilage. The function of this protein is believed to inhibit mineralization of cartilage as it has been shown to specifically inhibit hydroxyapatite crystal formation in growth plate cartilage.[115]

V. CONCLUDING REMARKS

Chondrocytes synthesize a diverse group of proteins and glycoproteins that are incorporated into or associated with the extracellular matrix of cartilage, some of these macromolecules being specific to cartilage. Our knowledge of the structure, metabolism, and function of these macromolecules varies considerably. This is most evident with the matrix proteins of noncollagen and nonproteoglycan origin, some of which are only described in terms of molecular mass. The understanding of the structure and metabolism of these macromolecules can be expected to have a far-reaching impact on the understanding of the functioning of normal and diseased cartilage. Even with the proteoglycan aggregate, a major structural component of cartilage, our understanding of the molecular events associated with the regulation of the synthesis of individual components of this structure, as well as the mechanism and regulation of the catabolism of the aggregate, awaits clarification. It is now clear that the chondrocyte is responsible for the metabolism of a growing list of macromolecules that are necessary for the normal functioning of cartilage and the integration of the individual metabolism of these matrix components in the growth, maintenance, pathology, and repair of cartilage will be important and also fascinating.

ACKNOWLEDGMENTS

This work was supported by the National Health and Medical Research Council of Australia and the Arthritis Foundation of Australia.

REFERENCES

1. **Kempson, G. E.,** The effects of proteoglycan and collagen degradation on the mechanical properties of adult human articular cartilage, in *Dynamics of Connective Tissue Macromolecules,* Burleigh, P. M. C. and Poole, A. R., Eds., North-Holland, Amsterdam, 1975, 277.
2. **Maroudas, A., Mizrahi, J., Katz, E. P., Wachtel, E. J., and Soudry, M.,** Physiochemical properties and functional behaviour of normal and osteoarthritic human cartilage, in *Articular Cartilage Biochemistry,* Kuettner, K., Schleyerbach, R., and Hascall, V. C., Eds., Raven Press, New York, 1986, 311.
3. **Handley, C. J., McQuillan, D. J., Campbell, M. A., and Bolis, S.,** Steady-state metabolism in cartilage explants, in *Articular Cartilage Biochemistry,* Kuettner, K., Schleyerbach, P., and Hascall, V. C., Eds., Raven Press, New York, 1986, 163.
4. **Handley, C. J., Ng, C. K., and Curtis, A. J.,** Short- and long-term explant culture of cartilage, in *Methods of Cartilage Research,* Maroudas, A. and Kuettner, K., Eds., Academic Press, London, 1990, 105.
5. **Doege, K., Sasaki, M., Horigan, E., Hassell, J. R., and Yamada, Y.,** Complete primary structure of the rat cartilage proteoglycan core protein deduced from cDNA clones, *J. Biol. Chem.,* 262, 17757, 1987.
6. **Doege, K. J., Sasaki, M., Kimura, T., and Yamada, Y.,** The complete coding sequence and deduced primary structure of the human cartilage large aggregating proteoglycan, aggrecan, *J. Biol. Chem.,* 266, 894, 1991.
7. **Doege, K., Sasaki, M., and Yamada, Y.,** Rat and human proteoglycan (aggrecan) gene structure, *Biochem. Soc. Trans.,* 18, 200, 1990.
8. **Paulsson, M., Mörgellin, M., Wiedemann, H., Beardmore-Gray, M., Dunham, D., Hardingham, T., Heinegård, D., Timpl, R., and Engel, J.,** Extended and globular protein domains in cartilage proteoglycans, *Biochem. J.,* 245, 763, 1987.
9. **Fosang, A. J. and Hardingham, T. E.,** Isolation of the N-terminal globular protein domain from cartilage proteoglycans. Identification of G2 domain and its lack of interaction with hyaluronate and link protein, *Biochem. J.,* 261, 801, 1989.
10. **Ratcliffe, A., Tyler, J. A., and Hardingham, T. E.,** Articular cartilage cultured with interleukin-1. Increased release of link protein, hyaluronate binding region and other proteoglycan fragments, *Biochem. J.,* 238, 571, 1986.
11. **Campbell, M. A., Handley, C. J., Hascall, V. C., Campbell, R. A., and Lowther, D. A.,** Turnover of proteoglycans in cultures of bovine articular cartilage, *Arch. Biochem. Biophys.,* 234, 275, 1984.
12. **Campbell, M. A., D'Souza, S. E., and Handley, C. J.,** Turnover of proteoglycans in articular-cartilage cultures, *Biochem. J.,* 259, 21, 1989.
13. **Lohmander, L. S., Nilsson, B., De Luca, S., and Hascall, V. C.,** Structures of N- and O-linked oligosaccharides from chondrosarcoma proteoglycan, *Semin. Arthritis Rheum.,* 11, 12, 1981.
14. **Oldberg, Å., Antonsson, P., and Heinegård, D.,** The partial amino acid sequence of bovine articular cartilage proteoglycan deduced from a cDNA clone, contains numerous Ser-Gly sequences arranged in homologous repeats, *Biochem. J.,* 243, 255, 1987.
15. **Antonsson, P., Heinegård, D., and Oldberg, Å.,** The keratan sulfate-enriched region of bovine cartilage proteoglycan consists of a consecutively repeated hexapeptide motif, *J. Biol. Chem.,* 264, 16170, 1989.
16. **Voss, B., Glössl, J., Cully, Z., and Kresse, H.,** Immunological investigation on the distribution of small chondroitin sulfate-dermatan sulfate proteoglycan in the human, *J. Histochem. Cytochem.,* 34, 1013, 1986.
17. **Rosenberg, L. C., Choi, H. U., Tang, L.-H., Johnsson, T.-L., Pal, S., Webber, C., Reiner, A., and Poole, A. R.,** Isolation of dermatan sulfated proteoglycans from mature bovine articular cartilage, *J. Biol. Chem.,* 260, 6304, 1985.

18. **Fisher, L. W., Termine, J. D., and Young, M. F.,** Deduced protein sequence of bone small proteoglycan I (biglycan) shows homology with PG II (decorin) and several nonconnective tissue proteins in a variety of species, *J. Biol. Chem.,* 264, 4571, 1989.

19. **Choi, H. U., Johnsson, T. L., Pal, S., Tang, L.-H., Rosenberg, L., and Neame, P. J.,** Characterisation of the DSPGs, DS-PG I and DS-PG II, from bovine articular cartilage and skin isolated by octyl-Sepharose chromatography, *J. Biol. Chem.,* 264, 2876, 1989.

20. **Vogel, K. G., Paulsson, M., and Heinegård, D.,** Specific inhibitor of Type I and II collagen fibrillogenesis by the small proteoglycan of tendon, *Biochem. J.,* 223, 587, 1984.

21. **Lewandowska, K., Choi, H. U., Rosenberg, L. C., Zardi, L., and Culp, L. A.,** Fibronectin-mediated adhesion of fibroblasts: inhibition by dermatan sulfate proteoglycans and evidence for cryptic glycosaminoglycan-binding domain, *J. Cell Biol.,* 105, 1443, 1987.

22. **Rosenberg, L. C., Choi, H. U., Poole, A. R., Lewandowska, K., and Culp, L. A.,** Biological roles of dermatan sulfate proteoglycans, in *Functions of the Proteoglycans,* Evered, D. and Whelan, J., Eds, Wiley, Chichester, 1986, 47.

23. **Scott, J. E.,** Proteoglycan-fibrillar collagen interactions, *Biochem. J.,* 252, 313, 1988.

24. **Oldberg, Å., Hayman, E. G., and Rouslahti, E.,** Isolation of a chondroitin sulfate proteoglycan from a rat yolk sac tumor and immunochemical demonstration of its cell surface localisation, *J. Biol. Chem.,* 256, 10847, 1988.

25. **Plaas, A. H. K., Neame, P. J., Nivens, C. M., and Reiss, L.,** Identification of the keratan sulfate attachment sites on bovine fibromodulin, *J. Biol. Chem.,* 265, 20634, 1990.

26. **Heinegård, D., Larsson, T., Sommarin, Y., Franzén, A., Paulsson, M., and Hedbom, E.,** Two novel matrix proteins isolated from articular cartilage show wide distribution among connective tissue, *J. Biol. Chem.,* 261, 13866, 1986.

27. **Hedbom, E. and Heinegård, D.,** Interactions of a 59-kDa connective tissue matrix protein with collagen I and collagen II, *J. Biol. Chem.,* 264, 6898, 1989.

28. **Sommarin, Y. and Heinegård, D.,** Four classes of cell-associated proteoglycans in suspension cultures of articular cartilage chondrocytes, *Biochem. J.,* 233, 809, 1986.

29. **Höök, M., Kjellén, L., Johansson, S., and Robinson, J.,** Cell-surface glycosaminoglycans, *Ann. Rev. Biochem.,* 53, 847, 1984.

30. **Koda, J. E., Rapraeger, A., and Bernfield, M.,** Heparan sulfate from mouse mammary epithelial cells: cell surface proteoglycans as a receptor for interstitial collagens, *J. Biol. Chem.,* 260, 8157, 1985.

31. **Andres, J. L., Stanley, K., Cheifez, S., and Massagué, J.,** Membrane-anchored and soluble forms of betaglycan, a polymorphic proteoglycan that binds transforming growth factor-ß, *J. Cell Biol.,* 109, 3137, 1989.

32. **Dyan, W. S.,** Promoters for housekeeping genes, *Trends Genet.,* 2, 196, 1986.

33. **Blobel, G.,** Intracellular protein topogenesis, *Proc. Natl. Acad. Sci. U.S.A.,* 77, 1496, 1980.

34. **Fellini, S. A., Kimura, J. H., and Hascall, V. C.,** Polydispersity of proteoglycans synthesized by chondrocytes from the Swarm rat chondrosarcoma, *J. Biol. Chem.,* 256, 7883, 1981.

35. **Lohmander, L. S. and Kimura, J. H.,** Biosynthesis of cartilage proteoglycans, in *Articular Cartilage Biochemistry,* Kuettner, K., Schleyerbach, R., and Hascall V. C., Eds., Raven Press, New York, 1986, 93.

36. **Fellini, S. A., Hascall, V. C., and Kimura, J. H.,** Localization of proteoglycan core protein in subcellular fractions isolated from rat chondrosarcoma, *J. Biol. Chem.,* 259, 4634, 1984.

37. **McQuillan, D. J., Handley, C. J., Robinson, H. C., Ng, K., Tzaicos, C., Brooks, P. R., and Lowther, D. A.,** The relation of protein synthesis to chondroitin sulfate biosynthesis in cultured bovine articular cartilage, *Biochem. J.,* 224, 977, 1984.

38. **Kimura, J. H. and Kuettner, K. E.,** Studies on the synthesis and assembly of cartilage proteoglycan, in *Articular Cartilage Biochemistry,* Kuettner, K., Schleyerbach, R., and Hascall, V. C., Eds., Raven Press, New York, 1986, 113.

39. **Oegema, T. R., Jr.,** Delayed formation of proteoglycan aggregate structures in human articular cartilage disease states, *Nature (London),* 288, 583, 1980.

40. **Plaas, A. H. K. and Sandy, J. D.,** The affinity of newly synthesized proteoglycans for hyaluronic acid can be enhanced by exposure to mild alkali, *Biochem. J.,* 234, 221, 1986.

41. **Bolis, S., Handley, C. J., and Comper, W. C.,** Passive loss of proteoglycan from articular cartilage, *Biochim. Biophys. Acta,* 993, 157, 1989.
42. **Slapolsky, A. I. and Howell, D. S.,** Further characterization of a neutral metalloprotease isolated from human articular cartilage, *Arthritis Rheum.,* 25, 981, 1982.
43. **Lowther, D. A., Sandy, J. B., Cartwright, E. C., and Brown, H. L. G.,** Isolation and secretion of proteolytic enzymes from articular cartilage in organ culture, *Semin. Arthritis Rheum.,* 11, 65, 1981.
44. **Galloway, W. A., Murphy, G., Sandy, J. D., Gavrilovic, J., Cawston, T. E., and Reynolds, J. J.,** Purification and characterisation of a rabbit bone metalloproteinase that degrades proteoglycan and other connective-tissue components, *Biochem. J.,* 209, 741, 1983.
45. **Murphy, G., Hemby, R. M., Hughes, C. E., Fosang, A. J., and Hardingham, T. E.,** Role and regulation of metalloproteinases in connective tissue turnover, *Biochem. Soc. Trans. (London),* 18, 812, 1990.
46. **Nagase, H., Enghild, J. J., Suzuki, K., and Salvesen, G.,** Stepwise activation mechanism of the precursor of matrix metalloproteinase-3 (stromelysin) by proteinases and (4-aminophenyl)mercuric acetate, *Biochemistry,* 29, 5783, 1990.
47. **Gavrilovic, J. and Murphy, G.,** The role of plasminogen in cell-mediated collagen degradation, *Cell Biol. Int. Rep.,* 13, 367, 1989.
48. **Campbell, M. A. and Handley, C. J.,** The effect of retinoic acid on proteoglycan biosynthesis in bovine articular cartilage cultures, *Arch. Biochem. Biophys.,* 253, 462, 1987.
49. **McQuillan, D. J., Handley, C. J., Campbell, M. A., Bolis, S., Milway, V. E., and Herington, A. C.,** Stimulation of proteoglycan biosynthesis by serum and insulin-like growth factor I in cultured bovine articular cartilage, *Biochem. J.,* 240, 423, 1986.
50. **Morales, T. I. and Roberts, A. B.,** Transforming growth factor-ß regulates the metabolism of proteoglycans in bovine cartilage organ cultures, *J. Biol. Chem.,* 263, 12828, 1988.
51. **Tyler, J. A.,** Insulin-like growth factor-I can decrease degradation and promote synthesis of proteoglycan in cartilage exposed to cytokines, *Biochem. J.,* 260, 543, 1989.
52. **McQuillan, D. J., Handley, C. J., Robinson, H. C., Ng, K., and Tzaicos, C.,** The relation of RNA synthesis to chondroitin sulfate biosynthesis in cultured bovine cartilage, *Biochem. J.,* 235, 499, 1986.
53. **McQuillan, D. J., Handley, C. J., and Robinson, H. C.,** Control of proteoglycan biosynthesis. Further studies on the effect of serum on cultured bovine articular cartilage, *Biochem. J.,* 237, 741, 1986.
54. **Curtis, A. J., Handley, C. J., and Devenish, R. J.,** unpublished data, 1991.
55. **Tesch, G. H., Handley, C. J., Cornell, H. J., and Herington, A. C.,** Effects of free and bound insulin-like growth factors-I and II on proteoglycan metabolism in bovine articular cartilage, *J. Orthop. Res.,* in press, 1992.
56. **Handley, C. J.,** unpublished data, 1991.
57. **Benton, H. P. and Tyler, J. A.,** Inhibition of cartilage proteoglycan synthesis by interleukin 1, *Biochim. Biophys. Res. Commun.,* 154, 421, 1988.
58. **Schynder, J., Payne, T., and Dinarello, C. A.,** Human monocyte or recombinant IL-1's are specific for the secretion of a metalloproteinase from chondrocytes, *J. Immunol.,* 138, 496, 1987.
59. **Meyer, K. and Palmer, J. W.,** The polysaccharide of the vitreous humor, *J. Biol. Chem.,* 107, 629, 1934.
60. **Christner, J. E., Brown, M. L., and Dziewiatkowski, D. D.,** Interactions of cartilage proteoglycans with hyaluronate. The role of the hyaluronate acetomido groups, *J. Biol. Chem.,* 254, 12303, 1979.
61. **Nieduszynski, I. A., Sheehan, J. K., Phelps, C. F., Hardingham, T. E., and Muir, H.,** Equilibrium-binding studies of pig laryngeal cartilage proteoglycans with hyaluronate oligosaccharides fractions, *Biochem. J.,* 185, 107, 1980.
62. **Larsson, S.-E., Ray, R. D., and Kuettner, K. E.,** Microchemical studies on acid glycosaminoglycans of the epiphyseal zones during endochondral calcification, *Calcif. Tissue Res.,* 13, 271, 1973.

63. **Toole B. P.,** Hyaluronan and its binding proteins, the hyaladherins, *Curr. Opin. Cell Biol.,* 2, 839, 1990.

64. **Toole, B. P., Munaim, S. I., Welles, S., and Knudson, C. B.,** Hyaluronate-cell interactions and growth factor regulation of hyaluronate synthesis during limb development, in *The Biology of Hyaluronan,* Evered, D. and Whelan, J., Eds., John Wiley & Sons, Chichester, U.K., 1989, 138.

65. **Skantze, F., Binkerhoff, C. E., and Collier, J.-P.,** Use of agarose culture to measure the effect of transforming growth beta and epidermal growth factor on rabbit articular chondrocytes, *Cancer Res.,* 45, 4416, 1985.

66. **Kleine, T. O.,** Hyaluronate-proteoglycan complex: Evidence for separate biosynthesis mechanisms of the macromolecules, *Connect. Tissue Res.,* 5, 195, 1978.

67. **Mason, R. M.,** Recent advances in the biochemistry of hyaluronic acid in cartilage, in *Connective Tissue Research: Chemistry, Biology and Physiology,* Alan R. Liss, New York, 1981, 87.

68. **Prehm, P.,** Synthesis of hyaluronate in differentiated teratocarcinoma cells. Characterisation of the synthase, *Biochem. J.,* 211, 181, 1983.

69. **Prehm, P.,** Synthesis of hyaluronate in differentiated teratocarcinoma cells. Mechanism of chain growth, *Biochem. J.,* 211, 191, 1983.

70. **Prehm, P.,** Hyaluronate is synthesized at plasma membrane, *Biochem. J.,* 220, 597, 1984.

71. **Prehm, P.,** Identification and regulation of the eukaryotic hyaluronate synthase, in *The Biology of Hyaluronan,* Evered, D. and Whelan, J., Eds., John Wiley & Sons, Chichester, U.K., 1989, 21.

72. **Prehm, P.,** Release of hyaluronate from eukaryotic cells, *Biochem. J.,* 267, 185, 1990.

73. **Ng, C. K., Handley, C. J., Mason, R. M., and Robinson, H. C.,** Synthesis of hyaluronate in cultured bovine articular cartilage, *Biochem. J.,* 263, 761, 1989.

74. **Ng, C. K., Handley, C. J., and Robinson, H. C.,** unpublished data, 1991.

75. **Holmes, M. W. A., Bayliss, M. T., and Muir, H.,** Hyaluronic acid in human articular cartilage. Age-related changes in content and size, *Biochem. J.,* 250, 435, 1988.

76. **Morales, T. I. and Hascall, V. C.,** Correlated metabolism of proteoglycans and hyaluronic acid in bovine cartilage organ cultures, *J. Biol. Chem.,* 263, 3632, 1988.

77. **Bansal, M. K. and Mason, R. M.,** Evidence for rapid metabolic turnover of hyaluronate synthetase in Swarm rat chondrosarcoma chondrocytes, *Biochem. J.,* 236, 515, 1986.

78. **Heinegård, D. and Oldberg, Å.,** Structure and biology of cartilage and bone matrix noncollagenous macromolecules, *FASEB J.,* 3, 2042, 1989.

79. **Hardingham, T. E.,** The role of link protein in the structure of cartilage proteoglycan aggregates, *Biochem. J.,* 177, 237, 1979.

80. **Tang, L.-H., Rosenberg, L. C., Reiner, A., and Poole, A. R.,** Proteoglycans from bovine nasal cartilage, properties of a soluble-form of link protein, *J. Biol. Chem.,* 254, 10523, 1979.

81. **Hering, T. M. and Sandell, L. J.,** Biosynthesis and processing of bovine cartilage link proteins, *J. Biol. Chem.,* 265, 2375, 1990.

82. **Baker, J. R. and Caterson, B.,** The isolation and characterisation of link proteins from proteoglycan aggregates of bovine nasal cartilage, *J. Biol. Chem.,* 254, 2387, 1979.

83. **Bonnet, F., Périn, J.-P., Lorenzo, F., Jollès, J., and Jollès, P.,** An unexpected sequence homology between link proteins of the proteoglycan complex and immunoglobulin-like proteins, *Biochim. Biophys. Acta,* 873, 152, 1986.

84. **Hering, T. H. and Sandell, L. J.,** Biosynthesis and cell-free translation of Swarm rat chondrosarcoma and bovine articular cartilage link protein, *J. Biol. Chem.,* 263, 1030, 1988.

85. **Mort, J. S., Caterson, B., Poole, A. R., and Roughley, P. J.,** The origin of human cartilage proteoglycan link-protein heterogeneity and fragmentation during ageing, *Biochem. J.,* 232, 805, 1985.

86. **Mort, J. S., Poole, A. R., and Roughley, P. J.,** Age-related changes in the structure of proteoglycan link protein present in normal human articular cartilage, *Biochem. J.,* 214, 269, 1983.

87. **Dudhia, J. and Hardingham, T. E.,** The primary structure of human cartilage link protein, *Nucleic Acids Res.,* 18, 1292, 1990.

88. **Perkins, S. J., Nealis, A. S., Dudhia, J., and Hardingham, T. E.,** Immunological fold and tandem repeat structures in proteoglycan N-terminal domains and link protein,*J. Mol. Biol.,* 206, 737, 1989.

89. **Doege, K., Fernandez, P., Hassell, J. R., Sasaki, M., and Yamada, Y.,** Partial cDNA sequence encoding a globular domain at the C-terminus of the rat cartilage proteoglycan,*J. Biol. Chem.,* 261, 8108, 1986.

90. **Kiss, I., Deák, F., Metri, S., Delius, H., Soos, J., Dékány, K., Argraves, W. S., Sparks, K. J., and Goetinck, P. F.,** Structure of the chicken link protein gene: exons correlate with the protein domains, *Proc. Natl. Acad. Sci. U.S.A.,* 84, 6399, 1987.

91. **Rhodes, C., Doege, K., Sasaki, M., and Yamada, Y.,** Alternative splicing generates two different mRNA species for rat link protein, *J. Biol. Chem.,* 263, 6063, 1988.

92. **Ratcliffe, A., Hughes, C., Fryer, P. R., Saed-Najed, F., and Hardingham, T. E.,** Immunological studies on the synthesis and secretion of link protein and aggregating proteoglycan by chondrocytes, *Collagen Rel. Res.,* 7, 407, 1987.

93. **Plaas, A. H. K., Sandy, J. D., and Kimura, J. H.,** Biosynthesis of cartilage proteoglycan and link protein by articular chondrocytes from immature and mature rabbits,*J. Biol. Chem.,* 263, 7560, 1988.

94. **Curtis, A. J., Ng, C. K., Handley, C. J., and Robinson, H. C.,** Effect of insulin-like growth factor-I on the synthesis and distribution of link protein and hyaluronate in explant cultures of articular cartilage, *Biochim. Biophys. Acta,* in press, 1992.

95. **Stirpe, N. S. and Goetinck, P. F.,** Gene regulation during cartilage differentiation: temporal and spatial expression of link protein and cartilage matrix protein in the developing limb, *Development,* 107, 23, 1989.

96. **Mollenhauer, J. and von der Mark, K.,** Isolation and characterization of a collagen-binding glycoprotein from chondrocyte membranes, *EMBO J.,* 2, 45, 1983.

97. **Von der Mark, K., Mollenhauer, J., Pfäffle, M., van Menxel, M., and Mller, P. K.,** Role of anchorin CII in the interaction of chondrocytes with extracellular collagen, in *Articular Cartilage Biochemistry,* Kuettner, K., Schleyerbach, R., and Hascall, V. C., Eds., Raven Press, New York, 1986, 125.

98. **Mollenhauer, J., Bee, J. A., Lizarbe, M. A., and von der Mark, K.,** Role of anchorin CII, a 31,000-mol wt membrane protein, in the interaction of chondrocytes with type II collagen, *J. Cell Biol.,* 98, 1572, 1984.

99. **Pfäffle, M., Ruggiero, F., Hofmann, H., Fernández, M. P., Selmin, O., Yamada, Y., Garrone, R., and von der Mark, K.,** Biosynthesis, secretion and extracellular localisation of anchorin CII, a collagen-binding protein of the calpactin family, *EMBO J.,* 7, 2338, 1988.

100. **Hewitt, A. T., Kleinman, H. K., Pennypecker, J. P., and Martin, G. R.,** Identification of an adhesion factor for chondrocytes, *Proc. Natl. Acad. Sci. U.S.A.,* 77, 385, 1980.

101. **Hewitt, A. T., Varner, H. H., Silver, M. H., Dessau, W., Wilkes, C. M., and Martin, G. R.,** The isolation and partial characterisation of chondronectin, an attachment factor for chondrocytes, *J. Biol. Chem.,* 259, 2330, 1982.

102. **Varner, H. H., Horn, V. J., Martin, G. R., and Hewitt, A. T.,** Chondronectin interactions with proteoglycan, *Arch. Biochem. Biophys.,* 244, 824, 1986.

103. **Burton-Wurster, N., Lust, G., and Wert, R.,** Expression of the ED-B fibronectin isoform in adult human articular cartilage, *Biochem. Biophys. Res. Commun.,* 165, 782, 1989.

104. **Burton-Wurster, N. and Lust, G.,** Molecular and immunologic difference in canine fibronectin from articular cartilage and plasma, *Arch. Biochem. Biophys.,* 269, 32, 1989.

105. **Hynes, R.,** Molecular biology of fibronectin, laminin and other basement membrane components, *Annu. Rev. Cell Biol.,* 1, 67, 1985.

106. **Brown, R. A. and Jones, K. L.,** The synthesis and accumulation of fibronectin by human articular cartilage, *J. Rheumatol.,* 17, 65, 1990.

107. **Fife, R. S. and Brandt, K. D.,** Identification of a high-molecular-weight (>400,000) protein in hyaline cartilage, *Biochim. Biophys. Acta,* 802, 506, 1984.

108. **Fife, R. S., Palmoski, M. J., and Brandt, K. D.,** Metabolism of a cartilage matrix glycoprotein in normal and osteoarthritic canine articular cartilage, *Arthritis Rheum.,* 1256, 1986.

109. **Sommarin, Y., Larsson, T., and Heinegård, D.,** Chondrocyte-matrix interactions. Attachment to proteins isolated from cartilage, *Exp. Cell Res.,* 184, 181, 1989.

110. **Melching, L. I. and Roughley, P. J.,** A matrix protein of M_r 55,000 that accumulates in human articular cartilage with age, *Biochim. Biophys. Acta,* 1036, 213, 1990.

111. **Paulsson, M., Sommarin, Y., and Heinegård, D.,** Metabolism of cartilage proteins in cultured tissue sections, *Biochem. J.,* 212, 659, 1983.

112. **Cancedda, F. D., Dozin, B., Rossi, R., Molina, F., Cancedda, R., Negri, A., and Ronchi, S.,** The Ch 21 protein, developmentally regulated in chick embryo belongs to the superfamily of lipophilic molecule carrier proteins, *J. Biol. Chem.,* 265, 19060, 1990.

113. **Cancedda, F. D., Manduca, P., Tacchetti, C., Fossa, P., Quarto, R., and Cancedda, R.,** Developmentally regulated synthesis of a new low molecular weight protein by differentiated chondrocytes, *J. Cell Biol.,* 107, 2455, 1988.

114. **Price, P. A. and Williamson, M. K.,** Primary structure of bone matrix Gla protein, a new vitamin K-dependent bone protein, *J. Biol. Chem.,* 260, 14971, 1985.

115. **Hale, J. E., Fraser, F. D., and Price, P. A.,** The identification of matrix Gla protein cartilage, *J. Biol. Chem.,* 263, 5820, 1988.

Chapter 4

DIFFERENT TYPES OF CULTURED CHONDROCYTES — THE *IN VITRO* APPROACH TO THE STUDY OF BIOLOGICAL REGULATION

Monique Adolphe and Paul Benya

TABLE OF CONTENTS

I. INTRODUCTION

Among the different connective tissues, cartilage plays a special role of skeletal protection by withstanding pressure and absorbing shocks. During embryogenisis, regulation of the growth cartilage phenotype is particularly specific. The proliferative capacity of growth cartilage is followed by a maturation phase, leading to hypertrophic cartilage, and finally to calcified cartilage. During postnatal life, cartilage is located in two sites. The first one corresponds to extraskeletal cartilaginous structures located in the larynx, nose, eustachian tube, and to costal cartilage. The second one is located in movable joints and is called "articular cartilage". The mechanical properties of cartilage are due to the physicochemical properties of the matrix. This matrix is produced by the cartilage cell or chondrocyte. The chondrocytes, by modulation of their metabolism, attempt to adapt the cartilaginous tissue to all physiological, pathological, or pharmacological modifications. Whatever the type of cartilage, cartilaginous tissue is always in a complex environment in direct or indirect contact with other types of cells (i.e., osteoblasts and synoviocytes from surrounding tissues). This circumstance leads to complex regulatory mechanisms mediated by factors derived from chondrocytes themselves, their different matrix components, and from neighboring cells. This situation becomes more complicated in pathology (i.e., arthritis) because of the interactions of lymphocytes and macrophages. Investigation of chondrocytes *in vitro* appears to be a particularly useful way to reduce the complexity of phenomena and gain a better understanding of the different regulatory mechanisms that are involved in cartilage maintenance and pathology. Schematically, the *in vivo* functions of chondrocytes can be divided into three parts: (1) the general functions, proliferation, and respiration; (2) the differentiative functions corresponding to the fabrication of matrix; and (3) the metabolic functions performed by various enzymes.

Even though cell proliferation is important in the early stages of growth of all cartilages, normal adult cartilage cell growth is infrequent, according to Mankin.[1] However, Kuntz et al.[2] showed proliferation of articular cartilage cells in the course of repair after damage. Haudrup and Telnag[3] also found mitoses of chondrocytes in the contralateral control joint. This phenomenon could be explained by paracrine regulation. Due to the avascular nature of articular cartilage, its metabolism is predominantly anaerobic. Mitochondrial content seems to correlate with the age of chondrocytes and their localization in the different zones of the joints. According to Ghadially,[4] mitochondria are more numerous and larger in the metabolically active cartilaginous zones. In growth plate, oxygen tension in the hypertrophic zone is quite low: 24.3 ± 2.4 mm, according to Brighton.[5] This low O_2 tension is also due to the avascularity of the zone. O_2 tension in the proliferative zone is high; aerobic metabolism occurs and mitochondria can form ATP or store calcium.

With matrix production being covered in detail in Chapters 2 and 3, only essential facts are presented here. The two major constituents of the matrix are collagen fibers and aggregates of proteoglycans. Type II collagen $[\alpha 1(II)]_3$ is the dominant type in cartilage. Ricard-Blum et al.[6] described the presence of a minor collagen (type IX) in hyaline cartilage which represents 5 to 10% in articular cartilaginous tissue. Mayne and Irwin[7] reported that type X collagen is predominantly synthetized by hypertrophic chondrocytes of the growth plate. Type XI collagen appears to be preferentially retained on the chondrocyte surface where it may be involved in the organization of the pericellular matrix. In normal cartilage matrix, proteoglycans consist of a central protein core along which chondroitin sulfate and keratan sulfate are situated. These subunits interact with hyaluronic acid to form aggregates stabilized by link proteins. In addition, several glycoproteins have been found in cartilage matrix: fibronectin, chondronectin, and more recently thrombospondin.[8] Interaction of cells with such extracellular matrix macromolecules particularly influences their migration, adhesion, growth, and differentiation. Various studies suggest that there are specific receptors for fibronectin and, more recently, for collagen. The structure of one of these, anchorin CII, has been described.[9,10]

Matrix degradation is due to numerous enzymes synthesized both by synovial cells and chondrocytes. Cathepsin B_1 has been found in cartilage and particularly high levels are found during osteoarthrosis. Glucosaminidase and hyaluronidase are also present in cartilage, but hyaluronidase and collagenase are not detectable due to the presence of metalloproteinase inhibitors. Important research concerns the immunolocalization of metalloproteinases and their inhibitors in the rabbit growth plate.[11] Chondrocytes located in resting and proximal proliferation zones were shown to synthesize and secrete all of the metalloproteinases and the tissue inhibitor of metalloproteinase. Synthesis of collagenase was also found in the remainder of the proliferative zone and in the most distal cells of the hypertrophic zone. Stromelysin was found to be synthesized in all zones, implying that it plays an important role in degradation.

One of the aims of chondrocyte culture is to reproduce, as far as possible, the *in vivo* situation. As described in this chapter, most methods of *in vitro* culture do not permit the maintenance of the different functional properties of chondrocytes. Despite this, much research has been carried out using cell culture, and the resulting data has implicated many mechanisms in the regulation of chondrocyte function.

The culture of chondrocytes has been performed since the beginning of tissue culture. Carrel and Burrows[12] described the *in vitro* survival of this cell type in the first historical paper on the cultivation of tissue outside the body. However, it was not until the 1960s that various experiments on cultured chondrocytes were performed. Chondrocytes have been cultured from a large variety of tissue sources. Many reports have been published on growth plate chondrocytes from chick or rabbit. However, cells from articular cartilage of mammalian origin have also been studied. Besides human and dog, rabbit chondrocytes have been used most often. The choice of rabbit is due to the fact that the articular tissue of young rabbits is particularly rich in cells. In addition, Webber et al.[13] concluded that rabbit chondrocytes grew more rapidly and incorporated sulfate into proteoglycans several times more rapidly than chondrocytes from cat, dog, sheep, or monkey.

II. ORGAN CULTURE

Organ culture of cartilaginous tissue does not present unusual problems when compared with organ culture of other tissues. The articular tissue is dissected, cut into slices 1 to 2 mm thick, and placed in a tissue culture dish containing media.[14] The medium level is adjusted to cover the tissue to allow efficient gas exchange. Cultures are generally fed every 1 to 2 d with fresh medium. A more recent technique consists of cultivating tissue at the interface of air and medium, using 0.45 µm-millipore filters cemented on to stainless steel grids.

Whatever the type of culture, the medium used is a rich synthetic basal medium. Indeed, chondrocytes have high nutritional requirements for survival and for production of matrix components. The media used are generally supplemented with fetal calf serum (10%), but Shurtz-Swirsk et al.[15] demonstrated that mouse mandibular cartilage maintained chondrogenic expression in serum-free medium.

The use of cartilage slices in organ culture allows chondrocytes to be studied in the presence of normal cell matrix interactions since the tissue remains intact. Several examples will be given in the following section and compared to other types of culture. Although less frequently used than monolayer cell culture, the demonstration of steady-state metabolism using adult bovine articular cartilage organ culture[16,17,100] has stimulated renewed interest in this technique.

III. MONOLAYER CULTURE

Monolayer is the most frequent type of culture referred to in the chondrocyte literature. This technique, and others, requires the release of cartilaginous cells which has been frequently achieved using the technique of Green.[18] Cartilage slices are washed and treated with trypsin to remove contaminating cells, and chondrocytes are subsequently released with collagenase digestion. In the original paper by Green, hyaluronidase was also used before trypsin to increase cartilage permeability by degrading proteoglycans.

More recently, Benya et al.[20] described a modification of this method using a short collagenase step, followed by overnight treatment with collagenase at a lower concentration in the presence of fetal calf serum. This procedure releases approximatively 90% of the chondrocytes present in the cartilage slices.

In most cases, cells are released without separation of the different subpopulations present in growth plate and articular cartilage. However, the separation of cartilage zones can be performed by manual dissection or subpopulations isolated by more sophisticated techniques (e.g., attachment kinetics, antibody selection, countercurrent centrifugal elutriation, or density gradient centrifugation).[21,22] These techniques permit chondrocytes that differ in location or physical/biochemical properties to be evaluated for variation in metabolic responsiveness and phenotype. For example, Aydelotte et al.[23] showed that the subpopulations of bovine articular chondrocytes from different depths of the cartilage varied in the amount of proteoglycan that they deposited. Chondrocytes from deep zones synthesized significantly higher amounts of proteoglycans than the cells from the superficial zone. Aydelotte and Kuettner[24] concluded that articular chondrocytes continue to express in culture metabolic differences which reflect their original anatomical location. Epiphysal chondrocytes separated by density gradient centrifugation showed differences between hypertrophic cells and proliferative cells. Fractions rich in hypertrophic cells contained larger cells which have a higher alkaline phophatase activity. After their release, cells are seeded into various dishes in a culture medium such as DMEM or Ham's F12. Articular chondrocyte behavior was studied comparatively by Green in these two media. If the synthesis of proteoglycans occurred in these two culture media, DMEM greatly increased the sulfate incorporation by rabbit chondrocytes, as compared with Ham's F12. Conversely, the average cloning efficiency was 29 and 3%, respectively, for Ham's F12 and DMEM. These media are generally used supplemented with 10% of fetal calf serum or serum substitutes such as ultroser G.[25] However, chondrocyte culture in serum-free medium (SFM) has also been attempted. The first synthetic SFM was proposed by Kato et al.[26] Supporting both proliferation and proteoglycan synthesis by chondrocytes in primary culture, this medium contains insulin, calcitonin, parathyroid hormone, and a somatomedin. Then, Jennings and Ham[27] established an SFM able to support clonal growth of rabbit ear chondrocytes. The medium was MCDB104, a lipid-enriched nutrient medium supplemented with insulin and fibroblast growth factor (FGF). The same authors described a similar medium for human hyaline chondrocytes.[28] However, no expression of cartilage-like differentiation occurred in this defined medium. In 1984, another medium was described for rabbit articular chondrocytes:[29] it contained insulin, FGF, hydrocortisone, bovine serum albumin, transferrin, selenium, and fibronectin. This medium permitted 75% of the growth rate obtained in the presence of FCS (10%) and maintained the expression of type II collagen detected using anti-type II collagen antibodies. From these different experiments, it is not clear whether the serum-free media described until now permit a better maintenance of differentiated functions in monolayer culture than media supplemented with FCS.

A. DIVISION

Monolayer culture is characterized by a high proliferative capacity. Ronot et al.[30] showed that rabbit articular chondrocytes cultivated in Ham's F12 + 10% FCS remain in suspension for the first 48 to 72 h. The growth curve contains a lag period until day 3 and an exponential growth phase from day 3 to day 5. In contrast, subcultured cells multiply exponentially until day 3 following a very short lag period. The length of the cell cycle and the doubling time are $T_c = 19$ h and $T_d = 20$ h, respectively. However, the time required to form a monolayer depends on cellular density. A cell culture is referred to as a "monolayer culture" when cell density varies between 5×10^3 to 16×10^3 per cm^2. Beyond this cellular concentration, the culture can be defined as high-density culture with characteristics very different from monolayer.

1. Growth Factors and Hormones

Since 1972, when Corvol et al.[31] described a pituitary growth factor in articular chondrocytes, later named chondrocyte growth factor (CGF), many other growth factors having a growth-promoting action on cultured chondrocytes have been isolated and later characterized. Chondrocyte growth regulation by various growth factors has been extensively studied in monolayer culture. In 1981, Kato et al.[32] showed that a polypeptide isolated from fetal bovine cartilage stimulated DNA synthesis in cultured rabbit costal chondrocyte. This polypeptide, "called cartilage-derived factor" (CDF), permits cell division in serum-free medium. The same year, Webber and Sokoloff[33] demonstrated that FGF and CGF increased DNA synthesis of rabbit growth plate. Prins et al.[34] described this phenomenon in rabbit articular chondrocytes with platelet-derived growth factor (PDGF), epidermal growth factor (EGF), and FGF. Insulin acted synergistically with these factors. On the other hand, FGF and EGF were continuously required in the culture medium during traverse of the entire G1 phase for stimulation of DNA synthesis. The mitogenic effects of FGF and EGF were stronger than those of CDF.[35]

Redini et al.[36] demonstrated that transforming growth factor-ß (TGF-ß) slightly enhanced the proliferation of cultured rabbit articular chondrocytes. However, Hiraki et al.[37] found a more complex result in rabbit growth plate chondrocytes. In serum-free medium, TGF-ß caused dose-dependent inhibition of DNA synthesis. The inhibitory effect was maximal at a dose of 1 ng/ml and extended for a duration of 16 to 42 h. TGF-ß potentiated the synthesis of DNA stimulated by fetal calf serum. Addition of TGF-ß to culture medium containing 10% FCS produced a sixfold increase of [^{3}H]-thymidine incorporation, as compared to culture medium with 10% FCS alone. TGF-ß also potentiated DNA synthesis stimulated by PDGF, EGF, and FGF. This TGF-ß-induced potentiation of DNA synthesis was associated with chondrocyte replication. Using countercurrent centrifugal elutriation, Rosier et al.[38] fractionated growth plate chondrocytes to obtain populations of cells in different stages of maturation. TGF-ß caused an increase in cell division and cellular maturation in the growth plate, with maximal stimulation in the proliferating and early hypertrophic cells. Somatomedins, a family of growth factors [insulin-like growth factors (IGF)-I and 2, multiplication stimulating activity (MSA), and somatomedine A and C] has been described as stimulating thymidine incorporation into DNA in chondrocytes. For example, Makower et al.[39] observed stimulation of DNA synthesis in resting, proliferative, and hypertrophic chondrocytes obtained by fractionation after treatment with FGF and IGF-I. In order to better define the role of somatomedin C insulin-like growth factor I (Sm-C/IGF-I) in skeletal development, Trippel et al.[40] studied the effects of Sm-C/IGF-I on bovine growth plate and rabbit articular and growth plate chondrocytes in primary culture. These cells were highly responsive with respect to [^{3}H]-thymidine incorporation. In articular chondrocytes cultured in serum-free medium, Froger-Gaillard et al.[41] have shown that there was a positive correlation between cell number and release into the medium of both IGF-I and IGF-II. Furthermore, the synthesis of binding proteins and IGF appeared to be coordinated. On the contrary, no significant mitogenic effect of growth hormone (GH) on rabbit articular chondrocytes has been found.[42] However, GH has been reported to exert direct effects on rat and rabbit epiphyseal chondrocytes *in vitro*[43] and growth hormone receptors have been recently demonstrated in cultured rat epiphyseal chondrocytes.[44]

Stimulation of DNA synthesis in primary cultures of chicken chondrocytes by parathyroid hormone (PTH) was studied by assaying [^{3}H]-thymidine incorporation into DNA. DNA synthesis was significantly stimulated by PTH and some of its fragments. Using a series of synthetic PTH peptides covering the central region of the hormone molecule, Schluter et al.[45] delimited the putative mitogenic functional domain. PTH action appears cAMP independent, but seems to involve calcium ions for signal transduction.

Dihydrotestosterone (DHT) and testosterone (T) significantly stimulated DNA synthesis by cultured human fetal epiphyseal chondrocytes.[46] In addition, DHT binding sites were present in these cells. Together, these observations support the idea that androgens elicit a biological response in epiphyseal chondrocytes. Similarly, estrogen receptors have been found by Dayani et al.[47] in pubertal and prepubertal rabbit articular cells.

2. Cytokines

The effects of various cytokines on matrix production have been intensively studied, while their action on division has received less attention. Ikebe et al.[48] have observed that human recombinant tumor necrosis factor α (rTNFα) stimulated [^{3}H]-thymidine incorporation in cultured rat costal chondrocytes in a dose-dependent manner, whereas interleukin-1α (IL-1α) and IL-1ß had no stimulatory activity on DNA synthesis. Fetal calf serum acted synergistically with rTNFα in increasing DNA synthesis. However, IL-1ß had a pronounced, but reversible cytostatic effect on rabbit articular chondrocytes.[49] This phenomenon has been confirmed by Iwamoto et al.[50] in two types of tridimensional culture. This suppression of chondrocyte replication by IL-1 may play an important role in cartilage destruction associated with chronic inflammatory joint diseases.

3. Vitamins

Several studies have suggested that vitamin D_3 is involved in cartilage metabolism. Physiological concentrations of 24,25-dihydroxyvitamin D_3 (24,25-[OH]$_2$D$_3$), a metabolite of vitamin D, were shown to be active in chondrocyte cultures, and specific nuclear binding was demonstrated in growth plate chondrocytes in culture.[51,52] The effects of 1,25-dihydroxyvitamin D_3 (1,25-[OH]$_2$D$_3$) and 24,25-[OH]$_2$D$_3$ were studied on third-passage chondrocytes derived from the resting zone and adjacent growth region of rat costochondral cartilage.[53] 1,25-$[OH]_2D_3$ inhibited growth cartilage increases in cell number at pharmacological concentrations and had no effect on resting cartilage cell number. In contrast, 24,25-$[OH]_2D_3$ appeared to stimulate resting cartilage cell number at physiological concentrations and inhibit these cells at pharmacological doses, but had no effect on growth cartilage chondrocytes.

Kristal et al.[54] showed that ascorbic acid increased the incorporation of [^{3}H]-thymidine in rabbit chondrocytes in monolayer and organ culture. However, ascorbate was cytotoxic at concentrations greater than 0.2 mM in the presence of some batches of serum.

An increasing number of studies have demonstrated that retinoids, a group of natural and synthetic vitamin A analogs, exert profound effects on cell growth and differentiation of normal and transformed cells. The DNA content of cultured embryonic chick sternal chondrocytes treated with retinoic acid for 6 d was 59% lower than in control cultures.[55] Two cell lines derived from human chondrosarcomas[56] exhibited changes in both cell growth and morphology following 4 d of retinoic acid treatment. These two cell lines exhibited different susceptibility to growth inhibitory effects (50% inhibition by 10^{-9} M or 2 × 10^{-7} M). This inhibition of proliferation has been confirmed by Hein et al.[57] When cultured rabbit articular chondrocytes were evaluated, a biphasic dose-dependent effect of retinoic acid was observed: at 10^{-7} M, a significant stimulation of proliferation occured; whereas at 10^{-5} M, growth inhibition occurred and was related to an accumulation of cells in the G0-G1 phase of the cell cycle.[58]

4. Slowing of Division after Subculture (Chondrocyte *in vitro* Aging)

When cells derived from articular cartilage are subcultured, chondrocytes undergo *in vitro* senescence similar to that described for other mesodermic diploid cells. Moskalewski et al.[59] demonstrated that chondrocytes isolated from rabbit articular cartilage exhibited the onset of phase III senescence after 10 to 14 population doublings. Evans and Georgescu[60] established that the population-doubling capacity of rabbit, dog, and human chondrocytes in culture is related to the life span of the donor and inversely related to the age of the donor. Using rabbit articular chondrocytes subcultured several-fold, Dominice et al.[61] described a slowing of the proliferative capacity at the forth passage followed by a complete loss of division after 8 ± 1 passages. Chondrocytes at the fourth subculture can be considered as senescent and are characterized by a decrease in proliferative capacity, a reduction of the proportion of cells in S and G2 + M phases of the cell cycle, and a concommitent enhancement in protein content related to the increase of cell size. Immunocytochemistry revealed a rigid cytoarchitecture with an increase in the number and organization of three cytoskeletal components: actin, tubulin, and vimentin. Tubulin content was modulated as a function of the protein content revealed by flow cytometry.[62] Addition of either acid or basic FGF to the culture medium of such senescent chondrocytes provoked proliferation and results in smaller cells with lower protein content.[63]

5. Division Modifications Induced by Drug Treatment

Several investigations have studied the effects of antirheumatic drugs on the proliferative capacity of chondrocytes in monolayer culture. Ronot et al.[64] showed that methylprednisolone (10^{-3} to 10^{-5} M) inhibited DNA synthesis in monolayer culture of rabbit articular chondrocytes. This growth inhibitory effect has also been found in organ culture of mouse neonatal condylar cartilage.[65] Dexamethasone at high doses induced slowing of the *in vitro* proliferation of rabbit articular chondrocytes. Even though glucocorticoid-specific receptors have been demonstrated in chondrocytes, this growth inhibition did not seem to be related to an interaction with the glucocorticoid receptor complex.[66] Among various nonsteroidal agents, Kirkpatrick et al.[67] showed that 10^{-4} M indomethacin markedly reduced proliferation of lapine articular chondrocytes. Jaffray et al.[63] demonstrated that a long-acting drug, D-penicillamine, also inhibited the growth of rabbit articular chondrocytes. This inhibitory effect was dose related between 5×10^{-4} and 5×10^{-3} M. Flow cytometry using propidium iodide DNA staining showed that drug exposure led to a slow-down in cell cycle progression, especially due to an accumulation of cells in the G0-G1 phase, and a slight decrease in cell transit through the G2M phase of the cell cycle. Sodium aurothiopropanolsulfonate also exerts a dose-dependent action on the *in vitro* proliferation kinetics of articular chondrocytes by reducing growth with a maximal and irreversible inhibitory effect at 5×10^{-4} M. Flow cytometry revealed a slight but significant cell arrest in G2 + M which, in fact, represents an increase in the proportion of binucleate cells.[69]

B. RESPIRATION

When transferred from the *in vivo* environment to conventional culture conditions, chondrocytes undergo many changes and, particularly, a modification in oxygen tension (the incubator usually contains 20% oxygen). Champagne et al.[70] showed that rabbit articular chondrocytes transferred from cartilage to culture conditions increased their stock of mtDNA. On the contrary, overall mitochondrial activity, estimated by rhodamine uptake by flow cytometry, is not modified. These data suggest the occurrence of a specific mitochondrial adaptation following the transfer of chondrocytes to *in vitro* conditions. This appears to begin with an increase of mtDNA synthesis without a correlative increase in respiratory activity.

When rabbit articular chondrocytes are cultivated in hypoxia (5% O_2), cell growth is stimulated as compared to normoxia (20% O_2) and the mitochondrial activity studied by rhodamine uptake decreases slightly (personal data). However, until now there have been few reports about mitochondria, metabolism in cell culture and many experiments remain to be done.

C. MATRIX PRODUCTION

The main disavantage of monolayer culture concerns the production of a matrix which differs from the one produced *in vivo*. In these *in vitro* culture conditions, the chondrocyte phenotype is labile with respect to collagen synthesis. It has been observed that there is a switch from type II to type I colla-gen[71-74] with the onset of synthesis of type I procollagen RNA accompanied by a loss of $\alpha 1(II)$ procollagen gene expression.[75] This phenomenon, which begins at the end of primary culture, is particularly evident in the first and subsequent subcultures.[74] In addition, proteoglycan production appears reduced, in spite of the experiments of Bjornsson and Heinegard[76] which demonstrated that proteoglycan synthesis was maintained during primary culture and character-ized the structure of the proteoglycans that were produced.

The kinetics of intracellular processing of chondroitin sulfate proteoglycan core protein have been studied[77] in chick embryonic chondrocytes. The core protein is processed very rapidly, exhibiting a $t_{1/2}$ less than 10 min, in both the rough endoplasmic reticulum and the Golgi regions. Link protein appears to be processed as rapidly as core protein, but type II collagen kinetics are three to four times slower.

A major part of this dedifferentiation is reversible before cellular senescence is well established. Benya and Shaffer[78] showed that chondrocytes dedifferen-tiated by serial subcultures can reexpress the differentiated phenotype during suspension culture in agarose gels. The rates of proteoglycan and collagen synthesis return to those of primary chondrocytes. Using SDS polyacrylamide gel electrophoresis of intact collagen chains and two-dimensional cyanogen bromide peptide mapping, they demonstrated a complete return to the differen-tiated collagen phenotype. This result emphasizes the important role of cell shape in the modulation and reexpression of chondrocyte phenotype.

Archer et al.[79] focused on cell shape and cartilage differentiation in early chick limb bud cells in culture and showed that these cells, maintained in a rounded configuration by culturing on a semiadhesive substratum (poly HEMA), synthesized more extracellular matrix than cells allowed to flatten on normal tissue culture plastic.

Takigawa et al.[80] studied the relationship between the cytoskeleton and differentiated properties in cultured rabbit costal chondrocytes. Cytochalasin B changed the shape of chondrocytes from polygonal to nearly spherical and stimulated glycosaminoglycan (GAG) synthesis, whereas colchicine changed them from polygonal to flattened, inhibited GAG synthesis. These authors suggested that the integrity of microtubules and disruption of microfilaments are involved in the regulation of the expression of the differentiated phenotype of chondrocytes in culture. The spreading of chondrocytes has been studied by fluorescence and interference reflection microscopy. When chondrocytes flat-ten, they develop stress fibers and show a diffuse system of vinculin-containing adhesion plaques scattered over the entire ventral side of the cells.[81] Newman and Watt[82] have explored the relationship between the morphology of articular chondrocytes in culture and the type of proteoglycan they synthesize, using cytochalasin D to induce reversible cell rounding. When chondrocytes were prevented from spreading or when spread cells were induced to round up, $^{35}SO_4$ incorporation into proteoglycan was stimulated. This stimulation reflected an increase in core protein synthesis rather than lengthening of GAG chains; but chondrocytes that have been passaged to stimulate dedifferentiation did not incorporate more $^{35}SO_4$ when treated with cytochalasin D, suggesting that increased proteoglycan synthesis in response to rounding may be, in itself, a differentiated property of chondrocytes.

The importance of cell shape has also been observed in hypertrophic chondrocytes regarding type X collagen synthesis.[83] Indeed, cells grown in monolayer and then resuspended exhibited a sixfold increase in type X collagen expression. It seems that cytoskeletal reorganization might be involved in the transduction of shape information into interpretable signals for gene expression. Recently, however, using *in situ* hybridization and immunocytochemistry, Mallein-Gerin et al.[84] have demonstrated that proteoglycan core protein and type II collagen are not correlated with cell shape changes during the culture of chicken embryo chondrocyte. According to these authors, cell shape changes do not have a causative role in chondrocyte phenotype expression, but rather are a secondary effect of the dedifferentiation process.

It should also be noted that the phenomenon of redifferentiation of chondrocytes dedifferentiated by serial monolayer culture has been observed *in vivo*.[85] Chondrocytes isolated from rabbit costal growth cartilage during successive passages lost their ability to proliferate, as previously described. This was accompanied by a decrease in GAG synthesis. However, subcutaneous transplantation of cells at the fifth passage into nude mice resulted in the formation of cartilage nodules which were able to actively produce GAG *in vitro*. Using rabbit articular chondrocytes modulated in primary culture by retinoic acid,[86] Brown and Benya[87] and Benya et al.[88] demonstrated that reexpression of differentiated collagen phenotype is possible in secondary culture following treatment with dihydrocytochalasin B (DHCB). Rhodamine-labeled phalloidin showed that treatment with DHCB resulted in rapid changes in microfilament stress fibers. These results identify microfilaments as the cytoskeletal element mainly affected and demonstrated that their modification, rather than their complete disruption, is sufficient for reexpression. Thus, chondrocytes need not become spherical to initiate the processes of reexpression.

1. Growth Factors and Hormones

The action of growth factors on proteoglycan synthesis has been particularly well studied. Prins et al.[89] studied the effects of FGF, PDGF, and EGF on rabbit articular chondrocytes while varying the concentrations of FCS. PDGF stimulated radiosulfate incorporation. The action of FGF was intermediate between that of EGF and PDGF. According to Kato and Gospodarowicz,[90] FGF effects were only observed when FGF was present during the cell logarithmic growth phase of rabbit costal chondrocytes, but not when it was added after chondrocyte cultures became confluent. Chondroitin sulfate proteoglycans synthesized in the presence of FGF were slightly larger in size than those produced in the absence of FGF.

TGF-ß caused dose-dependent stimulation of GAG synthesis in confluent cultures of growth-plate chondrocytes.[91] This stimulatory effect was greater than that of IGF-I or PDGF. Furthermore, TGF-ß stimulated GAG synthesis additively with IGF-I or PDGF. The effects of TGF-ß are enhanced by factors present in serum.[92] The results concerning collagen synthesis are divergent: a decrease in collagen synthesis was found in chick epiphyseal chondrocytes, and a stimulation of collagen synthesis was described in rabbit articular chondrocyte cultures.[93] More recently, Rosier et al.[94] have confirmed the inhibition of collagen synthesis of chick growth plate chondrocytes after TGF-ß treatment.

Early works[95-97] using purified somatomedin peptide with insulin-like activity have been followed by detailed studies concerning the effects of IGF-I on the synthesis and processing of GAG in cultured chick chondrocytes.[98] A stimulation of GAG synthesis was observed with a lag time of about 2 h, and the processing rate of proteoglycan core protein was investigated. The stimulatory effect of IGF-1 on proteoglycan synthesis was confirmed on bovine articular cartilage explants, indicating that serum IGFs are major regulatory factors of cartilage proteoglycan metabolism.[99,100] IGF-II appears less potent than IGF-I. Madsen et al.[101] have demonstrated that human growth hormone stimulates sulfate incorporation into proteoglycans of cultured chondrocytes from rat growth cartilage.

PTH has been found to increase GAG synthesis by rabbit costal chondrocytes, influencing chain elongation and termination of GAGs in PG.[102-104] This effect is different from the one observed with IGFs, which primarily is on the synthesis and secretion of PG. The stimulation of GAG synthesis by PTH has been confirmed on cultured chondrocytes of other origin.[105,106] The mechanism of action of PTH on proteoglycan synthesis has been approached by Ianotti et al.[107] using growth plate chondrocytes. The effect appears to be mediated by the degradation products of membrane phospho-inositides.

A direct effect of sex steroid hormones was demonstrated by Corvol et al.[108] They showed that testosterone and different derivatives provoked a stimulatory effect on [^{35}S] incorporation into newly synthetized PG of rabbit epiphyseal articular chondrocytes. The response of cells varied with animal age and sex.

Recently, Vukicevic et al.[109] described a bone inductive protein, osteogenin, which is able to increase the production of sulfated proteoglycans in fetal rat chondroblasts and in rabbit articular chondrocytes. This bone factor which was purified and partially sequenced appears to be able to stimulate the chondrogenic phenotype *in vitro*.

2. Vitamins
24,25[OH]$_2$D$_3$ stimulates proteoglycan synthesis by rabbit growth plate chondrocytes.[110] This phenomenon appears only in confluent culture when chondrocytes express their differentiated phenotype.[111] Chondrocytes derived from embryonic caudal sterna treated with 1,25[OH]$_2$D$_3$ showed a decrease in type II collagen synthesis which accompanied a decrease in its mRNA. In contrast, translational control of type I collagen synthesis was observed. An increase in proteoglycan synthesis and core protein mRNA was observed.[112]

Vitamin C stimulates matrix production in rabbit chondrocyte cultures.

McDevitt et al.[114] showed that ascorbate increased the amount of radiosulfate incorporated into GAGs and directed it into matrix deposition. Concerning collagen phenotype, the effects described by Sandell and Daniel[115] appear more complex. Ascorbic acid does not affect differentiated chondrocytes from the chicken embryo, while bovine articular chondrocytes begin to undergo a phenotypic change.

It has been shown that vitamin A inhibits chondrogenesis.[116,117] The effects of vitamin A and retinoids on matrix molecules appear complex. Sternal chondrocytes in culture, studied in the presence of vitamin A, produced smaller proteoglycans which were released and degraded more rapidly.[118] The profile of GAG produced by chondrocytes from rat cartilage, when grown in the presence of retinoids, resembled that of fibroblasts.[119] Retinoic acid completely inhibited the synthesis of "cartilage-specific" proteoglycans (PG-1), while the synthesis of the ubiquitous proteoglycan (PG-2) was only slightly affected.[120] Total collagen synthesis was reduced by retinol while α-2-chain synthesis was significantly increased. This suggested a switch of collagen synthesis in favor of type I and the presence of dedifferentiation.[121]

Retinoic acid treatment of rabbit articular chondrocytes led to the complete loss of the type II collagen synthesis and the induction of type I-trimer collagen, in parallel with eightfold decreases in collagen and proteoglycan synthesis.[86] Modulation of the phenotype is dose dependent and correlated with an alteration of cell morphology. Such cells have been described as fibroblast-like in appearance.[122] In summary, retinoids appear to contribute to cartilage destruction by increasing the turnover of matrix molecules and provoking the synthesis of abnormal matrix components.

3. Cytokines

Partially purified monocyte factor with IL-1 properties enhances proteoglycan synthesis by human neonatal articular chondrocytes in culture and changes the partition of these macromolecules between medium and cell layer.[123] In contrast, this factor inhibits [^{3}H]-proline incorporation into collagen synthesized by rabbit articular chondrocytes.[124] Goldring et al.[125,126] observed that the inhibition of type I, II, and III collagen synthesis was associated with the supression of the $\alpha 1$ (I), $\alpha 2$ (I), $\alpha 1$ (II), and $\alpha 1$ (III) procollagen mRNA levels after treatment of cultured human articular and costal chondrocytes with interferon γ. However, these authors observed an opposite effect with IL-1 when synthesis of prostaglandin E_2 was inhibited by indomethacin. They concluded that IL-1 and INF-γ could have an opposite role in the modulation of cartilage matrix turnover in joint disease by affecting repair as well as degradation.[127] Recently, Lefebvre et al.[128] have demonstrated that IL-1 and TNF-α profoundly influence collagen degradation, reduce type IX and XI collagen synthesis by differentiated chondrocytes, and type I and V by dedifferentiated chondrocytes.

Using human recombinant IL-1, a decrease in GAG synthesis was observed.[129,130] This phenomenon was also observed with tumor necrosis factor-α (TNF-α).[131] The mechanism of inhibition of cartilage proteoglycan synthesis by IL-1 has been investigated by Benton and Tyler;[132] they found a direct inhibitory effect on the synthesis of core protein without alteration in the rate of intracellular transport or secretion of completed PG.

4. Matrix Modifications Induced by Drugs

The effects of glucocorticoids on sulfated proteoglycan synthesis by rabbit costal chondrocytes have been studied by Kato and Gospodarowicz in serum-free conditions.[133] Chondrocytes maintained in presence of 10^{-7} M hydrocortisone reorganized at confluence into a homogeneous cartilage-like tissue. Cell ultrastructure and fibrils of the pericellular matrix were similar to those seen *in vivo*. The level of $^{35}SO_4$ incorporated into PG was 33-fold higher than that of glucocorticoid-free cultures. Dexamethasone also has been found to increase GAG synthesis.[134] On the contrary, Mitrovic et al.[135] have observed that between 10^{-5} and 10^{-7} M hydrocortisone, dexamethasone, and the nonsteroidal antiinflammatory drug, indomethacin, inhibited $^{35}SO_4$ incorporation into newly synthesized proteoglycan of bovine articular chondrocytes. In rabbit articular chondrocytes, several other nonsteroidal antiinflammatory drugs present the same effect.[136] This inhibitory effect has also been observed on the biosynthesis of type II collagen in cultured chondrocytes treated by indomethacin and aspirin. However, at lower doses, an increase in the biosynthesis of both collagen and noncollagen proteins has been found in rabbit articular chondrocytes.[137] Fujii et al.[138,139] have obtained similar results with several nonsteroidal antiinflammatory drugs tested on proteoglycan and collagen synthesis. In contrast, polysulphated polysaccharide, a potentially antiarthritic drug, improves proteoglycan incorporation into the extracellular matrix of cultured articular chondrocytes.[140] On the other hand, the long-acting drug D-penicillamine does not modify the chemical characteristics of rabbit articular chondrocytes proteoglycans.[141]

D. CATABOLIC ENZYME RELEASE

Normal rabbit articular chondrocytes in monolayer culture release latent metal-dependent neutral proteinases into the medium.[142-145] The neutral metal-dependent proteoglycanase predominates in the culture medium of human cartilage. This enzyme is mostly released in latent form.[146] Its activation results in the release of a 10,000-Da fragment. Chondrocytes also release inhibitory activity directed against proteoglycanase. The elaboration of neutral proteoglycanase has been confirmed by Mercier et al.[147] in cartilage growth plate in tissue culture. However, in cell culture, no protease activity could be detected. This might be due to the absence of hypertrophic chondrocytes which could be the origin of this neutral protease.

Solavagione et al.[148] have described the presence of N-acetyl ß-hexosaminidase, ß-galactosidase, and ß-glucuronidase in cultured rabbit articular chondrocytes. Secretion of enzyme activity seems to result preferentially in the accumulation of N-acetyl ß-hexosaminidase which may play a role in GAG catabolism.

When chondrocytes are grown in monolayer culture, both the intracellular pool of cathepsin B and its secretion are very low initially, but increase progressively by a factor of 110 after several weeks of culture. Transferring the cells into collagen gel reverses this change.[149] In contrast, collagenase is secreted in almost the same amounts. Furthermore, Morris has isolated a new inhibitor of collagenase from the medium of confluent differentiated rabbit chondrocytes.[150] This molecule is different from the low molecular weight collagenase inhibitor and similar to tissue inhibitor of metalloproteinase (TIMP).[151] Much attention has been devoted to soluble substances derived from monocytes and macrophages that are able to stimulate the production of collagenase and neutral proteases in cultured chondrocytes.[152-154] McGuire et al.[155] have emphasized that IL-1 could be responsible for these activities. It has also been shown that chondrocytes prepared from the articular cartilage of young rats synthesize and secrete an IL-1 activity into the medium.[156] The physical and biochemical properties of this IL-1-like activity are similar to macrophage-derived IL-1. Shinmei et al.[157] demonstrated using cartilage fragments in which sensitivity to IL-1 was higher in osteoarthritic and arthritic cartilage as compared with healthly cartilage. Hulkower et al.[158] suggested that signal transduction in chondrocytes responding to IL-1 involves the activation of one or more protein kinases.

It should be noted that the effects of IL-1 on the production of neutral proteases which cause degradation of the large aggregating proteoglycan were blocked by the pleiotropic growth factor, transforming growth factor-ß.[159]

In cultured human articular chondrocytes, the addition of TNF-α stimulates caseinase activity, whereas INF-γ has no significant effect.[160] Andrews et al.[161] have demonstrated that INF-γ inhibits metalloproteinase production and opposes the IL-1 stimulatory effect. This suggests a possible role for INF-γ in limiting cartilage degradation in inflammatory joint disease.

Alkaline phosphatase has been especially studied in cultured epiphyseal growth plate chondrocytes because it is a key enzyme involved in biomineralization.[162] Much research concerns the effects of vitamin D_3 and its metabolites on the expression of alkaline phosphatase which appears in relation to the stage of differentiation.[163-165] On the other hand, Leboy et al.[166] suggested that vitamin C plays an important role in endochondral bone formation by modulating gene expression of alkaline phosphatase and type X collagen in hypertrophic chondrocytes. Retinoic acid also modulates alkaline phosphatase activity.[167] In rabbit articular chondrocytes, $5 \times 10^{-5} M$ retinoic acid induced an increase and $10^{-7} M$ induced a decrease in this enzyme activity.

Kato and Iwamoto have studied the terminal differentiation of rabbit growth plate chondrocytes maintained as a pelleted mass in a centrifuge tube.[168] Their data show that FGF abolishes the increase in alkaline phosphatase and suggest that FGF inhibits chondrocyte terminal differentiation and calcification.

IV. THREE-DIMENSIONAL CELL CULTURE

Three-dimensional (3-D) cell culture may be divided in two main types: one consists of embedding the chondrocytes in a gel (collagen or agarose), and the other is the result of particular culture conditions (suspension or high-density culture) in which chondrocytes themselves produce their 3-D environment.

The first type of culture (chondrocytes embedded in a gel) was due to Horwitz and Dorfman in 1970;[169] however, it was not until 1982 that several reports were published about chondrocytes in collagen or agarose. Yasui et al.[170] and Gibson et al.[171] isolated chondrocytes from sternal embryonic cartilage and embedded them in collagen gels. Under these conditions, evidence for cell viability, growth, and deposition of extracellular matrix has been reported. Benya and Shaffer[172] have also observed that chondrocytes dedifferentiated by serial monolayer culture can reexpress their differentiated phenotype during suspension culture in firm gels of 0.5% low Tm agarose. Proteoglycan and collagen synthesis rates returned to those which existed in primary culture. Three-dimensional cell culture of limb mesenchyme, mesectoderm, and sternal chondrocytes have pointed out the importance of environmental regulation on the expression of specific collagen types.[173] This stability of chondrocyte phenotype in 3-D culture has been studied by various authors on different types of chondrocytes:[174] human,[175] swarm rat chondrosarcoma,[176] chick embryo sternal chondrocytes from the caudal and cephalic regions,[177] embryonic chick whole sterna,178 rabbit articular chondrocytes,[179] human epiphyseal,[180] and canine tumoral cells,[181] either in agarose or collagen gels. The attempts to investigate the mechanisms involved for the expression and reexpression of phenotype have been previously discussed. The expression of collagen mRNAs in human chondrocytes grown in agarose has recently been studied by Elima and Vuorio.[182] They observed the reappearance of type II collagen mRNA. Methods using 3-D culture in gels are particularly interesting because they permit the study of a fundamental problem in the regulation of chondrocyte differentiation. When most methods of gel culture are used, there is only a short period of cell proliferation. This limits the use of gel culture when growth factors and drugs are tested.

However, Skantze et al.,[183] noting that chondrocytes are the only normal diploid cells having the capability of growing in soft agar, tested the effects of TGF-ß and EGF. After treatment with these two growth factors, an increase in DNA content was observed along with a decrease in GAG and collagen synthesis. Stimulation of proliferation has also been found by Iwamoto et al.[184] using TGF-ß and FGF. The effects of vitamin D metabolites have been studied on chick embryo chondrocyte growth in soft agar. $1,25[OH]_2D_3$ induced colony formation, whereas $24-25[OH]_2D_3$ had little effect.[185] Kato et al.[186] observed a decrease in the efficiency of colony formation by chondrocytes in soft agar after treatment with indomethacin. Retinoic acid-treated chondrocytes in suspension in methylcellulose remained rounded, but synthesized proteins characteristic of flattened, dedifferentiated chondrocytes. This result suggests that retinoic acid has a direct effect on chondrocytes and that cell shape change is secondary.[187]

Chondrocyte culture in suspension has also been successful. Deshmukh and Kline[188] demonstrated that upon transfer from monolayer to suspension culture, the cells synthesize type II collagen in a medium devoid of $CaCl_2$. On the other hand, Wiebkin and Muir[189] found that chondrocytes isolated from larynges of adult pigs and cultured in suspension synthesized cartilage specific proteoglycan. Recently, the transfer of dedifferentiated chondrocytes to suspension culture was used to demonstrate that the resultant cellular accumulation of type II collagen mRNA is regulated not only at the transcriptional level, but also through stabilization of the RNA transcript.[190] Epiphyseal chondrocytes in suspension culture have been treated by GH or IGF-I.[191] GH potentiates the formation of particularly large colonies and suggests that GH promotes the differentiation of young chondrocytes. In contrast, IGF-I produces a higher proportion of small colonies and appears to stimulate chondrocytes at a later stage of differentiation. If normal or osteoarthritic articular chondrocytes are cultivated in suspension and treated with salicylate, the inhibitory effect on GAG synthesis is the same in the both cases.[192] Chondrocyte aggregates can be obtained by culturing cell suspensions in spinner bottles or in gyratory shakers. In spinner bottles, proliferation is reduced and metachromatic material is deposited about the aggregated cells.[193] The profile of GAGs formed corresponds closely to that of whole articular cartilage. Using a gyratory shaker, flaky aggregates appear after 5 d of culture and become more dense after 10 d. Chondrocytes are morphologically differentiated with a round shape; biochemical and immunocytochemical analysis of the extracellular matrix reveals the presence of type II collagen and cartilaginous PGs.[194] This last model has been used for different pharmacological experiments with human chondrocytes.[195-197] Pharmacological concentrations of calcitonin exerted no proliferative effect on chondrocytes, but displayed a dose-dependent and prolonged stimulatory effect on PG and type II collagen production.[198] Calcitonin may possess chondroprotective properties.

High-density cultures are obtained by seeding chondrocytes at a concentration of 10^5 cells/cm^2. After a few days, cells reestablish a territorial matrix which is rich in collagen fibrils and proteoglycans.[199] The number of cells decreases during the first 5 d and remains constant over a 4-week period.[201] The importance of seeding density on stability of the differentiated phenotype has been emphasized by Watt[200] using pig articular chondrocytes. Chondrocytes grown at high density for 21 d expressed predominantly large proteoglycans that aggregated with hyaluronic acid; whereas in low density cultures, a smaller, nonaggregating form was also present. By 21 d in culture, cells at both high and low density were expressing type I collagen, although the high-density cells also had an extensive extracellular matrix of type II collagen. Multilayer culture of canine articular chondrocytes on porous hydroxyapatite ceramic granules has been reported by Cheung.[202] Chondrocytes proliferate for 13 months and maintain their collagen phenotype for up to 11 months. Using high-density primary cultures of bovine articular chondrocytes, Watanabe et al.[203] have characterized a specific IGF-I receptor and studied the regulation of receptor concentration by somatomedin, insulin, and growth hormone. In both rabbit and human chondrocyte culture, chronic vitamin D$_3$ treatment inhibited chondrocyte proliferation and stimulated PG synthesis. Human chondrocytes derived from healthy and osteoarthritic cartilage and cultured in high-density display metabolic characteristics in relation to the histopathologic grade of osteoarthritis.[205] A comparative study of chick high-density cultures in medium with serum and defined medium appears to demonstrate that in serum-free medium PG production is modified.[206] This fact confirms the data obtained in monolayer with defined medium. It seems that serum-free medium does not improve expression of the chondrocyte phenotype; but the availability of this type of medium allows the exploration of bioactive factors which affect or modulate chondrocyte differentiation.[207]

Aggregate suspension culture and high-density culture appear more suitable for larger numbers of cells than cells suspended in gel. However, in most cases, cellular proliferation and extracellular matrix deposition are rapidly initiated and then decline substantially. This is perhaps due to matrix feedback on these processes.

In connection with these methods, culture on agarose-coated dishes or poly(2-hydroxyethylmethacrylate) (polyhema)-coated dishes has to be mentioned.[208,209] These products decrease the surface contact area of chick chondrocytes and induce differentiation.[210] Recently, Guo et al.[211] described chondrocyte culture in alginate beads which results in small cell clumps and large aggregates. The aggregates are surrounded by a dense alcian blue-positive halo. This methodology is applicable for chondrocytes cultured in single beads, in multiwell dishes, or mass culture.

V. IMMORTALIZED CELL LINES

Studies using cells in culture are often limited by the short life span and the instability of the differentiated properties of most cell types *in vitro*. The dream of obtaining cell lines possessing both the capacity for infinite proliferation and maintenance of differentiated properties has always stimulated researchers.

Historically, the concept of cellular immortality was born in 1910 with the first experiments of Carrel[212] who described the maintenance of fibroblasts in culture for 1 year and concluded that the cells, like microorganisms, multiply indefinitely. In fact, this was untrue, and in 1961 Hayflick and Moorhead[213] distinguished normal cell lines with limited life span from continuous cell lines possessing infinite growth capacity. This last category is the result of the establishment *in vitro,* often by accident, of various cell types from normal or pathological tissue. However, the idea of cellular immortality got a new start with the progress of virology in 1970 and then again with the use of recombinant DNA or viruses containing oncogenes with an immortalizing function. Thus, immortalized cell lines are continuous cell lines intentionally transfected with v and c oncogenes. These cell lines have acquired an infinite growth capacity and, on this point, the relationship with transformed precancerous cell lines is not clear. It seems that immortalized cell lines are maintained at a step preceding tumorigenic modification.

Attempts to immortalize various cell types have been made for three main reasons: immortalization permits the study of steps in progression to transformation, allows the establishment of cell lines producing biological products, and permits the generation of cell lines maintaining differentiated functions. Such cell lines will be useful tools for investigating cellular regulation and pharmacotoxicology.

An immortalized chondrocyte cell line would be particularly interesting because there are very few continuous cell lines of this type. The most famous one is a tumorigenic cell line, the Swarm rat chondrosarcoma. Takigawa et al.[214] have established a clonal cell line from a human chondrosarcoma with cartilage phenotype and tumorigenicity. However, in both cases it concerns transformed cancerous cell lines. Several attempts have been made to immortalize normal chondrocytes. Katoh and Takayama[215] described the first establishment of a clonal cell line of hamster sternal chondrocytes. These cells were obtained by treatment with 4-nitroquinoline 1-oxide, and maintained many of their differentiated properties. More recently, viral infection has been used to immortalize chondrocytes. However, the maintenance of differentiated properties has not always been studied. Nevertheless, Gionti et al.[216] used infection with the avian myelocytomatosis virus (MC29) carrying the myc oncogene to obtain immortalized quail embryo chondrocytes which still expressed type II collagen and cartilage proteoglycans. Primary fetal rat costal chondrocytes infected with a recombinant retrovirus carrying myc and raf oncogenes produced an immortalized cell line which synthesizes high levels of cartilage proteoglycan, but shows weak type II collagen expression.[217] Gionti et al.[218] generated a continuous cell line of chicken embryo cells from a culture of chondrocytes infected with Rous sarcoma virus. These cells exhibit reduced serum requirements, are able to grow in a semisolid medium, and are not tumorigenic. However, this cell line does not synthesize type II or type X collagen, which are the differentiation markers of normal chicken chondrocytes in culture. Immortalized rabbit articular chondrocytes were obtained by transfection with SV40 large T and little t encoding genes.[219] These cells have been maintained in culture for 2 years. Growth curves of normal and SV40 transfected chondrocytes were compared and displayed similar doubling times (approximately 20 h). These cells retained polygonal morphology, but the synthesis of alcian blue stainable matrix was 7- to 23-fold reduced. Northern Blot analysis of RNAs from monolayer culture of these immortalized chondrocytes did not permit the detection of any type II procollagen mRNAs. Two-dimensional cyanogen bromide peptide maps of labeled collagens from these cultures showed that the major collagen synthesized was type I collagen. No type II collagen was detectable. Several attempts to stimulate reexpression of collagen type II after culturing these immortalized cells in collagen gel or after treatment with cytochalasin B have been unsuccessful.[220]

Although immortalization of mammalian chondrocytes seems to be easily obtained by viral infection, by using recombinant retroviruses, or by transfection with plasmids encoding for SV40 early functions, conservation of all differentiated functions and their regulation have been imperfect. These results might be improved by transfecting chondrocytes before significant exposure to *in vitro* culture conditions that induce early modifications in their phenotype. Other immortalizing vectors more adapted to chondrocyte metabolism might also be used. Finally, transgenic animals carrying immortalizing oncogenes in their germ cells might permit the establishment of immortalized chondrocyte cell lines.

VI. CONCLUSION

Chondrocyte culture represents a large domain from a technical point of view and provides a useful model for studying cellular regulation in normal or pathological conditions. Since conventional monolayer culture is limited by phenotypic instability, numerous other types of culture have been described: culture in collagen or agarose, high density culture, suspension culture, etc. Each of them presents advantages and limitations. For this reason, the choice of a chondrocyte culture system should be based on careful evaluations of expected goals and the requirements of analytical methods chosen for each experimental design.

Monolayer culture at low cell density appears to be a good model to evaluate the proliferative capacity of chondrocytes, its regulation by growth factors and hormones, and its modifications by various drugs. However, matrix production is not similar to that *in vivo,* particularly at the end of the primary culture and during subcultures. Culture in gels, in suspension, or in high density maintain phenotypic stability with several exceptions and are good models to study the regulation of collagen, proteoglycan, and glycoprotein metabolism. Three-dimensional culture in agarose gel appears particularly fascinating to investigate the regulation of the chondrocyte phenotype. However, this technique is limited by a slow rate of cell proliferation.

Whatever the type of cultured chondrocyte, these various models have permitted the identification of numerous regulatory pathways which should be confirmed *in vivo.* A mitogenic effect has been described for EGF, FGF, and IGF-I. The action of TGF-ß is more complex; it potentiates DNA synthesis stimulated by other factors. On the contrary, IL-1 presents a cytostatic effect. Various growth factors and hormones have been found to increase GAG synthesis. TGF-ß stimulates GAG synthesis in chondrocyte culture and collagen synthesis appears stimulated or inhibited in parallel. Purified IL-1 provokes an inhibitory effect on core protein and on various types of collagen synthesis. On the other hand, articular chondrocytes are able to produce an IL-1 similar to that produced by macrophages.

The differentiation of epiphyseal growth plate chondrocytes has been extensively studied in culture in relation to the roles of vitamin D_3, vitamin C, and vitamin A on the modulation of alkaline phosphatase, a key enzyme in mineralization. The regulation of chondrocyte phenotype regarding type II collagen biosynthesis in 3-D agarose gel culture has underlined the importance of cytoskeletal reorganization as a regulatory signal.

However, cultured chondrocytes are not only useful for studying physiologically relevant regulation, but also for investigating regulation in pathological conditions and for studying the different effects of drugs or various modulators of chondrocyte function. The effect of pathological conditions could be evaluated by culturing chondrocytes from arthritic or osteoarthritic tissue or from infrequent diseases such as nanomelia.[221] Cartilage defect could be detected on these *in vivo-in vitro* models. For example, nanomelic chondrocytes have been shown to be deficient in the production of chondroitin sulfate PG and also synthesize a smaller glycoprotein related to core protein. Schuckett and Malemud[222] have demonstrated stable differences in proteoglycan synthesis, but not in intracellular processing between nonarthritic and osteoarthritic chondrocytes. A good correlation has been found between metabolic characteristics of *in vitro* cultured human osteoarthritic chondrocytes and the different histopathologic grades of the source cartilage.[223] Chondrocytes from the nasal septal cartilage of healthy adults and acromegalic patients have been used to determine differences in clonal proliferation.[224] Finally, tissue and cell studies of the growth plate promise to elucidate the cellular and molecular biology of chondrodysplasia.[225]

Alternatively, several attempts have been made to create pathological *in vitro-in vitro* models composed of normal cultured chondrocytes disturbed by various treatments such as exposure to IL-1, oxygen free-radicals, or senescence due to serial passage. For these *in vitro-in vitro* models, it is possible to study drugs interacting with the induced pathological state. For example, articular chondrocytes treated with oxygen free radicals produced by the enzymatic action of xanthine oxidase upon hypoxanthine grew more slowly. This was due to a pertubation in the cell cycle that led to an increase in the proportion of cells in G2M and in protein content.[226] The cell cycle arrest was associated with a decrease in c-myc and c-Ha-ras mRNA levels.[227] These different pertubations could be suppressed by catalase. Thus, this model may be useful for screening scavenger drugs.

An another application of cultured chondrocytes concerns the effects of electrical fields[228-230] or pulsed electromagnetic field.[231,232] Hiraki et al.[231] have shown that exposure of rabbit costal chondrocytes to a pulsed electromagnetic field resulted in functional differentiation of the cells. However, the study of cartilage response to mechanical force is more promising for the pathophysiology of cartilaginous tissue.[233-235] High-density cultures of chick embryonic chondrocytes were exposed to an intermittent compressive force of physiologic magnitude for 24 h. Proteoglycan synthesis was significantly increased in chondrocyte cultures exposed to the force.[236] Gray et al.[237] have shown, an organ culture system, that compressive loading produces physiochemical changes including a decreased water content and changes in local interstitial pH.

Cultured chondrocytes have also been used as grafts to fill defects in cartilage.[238-241] Itay et al.[242] described a graft with cultured homologous embryonic chick epiphyseal chondrocytes embedded in a biological resorbable immobilization vehicle. This graft was transplanted in mechanically induced defects of the tibiotarsal joint of 4-month-old roosters. Healing of the defects was observed for 6 months. The transplanted cells grew well and the resulting cartilage was structurally reorganized. The articular zone preserved its cartilaginous phenotype, whereas the subchondral regions were transformed into bone. Finally, cultured cells may be used to study the induction of chondrogenesis *in vitro*. Second passage of neonatal rat muscle can be induced to change to a chondrocyte-like mode of expression by the addition of materials prepared from decalcified bovine cortical bone.[243] Several experiments have described the differentiation of muscle, fat, cartilage, and bone from different clones under the influence of dexamethasone,[244] 5-azacytidine,[245] and vitamin D_3.[246] Atsumi et al.[247] used a differentiating culture of AT805 teratocarcinoma treated with 10 mg/ml insulin to obtain cell aggregates which appear similar to cartilage nodules and produce cartilage-specific proteoglycan and type II collagen. These data could be of great use in understanding the regulatory mechanisms of mesenchymal stem cell differentiation.

In conclusion, the firmly established experimental approach of chondrocyte culture facilitates and stimulates many research interests that will lead to a better understanding of the biological regulation of cartilaginous tissue.

VII. ACKNOWLEDGMENTS

Part of this original work was supported by INSERM grant CRE N°897001. The authors thank Mrs. B. Benoit, S. Demignot, S. Thenet, and F. Vincent for helpful suggestions and for critical reading of this article, and R. Frolleau for expert typing of the manuscript.

REFERENCES

1. **Mankin, H. J.,** Localization of tritiated thymidine in articular cartilage of rabbits. III. Mature articular cartilage, *J. Bone Joint Surg.,* 45, 529, 1963.
2. **Kunz, J., Wellmitz, G., Paul, U., and Fuhrmann, I.,** Histoautoradiographic studies on chondrocyte proliferation in healing of experimental cartilage injuries, *Zentralbl. Allg. Pathol.,* 123, 539, 1979.
3. **Havdrup, T. and Telhag, H.,** Mitosis of chondrocytes in normal adult joint cartilage, *Clin. Orthop.,* 153, 248, 1980.
4. **Ghadially, F. N.,** Fine structure of joints, in *The Joints and Synovial Fluid,* Sokolof, L., Ed., Academic Press, New York, 1978, 105.
5. **Brighton, C. T.,** The Growth Plate, *Orthop. Clin. North Am.,* 15, 571, 1984.
6. **Ricard-Blum, S., Tiollier, J., Garrone, R., and Herbage, D.,** Further biochemical and physiochemical characterization of minor disulfide bonded (type IX collagen, extract from foetal calf cartilage), *J. Cell Biochem.,* 27, 347, 1985.
7. **Mayne, I. R. and Irwin, M. H.,** Collagen types in cartilage, in *Articular Cartilage Biochemistry,* Kuettner, K. E., Schleyerbach, R., and Hascall, V. C., Eds., Raven Press, New York, 1985, 23.
8. **Miller, R. R. and McDevitt, C. A.,** Thrombospondin is present in articular cartilage and is synthesized by articular chondrocytes, *Biochem. Biophys. Res. Commun.,* 153, 708, 1988.

9. **Pilar-Fernandez, M., Selmin, O., Martin, G. R., and Yamada, Y.,** The structure of anchorin CII, a collagen binding protein isolated from chondrocyte membrane, *J. Biol. Chem.,* 263, 5921, 1988.

10. **Pfaffle, M., Ruggiero, F., Hofmann, H., Fernandez, M. P., Selmin, O., Yamada, Y., Garrone, R., and von der Mark, K.,** Biosynthesis, secretion and extracellular localization of anchorin CII, a collagen-binding protein of the calpactin family, *EMBO J.,* 7, 2335, 1988.

11. **Brown, C. C., Hembry, R. M., and Reynolds, J. J.,** Immunolocalization of metalloproteinases and their inhibitor in the rabbit growth plate, *J. Bone Joint Surg.,* 71, 580, 1989.

12. **Carrel, A. and Burrows, M. T.,** Cultivation of adult tissues and organs outside of the body, *JAMA,* 55, 1379, 1910.

13. **Webber, R. J., Malemud, C. J., and Sokoloff, L.,** Species differences in cell culture of mammalian articular chondrocytes, *Calcif. Tissue Res.,* 23, 61, 1977.

14. **McKenzie, L. S., Horsburgh, B. A., Ghosh, P., and Taylor, T. K. F.,** Organ culture of human articular cartilage: studies on sulfated glycosaminglycan synthesis, *In Vitro,* 13, 423, 1977.

15. **Shurtz-Swirski, R., Lewinsson, D., Shenzer, P., and Silbermann, M.,** Effects of different concentrations of serum on cartilage in an organ culture system, *In Vitro Cell Dev. Biol.,* 25, 995, 1989.

16. **Handley, C. J., Ng, C. K., and Curtis, A. J.,** Short and long-term explant culture of cartilage, introduction in *Methods in Cartilage Research,* Kuettner, K., and Maroudas, A., Eds., Academic Press, London, 1990, 105.

17. **Hascall, V. C., Luyton, F. P., Plaas, A. H. K., and Sandy, J. D.,** Steady-state metabolism of proteoglycans in bovine articular cartilage explants, in *Methods in Cartilage Research,* Kuettner, K. and Maroudas, A., Eds., Academic Press, London, 1990, 108.

18. **Green, W. T.,** Behavior of articular chondrocytes in cell culture, *Clin. Orthop.,* 75, 248, 1971.

19. **Kawiak, J., Moskalewski, S., and Darzynkiewicz, Z.,** Isolation of chondrocytes from calf cartilage, *Exp. Cell Res.,* 39, 59, 1965.

20. **Benya, P. D.,** Chondrocyte culture, introduction and survey of techniques for chondrocyte culture, in *Methods Cartilage Research,* Kuettner, K. and Maroudas, A., Eds., Academic Press, London, 1990, 85.

21. **O'Keefe, R. J., Crabb, I. D., Puzas, J. E., and Rosier, R. N.,** Countercurrent centrifugal elutriation. High-resolution method for the separation of growth-plate chondrocytes, *J. Bone Joint Surg.,* 71, 607, 1989.

22. **Ralphs, J. R., Evans, L., and Ali, S. Y.,** Separation of rabbit epiphyseal chondrocytes in various stages of differentiation, *Cell Tissue Res.,* 254, 393, 1988.

23. **Aydelotte, M. B., Greenhill, R. R., and Kuettner, K. E.,** Differences between sub-populations of cultured bovine articular chondrocytes. II. Proteoglycan metabolism, *Connect. Tissue Res.,* 18, 223, 1988.

24. **Aydelotte, M. B., and Kuettner, K. E.,** Differences between subpopulations of cultured bovine articular chondrocytes. I. Morphology and cartilage matrix production, *Connect. Tissue Res.,* 18, 205, 1988.

25. **Ronot, X., Sene, C., Boschetti, E., Hartmann, D. J., and Adolphe, M.,** Culture of chondrocytes in medium supplemented with fetal calf serum or a serum substitute: Ultroser G, *Biol. Cell,* 51, 307, 1984.

26. **Kato, Y. and Gospodarowicz, D.,** Effect of exogenous extracellular matrices on proteoglycan synthesis by cultured rabbit costal chondrocytes, *J. Cell Biol.,* 100, 486, 1985.

27. **Jennings, S. D. and Ham, R. G.,** Clonal growth of primary culture of rabbit ear chondrocytes in a lipid-supplemented defined medium, *Exp. Cell Res.,* 145, 415, 1983.

28. **Jennings, S. D. and Ham, R. G.,** Clonal growth of primary cultures of human hyaline chondrocytes in a defined medium, *Cell Biol. Int. Rep.,* 7, 149, 1983.

29. **Adolphe, M., Froger, B., Ronot, X., Corvol, M. T., and Forest, N.,** Cell multiplication and type II collagen production by rabbit articular chondrocytes cultivated in a defined medium, *Exp. Cell Res.,* 155, 527, 1984.

30. **Ronot, X., Hecquet, C., Jaffray, P., Guiguet, M., Adolphe, M., Fontagne, J., and Lechat, P.,** Proliferation kinetics of rabbit articular chondrocytes in primary culture and at the first passage, *Cell Tissue Kinet.,* 16, 531, 1983.

31. **Corvol, M. T., Malemud, C. J., and Sokoloff, L.,** A pituitary growth promoting factor for articular chondrocytes in monolayer culture, *Endocrinology,* 90, 263, 1972.

32. **Kato, Y., Nomura, Y., Tsuji, M., Ohmae, H., Nakazawa, T., and Suzuki, F.,** Multiplication-stimulating activity (MSA) and cartilage-derived factor (CDF): biological actions in cultured chondrocytes, *J. Biochem.,* 90, 1377, 1981.

33. **Webber, R. J. and Sokoloff, L.,** In vitro culture of rabbit growth plate chondrocytes. I. Age-dependence of response to fibroblast growth factor and "chondrocyte growth factor", *Growth,* 45, 252, 1981.

34. **Prins, A. P., Lipman, J. M., and Sokoloff, L.,** Effect of purified growth factor on rabbit articular chondrocytes in monolayer culture. I. DNA synthesis, *Arthritis Rheum.,* 25, 1217, 1982.

35. **Hiraki, Y., Kato, Y., Inoue, H., and Suzuki, F.,** Stimulation of DNA synthesis in quiescent rabbit chondrocytes in culture by limited exposure to somatomedin-like growth factors, *Eur. J. Biochem.,* 158, 333, 1986.

36. **Redini, F., Galera, P., Mauviel, A., Loyau, G., and Pujol, J. P.,** Transforming growth factor beta stimulates collagen and glycosaminoglycan biosynthesis in cultured rabbit articular chondrocytes, *FEBS Lett.,* 234, 172, 1988.

37. **Hiraki, Y., Inoue, H., Hirai, R., Kato, Y., and Suzuki, F.,** Effect of transforming growth factor beta on cell proliferation and glycosaminoglycan synthesis by rabbit growth-plate chondrocytes in culture, *Biochim. Biophys. Acta,* 969, 91, 1988.

38. **Rosier, R. N., O'Keefe, R. J., Crabb, I. D., and Puzas, J. E.,** Transforming growth factor beta: an autocrine regulator of chondrocytes, *Connect. Tissue Res.,* 20, 295, 1989.

39. **Makower, A. M., Wroblewski, J., and Pawlowski, A.,** Effects of IGF-1, rGH, FGF, EGF, and FCS on DNA-synthesis, cell proliferation and morphology of chondrocytes isolated from rat rib growth cartilage, *Cell Biol. Int. Rep.,* 13, 259, 1989.

40. **Trippel, S. B., Corvol, M. T., Dumontier, M. F., Rappaport, R., Hung, H. H., and Mankin, H. J.,** Effect of somatomedin-C/insulin-like growth factor 1 and growth hormone on cultured growth plate and articular chondrocytes, *Pediatr. Res.,* 25, 76, 1989.

41. **Froger-Gaillard, B., Hossenlopp, P., Binoux, M., and Adolphe, M.,** Production of insulin-like growth factors and their binding proteins by rabbit articular chondrocytes: relationships with cell multiplication, *Endocrinology,* 124, 2365, 1989.

42. **Jones, K. L., Villela, J. F., and Lewis, U. J.,** The growth of cultured rabbit articular chondrocytes is stimulated by pituitary growth factors but not by purified human growth hormone or ovine prolactin, *Endocrinology,* 118, 2588, 1986.

43. **Madsen, K., Friberg, U., Roos, P., Eden, S., and Isaksson, 0.,** Growth hormone stimulates the proliferation of cultured chondrocytes from rabbit ear and rat rib growth cartilage, *Nature (London),* 304, 545, 1983.

44. **Nilsson, A., Lindahl, A., Eden, S., and Isaksson, O. G.,** Demonstration of growth hormone receptors in cultured rat epiphyseal chondrocytes by specific binding of growth hormone and immunohistochemistry, *J. Endocrinol.,* 122, 69, 1989.

45. **Schluter, K. D., Hellstern, H., Wingender, E., and Mayer, H.,** The central part of parathyroid hormone stimulates thymidine incorporation of chondrocytes, *J. Biol Chem.,* 264, 11087, 1989.

46. **Carraccosa, A., Audi, L., Ferrandez, M. A., and Ballabriga, A.,** Biological effects of androgens and identification of specific dihydrotestosterone-binding sites in cultured human fetal epiphyseal chondrocytes, *J. Clin. Endocrinol. Metab.,* 70, 134, 1990.

47. **Dayani, N., Corvol, M. T., Robel, P., Eychenne, B., Moncharmont, B., Tsagris, L., and Rappaport, R.,** Estrogen receptors in cultured rabbit articular chondrocytes: influence of age, *J. Steroid Biochem.,* 31, 351, 1988.

48. **Ikebe, T., Hirata, M., and Koga, T.,** Effects of human recombinant tumor necrosis factor-alpha and interleukin-1 on the synthesis of glycosaminoglycan and DNA in cultured rat costal chondrocytes, *J. Immunol.,* 140, 827, 1988.

49. **Chin, J. E. and Lin, Y. A.,** Effects of recombinant human interleukin-1 beta on rabbit articular chondrocytes. Stimulation of prostanoid release and inhibition of cell growth, *Arthritis Rheum.,* 31, 1290, 1988.

50. **Iwamoto, M., Koike, T., Nakashima, K., Sato, K., and Kato, Y.,** Interleukin-1: a regulator of chondrocyte proliferation, *Immunol. Lett.,* 21, 153, 1989.

51. **Corvol, M. T., Dumontier, M. F., Garabedian, M. F., and Rappaport, R.,** Vitamin D and cartilage-II biological activity of 25-hydroxycholecalciferol and 24,25- and 1,25-dihydro cholecalciferol on cultured growth plate chondrocytes, *Endocrinology,* 102, 1269, 1978.

52. **Corvol, M. T., Ulman, A., and Garabedian, M. F.,** Specific nuclear uptake of 24,25-dihydroxycholecalciferol, a vitamin-D_3 metabolites biologically active in cartilage, *FEBS Lett.,* 116, 273, 1980.

53. **Boyan, B. D., Schwartz, Z., Carnes, D. L., and Ramirez, V.,** The effects of vitamin D metabolites on the plasma and matrix vesicle membranes of growth and resting cartilage cells in vitro, *Endocrinology,* 122, 2851, 1988.

54. **Krystal, G., Morris, C. M., and Sokoloff, L.,** Stimulation of DNA synthesis by ascorbate in cultures of articular chondrocytes, *Arthritis Rheum.,* 25, 318, 1982.

55. **Vasan, N. S.,** Proteoglycan synthesis by sternal chondrocytes pertubed vitamin A, *J. Embryol. Exp. Morphol.,* 63, 181, 1981.

56. **Thein, R. and Lotan, R.,** Sensitivity of cultured human osteosarcoma and chondrosarcoma cells to retinoic acid, *Cancer Res.,* 42, 4771, 1982.

57. **Hein, R., Krieg, T., Mueller, P. K., and Braun-Falco, O.,** Effect of retinoids on collagen production by chondrocytes in culture, *Biochem. Pharmacol.,* 33, 3263, 1984.

58. **Ronot, X., Nafziger, J., Hecquet, C., Dronne, N., Arock, M., Guillosson, J. J., and Adolphe, M.,** Retinoic acid: modulating effects on the proliferation kinetics of mastocytes and chondrocytes, *Biol. Cell,* 55, 5, 1985.

59. **Moskalewski, S., Adamiec, I., and Golaszewska, A.,** Maturation of rabbit articular chondrocytes grown in vitro in monolayer culture, *Am. J. Anat.,* 155, 339, 1979.

60. **Evans, C. H. and Georgescu, H. I.,** Observations on the senescence of cells derived from articular cartilage, *Mech. Ageing Dev.,* 22, 179, 1983.

61. **Dominice, J., Levasseur, C., Larno, S., Ronot, X., and Adolphe, M.,** Age-related changes in rabbit articular chondrocytes, *Mech. Ageing Dev.,* 37, 231, 1986.

62. **Ronot, X., Froger-Gaillard, B., Hainque, B., and Adolphe, M.,** In vitro ageing chondrocytes identified by analysis of DNA and tubulin content and relationship to cell size and protein content, *Cytometry,* 9, 436, 1988.

63. **Froger-Gaillard, B., Charrier, A. M., Thenet, S., Ronot, X., and Adolphe, M.,** Growth-promoting effects of acidic and basic fibroblast growth factor on rabbit articular chondrocytes. Ageing in culture, *Exp. Cell Res.,* 183, 388, 1989.

64. **Ronot, X., Blondelon, D., Perret, C., Adolphe, M., Fontagne, J., and Lechat, P.,** Effects of high doses of methylprednisolone on rabbit articular chondrocytes in culture, *Int. J. Tissue React.,* 2, 145, 1980.

65. **Maor, G. and Silbermann, M.,** In vitro effects of glucocorticoid hormones on the synthesis of DNA in cartilage of neonatal mice, *FEBS Lett.,* 129, 256, 1981.

66. **Hainque, B., Dominice, J., Jaffray, P., Ronot, X., and Adolphe, M.,** Effects of dexamethasone on the growth of cultured rabbit articular chondrocytes: relation with the nuclear glucocorticoid-receptor complex, *Ann. Rheum. Dis.,* 46, 146, 1987.

67. **Kirkpatrick, C. J., Mohr W., Wild Feller, A., and Haerkamp, O.,** Influence of nonsteroidal antiinflammatory agents on lapine articular chondrocyte growth in vitro, *Z. Rheumatol.,* 42, 58, 1983.

68. **Jaffray, P., Ronot, X., Adolphe, M., Fontagne, J., and Lechat, P.,** Effects of D-penicillamine on growth and cell cycle kinetics of cultured rabbit articular chondrocytes, *Ann. Rheum. Dis.,* 43, 333, 1984.

69. **Ronot, X., Hainque, B., Christen, M.-O., Froger, B., Hartmann, D. J., Adolphe, M., and Lechat, P.,** Rabbit articular chondrocytes: an in vitro model for studying the effect of sodium aurothiopropanolsulfonate on proliferation kinetics, type II collagen phenotype and mitochondrial activity, *Fund. Clin. Pharmacol.,* 2, 57, 1988.

70. **Champagne, A. M., Benel, L., Ronot, X., Mignotte, F., Adolphe, M., and Mounolou, J. C.,** Rhodamine 123 uptake and mitochondria DNA content in rabbit articular chondrocytes evolve differently upon transfer from cartilage to culture conditions, *Exp. Cell Res.,* 171, 404, 1987.

71. **Muller, P. K., Lemmen, G., Gay, S., and Kuhn, K.,** Immunochemical and biochemical study of collagen synthesis by chondrocytes in culture, *Exp. Cell Res.,* 108, 47, 1977.

72. **Von der Mark, K., Gauss, V., von der Mark, H., and Muller, P.,** Relationship between cell shape and type of collagen synthesized as chondrocytes lose their cartilage phenotype in culture, *Nature (London),* 267, 531, 1977.

73. **Benya, P. D., Padilla, S. R., and Nimni, M. E.,** The progeny of rabbit articular chondrocytes synthesize collagens type I and III and type I trimer, but not type II. Verifications by cyanogen bromide peptide analysis, *Biochemistry,* 16, 865, 1977.

74. **Benya, P. D., Padilla, S. R., and Nimni, M. E.,** Independent regulation of collagen types by chondrocytes during the loss of differentiated function in culture, *Cell,* 15, 1313, 1978.

75. **Duchene, M., Sobel, M. E., and Muller, P. K.,** Levels of collagen mRNA in dedifferentiating chondrocytes, *Exp. Cell Res.,* 142, 317, 1982.

76. **Bjornsson, S. and Heinegard, D.,** Fractionation and characterization of proteoglycans isolated from chondrocyte cell cultures, *Biochem. J.,* 197, 249, 1981.

77. **Campbell, S. C. and Schwartz, N. B.,** Kinetics of intracellular processing of chondroitin sulfate proteoglycan core protein and other matrix components, *J. Cell Biol.,* 106, 2191, 1988.

78. **Benya, P. D. and Shaffer, J. D.,** Dedifferentiated chondrocytes reexpress the differentiated collagen phenotype when cultured in agarose gels, *Cell,* 30, 215, 1982.

79. **Archer, C. W., Rooney, P., and Wolpert, L.,** Cell shape and cartilage differentiation of early chick limb bud cells in culture, *Cell Differ.,* 11, 245, 1982.

80. **Takigawa, M., Takano, T., Shirai, E., and Suzuki, F.,** Cytoskeleton and differentiation: effects of cytochalasin B and colchicine on expression of the differentiated phenotype of rabbit costal chondrocytes in culture, *Cell Differ.,* 14, 197, 1984.

81. **Marchisio, P. C., Capasso, O., Nitsch, L., Cancedda, R., and Gionti, E.,** Cytoskeleton and adhesion patterns of cultured chick embryo chondrocytes during cell spreading and rous sarcoma virus transformation, *Exp. Cell Res.,* 151, 332, 1984.

82. **Newman, P. and Watt, F. M.,** Influence of cytochalasin D-induced changes in cell shape on proteoglycan synthesis by cultured articular chondrocytes, *Exp. Cell Res.,* 178, 199, 1988.

83. **Adams, S. L., Pallante, K. M., and Pacifici, M.,** Effects of cell shape on type X collagen gene expression in hypertrophic chondrocytes, *Connect. Tissue Res.,* 20, 223, 1989.

84. **Mallein-Gerin, F., Ruggiero, F., and Garrone, R.,** Proteoglycan core protein and type II collagen gene expressions are not correlated with cell shape changes during low density chondrocyte cultures, *Differentiation,* 43, 204, 1990.

85. **Takigawa, M., Shirai, E., Fukuo, K., Tajima, K., Mori, Y., and Suzuki, F.,** Chondrocytes dedifferentiated by serial monolayer culture form cartilage nodules in nude mice, *Bone Miner.,* 2, 449, 1987.

86. **Benya, P. D. and Padilla, S. R.,** Modulation of the rabbit chondrocyte phenotype by retinoic acid terminates type II collagen synthesis without inducing type I collagen: the modulated phenotype differs from that produced by subculture, *Dev. Biol.,* 118, 296, 1986.

87. **Brown, P. D. and Benya, P. D.,** Alterations in chondrocyte cytoskeletal architecture during phenotypic modulation by retinoic acid and dihydrocytochalasin-B-induced reexpression, *J. Cell Biol.,* 106, 171, 1988.

88. **Benya, P. D., Brown, P. D., and Padilla, S. R.,** Microfilament modification by dihydrocytochalasin-B causes retinoic acid-modulated chondrocytes to reexpress the differentiated collagen phenotype without a change in shape, *J. Cell Biol.,* 106, 161, 1988.

89. **Prins, A. P., Lipman, J. M., McDevitt, C. A., and Sokoloff, L.,** Effect of purified factors on rabbit articular chondrocytes in monolayer culture. II. Sulfated proteoglycan synthesis, *Arthritis Rheum.,* 25, 1228, 1982.

90. **Kato, Y. and Gospodarowicz, D.,** Sulfated proteoglycan synthesis by confluent cultures of rabbit costal chondrocytes grown in the presence of fibroblast growth factor, *J. Cell Biol.,* 100, 477, 1985.

91. **Hiraki, Y., Inoue, H., Hirai, R., Kato, Y., and Suzuki, F.,** Effect of transforming growth factor beta on cell proliferation and glycosaminoglycan synthesis by rabbit growth-plate chondrocytes in culture, *Biochim. Biophys. Acta,* 969, 91, 1988.

92. **O'Keefe, R. J., Puzas, J. E., Brand, J. S., and Rosier, R. N.,** Effects of transforming growth factor-beta on matrix synthesis by chick growth plate chondrocytes, *Endocrinology,* 122, 2953, 1988.

93. **Redini, F., Galera, P., Mauviel, A., Loyau, G., and Pujol, J. P.,** Transforming growth factor beta stimulates collagen and glycosaminoglycan biosynthesis in cultured rabbit articular chondrocytes, *FEBS Lett.,* 234, 172, 1988.

94. **Rosier, R. N., O'Keefe, R. J., Crabb, I. D., and Puzas, J. E.,** Transforming growth factor beta: an autocrine regulator of chondrocytes, *Connect. Tissue Res.,* 20, 295, 1989.

95. **Corvol, M. T., Dumontier, M. F., Rappaport, R., Guyda, H., and Posner, B. I.,** The effect of a slightly acidic somatomedin peptide (ILAs) on the sulphation of proteoglycans from articular and growth plate chondrocytes in culture, *Acta Endocrinol.,* 89, 263, 1978.

96. **Kemp, S. F. and Hintz, R. L.,** The action of somatomedin on glycosaminoglycan synthesis in cultured chick chondrocytes, *Endocrinology,* 106, 744, 1980.

97. **Asakawa, K., Takano, K., Hizuka, N., Kogawa, M., and Shizume, K.,** Effects of somatomedin and other factors on glycosaminoglycan synthesis in cultured rat chondrocytes, *Endocrinol. Jpn.,* 30, 701, 1983.

98. **Kemp, S. F., Kearns, G. L., Smith, W. G., and Elders, M. J.,** Effects of IGF-1 on the synthesis and processing of glycosaminoglycan in cultured chick chondrocytes, *Acta Endocrinol.,* 119, 245, 1988.

99. **Luyten, F. P., Hascall, V. C., Nissley, S. P., Morales, T. I., and Reddi, A. H.,** Insulin-like growth factors maintain steady-state metabolism of proteoglycans in bovine articular cartilage explants, *Arch. Biochem. Biophys.,* 267, 416, 1988.

100. **Schalkwijk, J., Joosten, L. A., Van der Berg, W. B., Van Wyk, J. J., and Van de Putte, L. B.,** Insulin-like growth factor stimulation of chondrocyte proteoglycan synthesis by human synovial fluid, *Arthritis Rheum.,* 32, 66, 1989.

101. **Madsen, K., Makower, A. M., Friberg, U., Eden, S., and Isaksson, O.,** Effect of human growth hormone on proteoglycan synthesis in cultured rat chondrocytes, *Acta Endocrinol.,* 108, 338, 1985.

102. **Takigawa, M., Takano, T., and Suzuki, F.,** Effects of parathyroid hormone and cyclic AMP analogues on the activity of ornithine decarboxylase and expression of the differentiated phenotype of chondrocytes in culture, *J. Cell. Physiol.,* 106, 259, 1981.

103. **Takano, T., Takigawa, M., Shirai, E., Suzuki, F., and Rosenblatt, M.,** Effects of synthetic analogs and fragments of bovine parathyroid hormone on adenosine 3′,5′-monophosphate level, ornithine decarboxylase activity, and glycosaminoglycan synthesis in rabbit costal chondrocytes in culture, structure-activity relations, *Endocrinology,* 116, 2536, 1985.

104. **Hiraki, Y., Yutani, Y., Takigawa, M., Kato, Y., and Suzuki, F.,** Differential effects of parathyroid hormone and somatomedin-like growth factors on the sizes of proteoglycan monomers and their synthesis in rabbit costal chondrocytes in culture, *Biochim. Biophys. Acta,* 845, 445, 1985.

105. **Takano, T., Takigawa, M., Shirai, E., Nakagawa, K., Sakuda, M., and Suzuki, F.,** The effect of parathyroid hormone (1-34) on cyclic AMP level, ornithine decarboxylase activity, and glycosaminoglycan synthesis of chondrocytes from mandibular condylar cartilage, nasal septal cartilage, and spheno-occipital synchondrosis in culture, *J. Dent. Res.,* 66, 84, 1987.

106. **Kato, Y., Koike, T., Iwamoto, M., Kinoshita, M., Sato, K., Hiraki, Y., and Suzuki, F.,** Effects of limited exposure of rabbit chondrocyte cultures to parathyroid hormone and dibutyryl adenosine 3′,5′-monophosphate on cartilage-characteristic proteoglycan synthesis, *Endocrinology,* 122, 1991, 1988.

107. **Iannotti, J. P., Brighton, C. T., Iannotti, V., and Ohishi, T.,** Mechanism of action of parathyroid hormone-induced proteoglycan synthesis in the growth plate chondrocyte, *J. Orthop. Res.,* 8, 136, 1990.

108. **Corvol, M. T., Carrascosa, A., Tsagris, L., Blanchard, O., and Rappaport, R.,** Evidence for a direct in vitro action of sex steroids on rabbit cartilage cells during skeletal growth: influence of age and sex, *Endocrinology,* 120, 1422, 1987.

109. **Vukicevic, S., Luyten, F. P., and Reddi, A. H.,** Stimulation of the expression of osteogenic and chondrogenic phenotypes in vitro by osteogenin, *Proc. Natl. Acad. Sci. U.S.A.,* 86, 8793, 1989.

110. **Corvol, M. T., Dumontier, M. F., Tsagris, L., Lang, F., and Bourguignon, J.,** Cartilage and vitamin D in vitro, *Ann. Endocrinol.,* 42, 482, 1981.

111. **Takigawa, M., Enomoto, M., Shirai, E., Nishii, Y., and Suzuki, F.,** Differential effects of 1-alpha, 25-dihydroxycholecalciferol and 24,25-dihydroxycholecalciferol on the proliferation and the differentiated phenotype of rabbit costal chondrocytes in culture, *Endocrinology,* 122, 831, 1988.

112. **Gerstenfeld, L. C., Kelly, C. M., Von Deck, M., and Lian, J. B.,** Effect of 1,25-dihydroxyvitamin D_3 on induction of chondrocyte maturation in culture: extracellular matrix gene expression and morphology, *Endocrinology,* 126, 1599, 1990.

113. **Wright, G. C., Wei, X. Q., McDevitt, C. A., Lane, B. P., and Sokoloff, L.,** Stimulation of matrix formation in rabbit chondrocyte cultures by ascorbate. I. Effect of ascorbate analogs and beta-aminopropionitrile, *J. Orthop. Res.,* 6, 397, 1988.

114. **McDevitt, C. A., Lipman, J. M., Ruemer, R. J., and Sokoloff, L.,** Stimulation of matrix formation in rabbit chondrocyte cultures by ascorbate. II. Characterization of proteoglycans, *J. Orthop. Res.,* 6, 518, 1988.

115. **Sandell, L. J. and Daniel, J. C.,** Effects of ascorbic acid on collagen mRNA levels in short term chondrocyte cultures, *Connect. Tissue Res.,* 17, 11, 1988.

116. **Hassel, J. R., Pennypacker, J. P., and Lewis, O. A.,** Chondrogenesis and cell proliferation in limb bud cell cultures treated with cytosine arabinoside and vitamin A, *Exp. Cell Res.,* 112, 409, 1978.

117. **Pacifici, M., Cossu, G., Molinaro, M., and Tato, F.,** Vitamin A inhibits chondrogenesis but not myogenesis, *Exp. Cell Res.,* 129, 469, 1980.

118. **Vasan, N. S.,** Proteoglycan synthesis by sternal chondrocytes perturbed with vitamin A, *J. Embryol. Exp. Morphol.,* 63, 181, 1981.

119. **Shapiro, S. S., Mott, D. J., and Nutley, N. J.,** Modulation of glycosaminoglycan biosynthesis by retinoids, *Ann. NY Acad. Sci.,* 359, 306, 1981.

120. **Hiraki, Y., Yutani, Y., Fukura, M., Takigawa, M., and Suzuki, F.,** Differentiation and de-differentiation of cultured chondrocytes: increase in monomeric size of "cartilage-specific" proteoglycans by dibutyryl cyclic AMP and complete inhibition of their synthesis by retinoic acid, *Biochem. Int.,* 10, 267, 1985.

121. **Trechsel, U., Dew, G., Murphy, G., and Reynolds, J. J.,** Effects of products from macrophages, blood mononuclear cells and of retinol on collagenase secretion and collagen synthesis in chondrocyte culture, *Biochim. Biophys. Acta,* 720, 364, 1982.

122. **Hein, R., Xrieg, T., Muellner, P. K., and Braun Falco, O.,** Vitamin A analogs alter the regulation of collagen synthesis in connective tissue cells, *Arch. Dermatol. Res.,* 275, 280, 1983.

123. **Bocquet, J., Daireaux, M., Langris, M., Jouis, V., Pujol, J. P., Beliard, R., and Loyau, G.,** Effect of a interleukin-1 like factor (mononuclear cell factor) on proteoglycan synthesis in cultured human articular chondrocytes, *Biochem. Biophys. Res. Commun.,* 134, 539, 1986.

124. **Pujol, J. P., Brisset, M., Jourdan, C., Bocquet, J., Jouis, V., Beliard, R., and Loyau, G.,** Effect of a monocyte cell factor (MCF) on collagen production in cultured articular chondrocytes: role of prostaglandin E2, *Biochem. Biophys. Res. Commun.,* 119, 499, 1984.

125. **Goldring, M. B., Sandell, L. J., Stephenson, M. L., and Krane, S. M.,** Immune interferon suppresses levels of procollagen mRNA and type II collagen synthesis in cultured human articular and costal chondrocytes, *J. Biol. Chem.,* 261, 9049, 1986.

126. **Goldring, M. B.,** Control of collagen synthesis in human chondrocyte cultures by immune interferon and interleukin-1, *J. Rheumatol.,* 14, 64, 1987.

127. **Goldring, M. B. and Krane, S. M.,** Modulation by recombinant interleukin-1 of synthesis of types I and III collagens and associated procollagen mRNA levels in cultured human cells, *J. Biol. Chem.,* 262, 16724, 1987.

128. **Lefebvre, V., Peeters-Joris, C., and Vaes, G.,** Modulation by interleukin-1 and tumor necrosis factor-alpha of production of collagenase, tissue inhibitor of metalloproteinases and collagen types in differentiated and dedifferentiated articular chondrocytes, *Biochim. Biophys. Acta,* 1052, 366, 1990.

129. **Ikebe, T., Hirata, M., and Koga, T.,** Human recombinant interleukin 1-mediated suppression of glycosaminoglycan synthesis in cultured rat costal chondrocytes, *Biochem. Biophys. Res. Commun.,* 140, 386, 1986.

130. **Hubbard, J. R., Steinberg, J. J., Bednar, M. S., and Sledge, C. B.,** Effect of purified human interleukin-1 on cartilage degradation, *J. Orthop. Res.,* 6, 180, 1988.

131. **Ikebe, T., Hirata, M., and Koga, T.,** Effects of human recombinant tumor necrosis factor-alpha and interleukin-1 on the synthesis of glycosaminoglycan and DNA in cultured rat costal chondrocytes, *J. Immunol.,* 140, 827, 1988.

132. **Benton, H. P. and Tyler, J. A.,** Inhibition of cartilage proteoglycan synthesis by interleukin-1, *Biochem. Biophys. Res. Commun.,* 154, 421, 1988.

133. **Kato, Y. and Gospodarowicz, D.,** Stimulation by glucocorticoid of the synthesis of cartilage-matrix proteoglycans produced by rabbit costal chondrocytes in vitro, *J. Biol. Chem.,* 260, 2364, 1985.

134. **Takano, T., Takigawa, M., and Suzukl, F.,** Stimulation by glucocorticoids of the differentiated phenotype of chondrocytes and the proliferation of rabbit costal chondrocytes in culture, *J. Biochem.,* 97, 1093, 1985.

135. **Mitrovic, D., McCall, E., Front, P., Aprile, F., Darmon, N., and Dray, F.,** Anti-inflammatory drugs, prostanoid and proteoglycan production by cultured bovine articular chondrocytes, *Prostaglandins,* 28, 417, 1984.

136. **Collier, S. and Ghosh, P.,** Evaluation of the effect of antiarthritic drugs on the secretion of proteoglycans by lapine chondrocytes using a novel assay procedure, *Ann. Rheum. Dis.,* 48, 372, 1989.

137. **Fontagne, J., Loizeau, M., Adolphe, M., and Lechat, P.,** Effect of indomethacin on collagen biosynthesis by rabbit articular chondrocytes in monolayer culture, *Int. J. Tissue React.,* 6, 233, 1984.

138. **Fujii, K., Tajiri, K., Sai, S., Tanaka, T., and Murota, K.,** Effects of nonsteroidal antinflammatory drugs on collagen biosynthesis of cultured chondrocytes, *Semin. Arthritis Rheum.,* 18, 16, 1989.

139. **Fujii, K., Tajiri, K., Kajiwara, T., Tanaka, T., and Murota, K.,** Effects of NSAID on collagen and proteoglycan synthesis of cultured chondrocytes, *J. Rheumatol.,* 18, 28, 1989.

140. **Costeseque, R., Edmonds-alt, X., Breliere, J. C., and Roncucci, R.,** Polysulphated polysaccharides: an in vitro study of their effects on proteoglycan biosynthesis by articular chondrocytes, *Arch. Int. Pharmacodyn. Ther.,* 282, 196, 1986.

141. **Legendre, P., Bouakka, M., Langris, M., Pujol, J. P., Beliard, R., Loyau, G., and Bocquet, J.,** Proteoglycan biosynthesis by rabbit articular chondrocytes treated with D-penicillamine, *Agents Actions,* 25, 171, 1988.

142. **Malemud, C. J., Weitzman, G. A., Norby, D. P., Sapolsky, A. I., and Howell, D. S.,** Metal-dependent neutral proteoglycanase activity from monolayer cultured lapine articular chondrocytes, *J. Lab. Clin. Med.,* 93, 1018, 1979.

143. **Morales, T. I. and Kuettner, K. E.,** The properties of the neutral proteinase released by primary chondrocyte cultures and its action on proteoglycan aggregate, *Biochim. Biophys. Acta,* 705, 92, 1982.

144. **Sapolski, A. I., Malemud, C. J., Norby, D. P., Moskowitz, R. W., Matsuta, K., and Howell, D. S.,** Neutral proteinases from articular chondrocytes in culture. II. Metal-dependent latent neutral proteoglycanase, and inhibitory activity, *Biochim. Biophys. Acta,* 658, 138, 1981.

145. **Malemud, C. J., Norby, D. P., Sapolsky, A. I., Matsuta, K., Howell, D. S., and Moskowitz, R. W.,** Neutral proteinases from articular chondrocytes in culture. I. A latent collagenase that degrades human cartilage type II collagen, *Biochim. Biophys. Acta,* 657, 517, 1981.

146. **Sapolsky, S., Malemud, C., and Sheff, M.,** The proteoglycanase from human cartilage and cultured rabbit chondrocytes and its relation to osteoarthritis, *J. Rheumatol.,* 14, 33, 1987.

147. **Mercier, P., Ehrlich, M. G., Armstrong, A., and Mankin, H. J.,** Elaboration of neutral proteoglycanase by growth-plate tissue cultures, *J. Bone Joint Surg.,* 69, 76, 1987.

148. **Solavagione, E., Bourbouze, R., Percheron, F., Hecquet, C., and Adolphe, M.,** Partial characterization of intracellular and secreted glycosidases from rabbit articular chondrocytes in culture, *Biochimie,* 69, 239, 1987.

149. **Baici, A., Lang, A., Horler, D., and Knopfel, M.,** Cathepsin B as a marker of the dedifferentiated chondrocyte phenotype, *Ann. Rheum. Dis.,* 47, 684, 1988.

150. **Morris, G. M.,** A high molecular weight collagenase inhibitor made by rabbit chondrocytes in cell culture, *Matrix,* 9, 127, 1989.

151. **Lefebvre, V., Peeters-Joris, C., and Vaes, G.,** Production of collagens, collagenase and collagenase inhibitor during the dedifferentiation of articular chondrocytes by serial subcultures, *Biochim. Biophys. Acta,* 1051, 266, 1990.

152. **Deshmuth-Phadke, K., Nauda, S., and Lee, K.,** Macrophage factor that induces neutral protease secretion by normal rabbit chondrocytes. Studies of some properties and effects on metabolism of chondrocytes, *Eur. J. Biochem.,* 104, 175, 1980.

153. **Kerwar, S. S., Ridge, S. C., Landes, M. J., Nolan, J. C., and Oronsky, A. L.,** Induction of collagenase synthesis in chondrocytes by a factor synthesized by inflamed synovial tissue, *Agents Actions,* 14, 54, 1984.

154. **Gowen, M., Nood, D. D., Ihrie, E. J., Meats, J. E., and Russell, R. G.,** Stimulation by human interleukin-1 of cartilage breakdown and production of collagenase and proteoglycanase by human chondrocytes but not by human osteoblasts in vitro, *Biochim. Biophys. Acta,* 797, 186, 1984.

155. **McGuire-Goldring, M. B., Meats, J. E., Wood, D. D., Ihrie, E. J., Ebsworth, N. M., and Russell, R. G.,** In vitro activation of human chondrocytes and synoviocytes by a human interleukin-1-like factor, *Arthritis Rheum.,* 27, 654, 1984.

156. **Rath, N. C., Oronsky, A. L., and Kerwar, S. S.,** Synthesis of interleukin-1-like activity by normal rat chondrocytes in culture, *Clin. Immunol. Immunopathol.,* 47, 39, 1988.

157. **Shimei, M., Kikuchi, T., Masuda, K., and Shimomura, Y.,** Effects of interleukin-1 and anti-inflammatory drugs on the degradation of human articular cartilage, *Drugs,* 35, 33, 1988.

158. **Hulkower, K. I., Georgescu, H. I., and Evans, C. H.,** Altered patterns of protein phosphorylation in articular chondrocytes treated with interleukin-1 or synovial cytokines, *FEBS Lett.,* 257, 228, 1989.

159. **Chandrasekhar, S. and Harvey, A. K.,** Transforming growth factor-beta is a potent inhibitor of IL-1 induced protease activity and cartilage proteoglycan degradation, *Biochem. Biophys. Res. Commun.,* 157, 1352, 1988.

160. **Bunning, R. A. and Russell, R. G.,** The effect of tumor necrosis factor alpha and gamma-interferon on the resorption of human articular cartilage and on the production of prostaglandin E and of caseinase activity by human articular chondrocytes, *Arthritis Rheum.,* 32, 780, 1989.

161. **Andrews, H. J., Bunning, R. A., Dinarello, C. A., and Russell, R. G.,** Modulation of human chondrocyte metabolism by recombinant human interferon gamma: in vitro effects on basal and IL-1-stimulated proteinase production, cartilage degradation and DNA synthesis, *Biochim. Biophys. Acta,* 1012, 128, 1989.

162. **Boyan, B. D., Schwartz, Z., Bonewald, L. F., and Swain, L. D.,** Localization of 1,25-$(OH)_2D_3$ responsive alkaline phosphatase in osteoblast-like cells (ROS 17/2. 8, MG63 and MC3T3) and growth cartilage cells in culture, *J. Biol. Chem.,* 264, 11879, 1989.

163. **Schwartz, Z., Knight, G., Swain, L. D., and Boyan, B. D.,** Localization of vitamin D_3-responsive alkaline phosphatase in cultured chondrocytes, *J. Biol. Chem.,* 263, 6023, 1988.

164. **Hale, L. V., Kemick, M. L., and Wuthier, R. E.,** Effect of vitamin D metabolites on the expression of alkaline phosphatase activity by epiphyseal hypertrophic chondrocytes in primary cell culture, *J. Bone Miner. Res.,* 1, 489, 1986.

165. **Schwartz, Z., Schlader, D. L., Swain, L. D., and Boyan, B. D.,** Direct effects of 1,25-dihydroxyvitamin D_3 and 24,25-dihydroxyvitamin D_3 on growth zone and resting zone chondrocyte membrane alkaline phosphatase and phospholipase-A_2 specific activities, *Endocrinology,* 123, 2878, 1988.

166. **Leboy, P. S., Vaias, L., Uschmann, B., Golub, E., Adams, S. L., and Pacifici, M.,** Ascorbic acid induces alkaline phosphatase, type X collagen, and calcium deposition in cultured chick chondrocytes, *J. Biol. Chem.,* 264, 17281, 1989.

167. **Dronne, N., Benel, L., Thenet, S., Larno, S., Mokondjimobe, E., Bourbouze, R., and Adolphe, M.,** Effects of retinoic acid on the growth of cultured rabbit articular chondrocytes: relation with alkaline phosphatase activity and beta receptor, *Cytotechnology,* 2, 233, 1989.

168. **Kato, Y. and Iwamoto, M.,** Fibroblast growth factor is an inhibitor of chondrocyte terminal differentiation, *J. Biol. Chem.,* 265, 5903, 1990.

169. **Horwitz, A. I. and Dorfman, A.,** The growth of cartilage cells in soft agar and liquid suspension, *J. Cell Biol.,* 45, 434, 1970.

170. **Yasui, N., Osawa, S., Ochi, T., Nakashima, H., and Ono, K.,** Primary culture of chondrocytes embedded in collagen gels, *Exp. Cell Biol.,* 50, 92, 1982.

171. **Gibson, G. J., Schor, S. L., and Grant, M. E.,** Effects of matrix macromolecules on chondrocyte gene expression: synthesis of a low molecular weight collagen species by cells cultured within collagen gels, *J. Cell Biol.,* 93, 767, 1982.

172. **Benya, P. D. and Shaffer, J. D.,** Dedifferentiated chondrocytes reexpress the differentiated collagen phenotype when cultured in agarose gels, *Cell,* 30, 215, 1982.

173. **Solursh, M., Jensen, K. L., Reiter, R. S., Schmid, T. M., and Linsenmayer, T. F.,** Environmental regulation of type X collagen production by cultures of limb mesenchyme, mesectoderm and sternal chondrocytes, *Dev. Biol.,* 117, 90, 1986.

174. **Kimura, T., Yasui, N., Ohsawa, S., and Ono, K.,** Chondrocytes embedded in collagen gels maintain cartilage phenotype during long-term cultures, *Clin. Orthop.,* 186, 231, 1984.

175. **Delbruck, A., Dresow, B., Gurr, E., Reale, E., and Schroder, H.,** In vitro culture of human chondrocytes from adult subjects, *Connect. Tissue Res.,* 15, 155, 1986.

176. **Sun, D., Aydelotte, M. B., Maldonado, B., Kuettner, K. E., and Kimura, J. H.,** Clonal analysis of the population of chondrocytes from the swarm rat chondrosarcoma in agarose culture, *J. Orthop. Res.,* 4, 427, 1986.

177. **Thomas, J. T. and Grant, M. E.,** Cartilage proteoglycan aggregate and fibronectin can modulate the expression of type X collagen by embryonic chick chondrocytes cultured in collagen gels, *Biosci. Rep.,* 8, 163, 1988.

178. **McClure, J., Bates, G. P., Rowston, H., and Grant, M. E.,** Comparison of the morphological, histochemical and biochemical features of embryonic chick sternal chondrocytes in vivo with chondrocytes cultured in three-dimensional collagen gels, *Bone Miner.,* 3, 235, 1988.

179. **Dewilde, B., Benel, L., Hartmann, D. J., and Adolphe, M.,** Subculture of rabbit articular chondrocytes within a collagen gel: growth and analysis of differentiation, *Cytotechnology,* 1, 123, 1988.

180. **Aulthouse, A. L., Beck, M., Griffey, E., Sanford, J., Arden, K., Machado, M. A., and Horton, W. A.,** Expression of the human chondrocyte phenotype in vitro, *In Vitro Cell Dev. Biol.,* 25, 659, 1989.

181. **Arai, K., Uehara, K., and Nagai, Y.,** Expression of type II and type XI collagens in canine mammary mixed tumors and demonstration of collagen production by tumor cells in collagen gel culture, *Jpn. J. Cancer Res.,* 80, 840, 1989.

182. **Elima, K. and Vuorio, E.,** Expression of mRNAs for collagens and other matrix components in dedifferentiating and redifferentiating human chondrocytes in culture, *FEBS Lett.,* 258, 195, 1989.

183. **Skantze, K. A., Brinckerhoff, C. E., and Collier, J. P.,** Use of agarose culture to measure the effect of transforming growth factor-beta and epidermal growth factor on rabbit articular chondrocytes, *Cancer Res.,* 45, 4416, 1985.

184. **Iwamoto, M., Sato, K., Nakashima, K., Fuchihata, H., Suzuki, F., and Kato, Y.,** Regulation of colony formation of differentiated chondrocytes in soft agar by transforming growth factor-beta, *Biochem. Biophys. Res. Commun.,* 159, 1006, 1989.

185. **Sato, K., Iwamoto, M., Nakashima, K., Suzuki, F., and Kato, Y.,** 1-alpha,25-dihydroxyvitamin D_3 stimulates colony formation of chick embryo chondrocytes in soft agar, *Exp. Cell Res.,* 187, 335, 1990 .

186. **Kato, Y., Iwamoto, M., Nakashima, K., Sato, K., Yan, W., and Koike, T.,** Rabbit articular chondrocytes in soft agar are useful to test the effect of nonsteroidal anti-inflammatory drugs on chondrocyte replication, *Scand. J. Rheumatol.,* 77, 3, 1988.

187. **Horton, W. and Hassell, J. R.,** Independence of cell shape and loss of cartilage matrix production during retinoic acid treatment of cultured chondrocytes, *Dev. Biol.,* 115, 392, 1986.

188. **Deshmukh, K. and Kline, W. H.,** Characterization of collagen and its precursors synthesized by rabbit articular cartilage cells in various culture systems, *Eur. J. Biochem.,* 69, 117, 1976.

189. **Wiebkin, O. W. and Muir, H.,** Synthesis of cartilage specific proteoglycan by suspension cultures of adult chondrocytes, *Biochem. J.,* 164, 269, 1977.

190. **Dozin, B., Quarto, R., Rossi, F., and Cancedda, R.,** Stabilization of the mRNA follows transcriptional activation of type II collagen in differentiating chicken chondrocytes, *J. Biol. Chem.,* 265, 7216, 1990.

191. **Lindahl, A., Isgaard, J., Carlsson, L., and Isaksson, O. G.,** Differential effects of growth hormone and insulin-like growth factor 1 on colony formation of epiphiseal chondrocytes in suspension culture in rats of different ages, *Endocrinology,* 121, 1061, 1987.

192. **Slowman-Kovacs, S. D., Albrecht, M. E., and Brandt, K. D.,** Effects of salycilate on chondrocytes from osteoarthritic and contralateral knees of dogs with unilateral anterior cruciate ligament transection, *Arthritis Rheum.,* 32, 486, 1989.

193. **Srivastava, V. M., Malemud, C. J., and Sokoloff, L.,** Chondroid expression by lapine articular chondrocytes in spinner culture following monolayer growth, *Connect. Tissue Res.,* 2, 127, 1974.

194. **Bassleer, C., Gysen, P., Foidary, J. M., Bassleer, R. and Franchimont, P.,** Human chondrocytes in three-dimensional culture, *In Vitro Cell Dev. Biol.,* 22, 113, 1986.

195. **Bassleer, C., Gysen, P., Bassleer, R., and Franchimont, P.,** Effects of peptidic glycosaminoglycans complex on human chondrocytes cultivated in three dimensions, *Biochem. Pharmacol.,* 37, 1939, 1988.

196. **Henrotin, Y., Bassleer, C., Reginster, J. Y., and Franchimont, P.,** Effects of etodolac on human chondrocytes cultivated in three-dimensional culture, *Clin. Rheumatol.,* 8, 36, 1989.

197. **Franchimont, P., Bassleer, C., and Henrotin, Y.,** Effects of hormones and drugs on cartilage repair, *J. Rheumatol.,* 18, 5, 1989.

198. **Franchimont, P., Bassleer, C., Henrotin, Y., Gysen, P., and Bassleer, R.,** Effects of human and salmon calcitonin on human articular chondrocytes cultivated in clusters, *J. Clin. Endocrinol. Metab.,* 69, 259, 1989.

199. **Handley, C. J., and Lowther, D. A.,** Extracellular matrix metabolism by chondrocytes. V. The proteoglycans and glycosaminoglycans synthesized by chondrocytes in high-density cultures, *Biochim. Biophys. Acta,* 582, 234, 1979.

200. **Watt, F. M.,** Effect of seeding density on stability of the differentiated phenotype of pig articular chondrocytes in culture, *J. Cell Sci.,* 89, 373, 1988.

201. **Kuettner, K. E., Pauli, B. U., Gall, G., Memoli, V. A., and Schenk, R. K.,** Synthesis of cartilage matrix by mammalian chondrocytes in vitro. I. Isolation, culture characteristics and morphology, *J. Cell Biol.,* 93, 743, 1982.

202. **Cheung, H. S.,** In vitro cartilage formation on porous hydroxyapatite ceramic granules, *In vitro Cell Dev. Biol.,* 21, 353, 1985.

203. **Watanabe, N., Rosenfeld, R. G., Hintz, R. L., Dollar, L. A., and Smith, R. L.,** Characterization of a specific insulin-like growth factor-1/somatomedin-C receptor on high density, primary monolayer cultures of bovine articular chondrocytes: regulation of receptor concentration by somatomedin, insulin and growth hormone, *J. Endocrinol.,* 107, 275, 1985.

204. **Harmand, M. F., Thomasset, M., Rovars, F., and Ducassou, D.,** In vitro stimulation of articular chondrocytes differentiated function by 1,25-dihydroxicholecalciferol or 24,25-dihydroxycholecalciferol, *J. Cell. Physiol.,* 119, 359, 1984.

205. **Bulstra, S. K., Buurman, W. A., Walenkamp, G. H., and Van der Linden, A. J.,** Metabolic characteristics of in vitro cultured human chondrocytes in relation to the histopathologic grade of osteoarthritis, *Clin. Orthop.,* 242, 294, 1989.

206. **Carrino, D. A., Kujawa, M. J., Lennon, D. P., and Caplan, A. I.,** Altered cartilage proteoglycans synthesized by chick limb bud chondrocytes cultured in serum-free defined medium, *Exp. Cell Res.,* 183, 62, 1989.

207. **Kujawa, M. J., Lennon, D. P., and Caplan, A. I.,** Growth and differentiation of stage 24 limb mesenchyme cells in a serum free chemically defined medium, *Exp. Cell Res.,* 183, 45, 1989.

208. **Castagnola, P., Torella, G., and Cancedda, R.,** Type X collagen synthesis by cultured chondrocytes derived from the permanent cartilaginous region of chick embryo sternum, *Dev. Biol.,* 123, 332, 1987.

209. **Giaretti, W., Moro, G., Quarto, R., Bruno, S., Di Vinci, A., Geido, E., and Cancedda, R.,** Flow cytometric evaluation of cell cycle characteristics during in vitro differentiation of chick embryo chondrocytes, *Cytometry,* 9, 281, 1988.

210. **Glowacki, J., Trepman, E., and Folkman, J.,** Cell shape and phenotypic expression in chondrocytes, *Proc. Soc. Exp. Biol. Med.,* 172, 93, 1983.

211. **Guo, J., Jourdian, G. W., and MacCallum, D. K.,** Culture and growth characteristics of chondrocytes encapsulated in alginate beads, *Connect. Tissue Res.,* 19, 277, 1989.

212. **Carrel, A. and Burrows, M. T.,** Cultivation of adult tissues and organs outside of the body, *Arthritis Rheum.,* 22, 220, 1910.

213. **Hayflick, L. and Moorhead, P. S.,** The serial cultivation of human diploid cell strains, *Exp. Cell Res.,* 25, 585, 1961.

214. **Takigawa, M., Tajima, K., Pan, H. O., Enomoto, M., Kinoshita, A., Suzuki, F., Takano, Y., and Mori, Y.,** Establishment of a clonal human chondrosarcoma cell line with cartilage phenotypes, *Cancer Res.,* 49, 3996, 1989.

215. **Katoh, Y. and Takayama, S.,** Establishment of clonal cell lines of hamster chondrocytes maintaining their phenotypic traits, *Exp. Cell Res.,* 106, 285, 1977.

216. **Gionti, E., Pontarelli, G., and Cancedda, R.,** Avian myelocytomatosis virus immortalizes differentiated quail chondrocytes, *Proc. Natl. Acad. Sci. U.S.A.,* 82, 2756, 1985.

217. **Horton, W. E., Cleveland, J., Rapp, U., Nemuth, G., Bolander, M., Doege, K., Yamada, Y., and Massel, J. R.,** An established rat cell line expressing chondrocyte properties, *Exp. Cell Res.,* 178, 457, 1988.

218. **Gionti, E., Jullien, P., Pontarelli, G., and Sanchez, M.,** A continuous line of chicken embryo cells derived from a chondrocyte culture infected with RSV, *Cell Differ. Dev.,* 27, 215, 1989.

219. **Adolphe, M., and Thenet, S.,** Le concept d'immortalité cellulaire, un mythe ou une réalité. Exemple de chondrocytes articulaires "immortalisés", *Bull. Acad. Natl. Med.,* 174, 139, 1990.

220. **Thenet, S., Benya, P. D., Demignot S., Feunteun, J., and Adolphe M.,** SV40-immortalization of rabbit articular chondrocytes. Alteration of differentiated functions, *J. Cell. Physiol.,* 1992, in press.

221. **O'Donnell, C. M., Kaczman-Daniel, K., Goetinck, P. F., and Vertel, B. M.,** Nanomelic chondrocytes synthesize a glycoprotein related to chondroitin sulfate proteoglycan core protein, *J. Biol. Chem.,* 263, 17749, 1988.

222. **Shuckett, R. and Malemud, C. J.,** Proteoglycans synthesized by chondrocytes of human nonarthritic and osteoarthritic cartilage, *Proc. Soc. Exp. Biol. Med.,* 190, 275, 1989.

223. **Bulstra, S. K., Buurman, W. A., Walenkamp, G. H., and Van der Linden, A. J.,** Metabolic characteristics of in vitro cultured human chondrocytes in relation to the histopathologic grade of osteoarthritis, *Clin. Orthop.,* 242, 294, 1989.

224. **Vetter, U., Zapf, J., Henrichs, I., Gammery, C., Heinze, E., and Pirsig, W.,** Human nasal septal cartilage: analysis of intracellular enzyme activities, glycogen content, cell density and clonal proliferation of septal chondrocytes of healthy adults and acromegalic patients, *Connect. Tissue Res.,* 18, 243, 1989.

225. **Horton, W. A., Campbell, D., Machado, M. A., Aulthouse, A. L., Ahmed, S., and Ellard, J. T.,** Tissue and cell studies of the growth plate in the chondrodysplasias, *Am. J. Med. Genet.,* 34, 91, 1989.

226. **Vincent, F., Brun, H., Clain, E., Ronot, X., and Adolphe, M.,** Effects of oxygen-free radicals on proliferation kinetics of cultured rabbit articular chondrocytes, *J. Cell. Physiol.,* 141, 262, 1989.

227. **Vincent, F., Corral, M., Defer, N., and Adolphe, M.,** Effects of oxygen free radicals in culture: C-myc and C-HA-ras messenger RNAs and proliferation kinetics, *Exp. Cell Res.,* 192, 333, 1991.

228. **Brighton, C. T., Unger, A. S., and Stambough, J. L.,** In vitro growth of bovine articular cartilage chondrocytes in various capacitively coupled electrical fields, *J. Orthop. Res.,* 2, 15, 1984.

229. **Armstrong, P. F., Brighton, C. T., and Star, A. M.,** Capacitively coupled electrical stimulation of bovine growth plate chondrocytes grown in pellet form, *J. Orthop. Res.,* 6, 265, 1988.

230. **Brighton, C. T., Jensen, L., Pollack, S. R., Tolin, B. S., and Clark, C. C.,** Proliferative and synthetic response of bovine growth plate chondrocytes to various capacitively coupled electrical fields, *J. Orthop. Res.,* 7, 759, 1989.

231. **Hiraki, Y., Endo, N., Takigawa, M., Asada, A., Takahashi, H., and Suzuki, F.,** Enhanced responsiveness to parathyroid hormone and induction of functional differentiation of cultured rabbit costal chondrocytes by a pulsed electromagnetic field, *Biochim. Biophys. Acta,* 931, 94, 1987.

232. **Elliott, J. P., Smith, R. L., and Block, C. A.,** Time-varying magnetic fields: effects of orientation on chondrocyte proliferation, *J. Orthop. Res.,* 6, 259, 1988.

233. **De Witt, M. T., Handley, C. J., Oakes, B. W., and Lowther, D. A.,** In vitro response of chondrocytes to mechanical loading. The effect of short term mechanical tension, *Connect. Tissue Res.,* 12, 97, 1984.

234. **Van Xampen, G. P., Veldhuijzen, J. P., Kuijer, R., Van de Stadt, R. J., and Schipper, C. A.,** Cartilage response to mechanical force in high-density chondrocyte cultures, *Arthritis Rheum.,* 28, 419, 1985.

235. **Veldhuijzen, J. P., Huisman, A. H., Vermeiden, J. P., and Prahl-Andersen, B.,** The growth of cartilage cells in vitro and the effect of intermittent compressive force. A histological evaluation, *Connect. Tissue Res.,* 16, 187, 1987.

236. **Uchida, A., Yamashita, K., Hashimoto, K., and Shimomura, Y.,** The effect of mechanical stress on cultured growth cartilage cells, *Connect. Tissue Res.,* 17, 305, 1988.

237. **Gray, M. L., Pizzanelli, A. M., Grodzinsky, A. J., and Lee, R. C.,** Mechanical and physiochemical determinants of the chondrocyte biosynthetic response, *J. Orthop. Res.,* 6, 777, 1988.

238. **Green, W. T.,** Articular cartilage repair. Behavior of rabbit chondrocytes during tissue culture and subsequent allografting, *Clin. Orthop.,* 124, 237, 1977.

239. **Tsuge, H., Sasaki, T., Susuda, K., and Abe, K.,** Allograft of cultured chondrocytes into articular cartilage defects in rabbits. Experimental study of the repair of cartilage injuries, *Nippon Seikeigeka gakkai Zasshi.,* 57, 837, 1983.

240. **Foster, B. K., Hansen, A. L., Gibson, G. J., Hopwood, J. J., Binns, G. F., and Wiebkin, O. W.,** Reimplantation of growth plate chondrocytes into growth plate defects in sheep, *J. Orthop. Res.,* 8, 555, 1990.

241. **Robinson D., Halpering N., and Nevo Z.,** Regenerating hyaline cartilage in articular defects of old chickens using implants of embryonal chick chondrocytes embedded in a new natural delivery substance, *Calcif. Tissue Int.,* 46, 246, 1990.

242. **Itay, S., Abramovici, A., and Nevo, Z.,** Use of cultured embryonal chick epiphyseal chondrocytes as grafts for defects in chick articular cartilage, *Clin. Orthop.,* 220, 284, 1987.

243. **Koskinen, K. P., Kanwar, Y. S., Sires, B., and Veis, A.,** An electron microscopic demonstration of induction of chondrogenesis in neonatal rat muscle outgrowth cells in monolayer cultures, *Connect. Tissue Res.,* 14, 141, 1985.

244. **Grigoriadis, A. E., Heersche, J. N., and Aubin, J. E.,** Differentiation of muscle, fat, cartilage and bone from progenitor cells present in bone-derived clonal cell populations: effect of dexamethasone, *J. Cell Biol.,* 106, 2139, 1988.

245. **Davis, C. M., Constantinides, P. G., Van der Riet, F., Van Schalkwyk, L., Gevers, N. and Parker, M. I.,** Activation and demethylation of the intracisternal a particle genes by 5-azacytidine, *Cell Differ. Dev.,* 27, 83, 1989.

246. **Azuma, M., Kawamata, H., Kasai, Y., Yanagawa, T., Yoshida, H., and Sato, M.,** Induction of cells with a chondrocyte-like phenotype by treatment with 1-alpha, 25-dihydroxyvitamin D_3 in a human salivary acinar cell line, *Cancer Res.,* 49, 5435, 1989.

247. **Atsumi, T., Miwa, Y., Kimata, K., and Ikawa, Y.,** A chondrogenic cell line derived from a differentiating culture of AT805 teratocarcinoma cells, *Cell Differ. Dev.,* 30, 109, 1990.

Chapter 5

ROLES OF FIBROBLAST GROWTH FACTOR AND TRANSFORMING GROWTH FACTOR-ß FAMILIES IN CARTILAGE FORMATION

Yukio Kato

TABLE OF CONTENTS

I. INTRODUCTION

The purpose of this chapter is to review recent studies on roles of fibroblast growth factor (FGF) and transforming growth factor-ß (TGF-ß) families in development, growth, and remodeling of cartilage. FGF and TGF-ß are mitogenic for chondrocytes and have a profound effect on their cytodifferentiation. FGF and TGF-ß synthesized by chondrocytes may serve as autocrine regulators. Although FGF and TGF-ß have a number of similar actions on chondrocytes, the two factors will be considered separately here in order to emphasize that they belong to different gene families.

II. FGF

A. FGF GENE FAMILY

Basic FGF (bFGF) and acidic FGF (aFGF) belong to a multigene family of heparin-binding peptides. Their prepropeptides lack a signal peptide domain and are not efficiently secreted, whereas other members of the FGF family [e.g., kFGF (*KS*FGF, *hst*), *int*-2, FGF-5, and keratinocyte growth factor] have a conventional hydrophobic leader sequence. bFGF, aFGF, and kFGF induce a similar spectrum of biological actions and bind to a common receptor on fibroblasts.[1,2]

Patterns of expressions of bFGF, aFGF, kFGF, FGF-5, and *int*-2 during mouse embryogenesis have been examined using their respective cDNA.[3] The expression of bFGF transcripts showed the greatest tissue-specific variability, with the highest levels in developing long bone (cartilage and bone) and tail. aFGF transcripts showed the least variable pattern of expression: transcripts were detected at similar levels in almost all tissues tested, including limbs. FGF-5 transcripts were detected at every stage in every tissue tested. kFGF transcripts were detected only in undifferentiated teratocarcinoma stem cells. *int*-2 was expressed in parietal endoderm, early-migrating mesoderm, Purkinje cells, developing retina, sensory regions of the inner ear, and developing teeth.[4] These different patterns of expressions suggest that FGF-related peptides have different roles in embryogenesis.

B. bFGF IS PRESENT IN CARTILAGE AND BONE

bFGF was purified from various cartilages of human, chick, and bovine origin, and a rat chondrosarcoma cell line.[5] Immunohistochemical analyses showed that it is present in premature cartilage, the proliferating and maturing zones of growth plate, and the perichondrium.[6] However, hypertrophic cartilage had no detectable bFGF. The content of bFGF in the resting cartilage, proliferating zone, maturing zone, and the hypertrophic zone of 4-week-old rabbit growth plates, was 11 ± 2, 9 ± 2, 7 ± 2, and 3 ± 1 pg/0.1 μg soluble protein, respectively, as determined by radioimmunoassay.[7] The hypertrophic zone might be contaminated by capillaries.

C. RELEASE OF bFGF FROM MATRIX

Various forms of bFGF are reported to be in nuclei (22 and 24 kDa), cytoplasm (18 kDa), and the extracellular matrix (18 kDa) of cells.[8] In the extracellular matrix and on the cell surface, FGF is associated with heparan sulfate proteoglycans.[9] FGF cannot bind to its cell surface receptor unless it is released from the matrix. Treatment of cultured cells with heparitinase, plasmin, or trypsin released a bioactive form of bFGF which is less stable ($t_{1/2}$ = 24 h).[10,11]

D. MITOGENIC ACTIONS OF FGF

In 4-week-old rabbit chondrocyte cultures, FGF increased [^{3}H]thymidine incorporation into DNA in the presence of 0.3 to 2% serum.[12-14] The effect of FGF on [^{3}H]thymidine incorporation into DNA was greater than that of insulin-like growth factor (IGF)-I, IGF-II or insulin, and was similar to that of epidermal growth factor (EGF) or platelet-derived growth factor (PDGF).[12,13] Combination of FGF with IGF-II produced a synergistic increase in [^{3}H]thymidine incorporation into DNA.[13] In chondrocyte cultures, continuous presence of FGF in the medium was necessary in order to traverse the entire G_1 phase prior to stimulation of DNA synthesis. On the other hand, exposure of chondrocytes to IGF-II for only the initial 5 h was sufficient for transmission of their full stimulatory effect.[15] FGF appears to act throughout the G_1 phase, whereas IGF-II acts only during early G_1. Accordingly, FGF had synergistic effects on ^{3}H-thymidine incorporation into DNA with IGF-II even when added 10 h after the latter in rabbit chondrocyte cultures. The synergism was also observed in cells pretreated with IGF-II; that is, the cultures were treated for 5 h with IGF-II and washed and then treated with FGF.[16] However, when IGF-II was added more than 5 h after FGF, no synergism of effects occurred. These observations suggest that chondrocytes become activated to interact with FGF for commitment to DNA synthesis when they are exposed to IGF-II during an early stage of the G_1 phase.

Chondrocytes proliferate actively in the absence of serum when cultured in medium supplemented with high-density lipoprotein, transferrin, hydrocortisone, insulin (or IGF-I), and bFGF.[17] Optimal proliferation was observed when dished were coated with an extracellular matrix produced by cultured endothelial cells. Omission of FGF from the complete medium had the most deleterious effect on cell growth.

In suspension cultures in liquid medium or soft agar, chondrocytes maintain the ability to synthesized cartilage-matrix proteoglycan for more than a month. Thus, it was of interest to examine the effect of FGF on proliferation of chondrocytes in the absence of the artificial substrate. Addition of bFGF (at 0.1 ng/ml) increased colony formation 50- to 100-fold by chick embryo and rabbit chondrocytes in soft agar in the presence of 10% serum (Figure 1).[18] However, other growth factors such as EGF, insulin, IGF-I, and PDGF supported only very low levels of colony formation. bFGF did not induce soft agar growth of chondrocytes whose phenotypic expression was suppressed by retinoic acid or

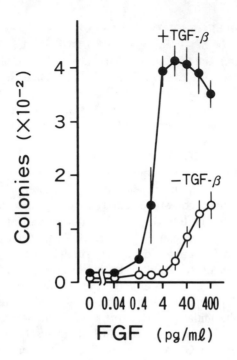

FIGURE 1.   Effects of increasing concentrations of bFGF on colony formation in soft agar by chondrocytes in the presence or absence of TGF-ß. Rabbit chondrocytes were seeded at 1000 cells per dish and maintained in soft agar for 2 weeks in the presence of various concentrations of bFGF with or without an optimal dose (3 ng/ml) of TGF-ß1.

5-bromodeoxyuridine.[18] Thus, FGF selectively stimulated growth in soft agar of overtly differentiated chondrocytes. Perhaps, the extracellular matrix produced by chondrocytes serves as an anchoring support for the growth of chondrocytes in soft agar in response to FGF.

E. STABILIZATION OF CHONDROCYTE PHENOTYPIC EXPRESSION

Chondrocytes grown on plastic or glass dishes progressively transform into motile cells that are morphologically indistinguishable from fibroblasts.[19] More than 50% of rabbit chondrocytes became fibroblastic after proliferation for four generations on plastic culture dishes.[20] However, 50% of bFGF-exposed chondrocytes maintained the differentiated state after proliferation for 10 to 12 generations.[20,21] EGF, insulin, and IGF-I did not stabilize chondrocyte phenotypic expression in long term cultures.

F. EXTRACELLULAR MATRIX SYNTHESIS

In the proliferative stage of rabbit chondrocytes in culture, bFGF decreased the incorporation of [^{35}S]sulfate incorporation into proteoglycans by 30 to 50%.[20] This inhibition of proteoglycan synthesis may be secondary to the FGF stimu-

TABLE 1
Effect of Length of Exposure to bFGF on the Rate of Proteoglycan
Synthesis by 18-day Cultures of Rabbit Chondrocytes

Addition of bFGF (period)	Cells/dish ($\times 10^{-5}$)	^{35}S-Sulfate uptake[a] (cpm $\times 10^3$/cell)
None	10.5	9
d0—d8	19.2	58
d0—d12	19.5	68
d0—d18	19.6	77
d12—d18	12.9	6

[a] Incorporation of [^{35}S]sulfate into proteoglycans was measured on day 18. Rabbit chondrocytes were seeded at 3000 cells per 35-mm culture dishes and exposed to Dulbecco's modified Eagle's medium with 10% fetal bovine serum. These cells became confluent on day 8. bFGF (0.6 ng/ml) was added every other day during the indicated culture periods. After 18 d in culture, triplicate plates were exposed for 3 h to 2 μCi/ml of [^{35}S]sulfate in 0.8 ml of Dulbecco's modified Eagle's medium. Values are averages for triplicate cultures.

lation of DNA synthesis; rapid cell division is generally incompatible with increase in proteoglycan synthesis. bFGF treatment during the proliferative stage also resulted in sevenfold increase in [^{35}S]sulfate incorporation into proteoglycans in confluent cultures (Table 1). Since bFGF treatment initiating after cessation of cell division had little effect on [^{35}S]sulfate incorporation into proteoglycans (Table 1), bFGF does not seem to have a direct effect on proteoglycan synthesis.[20] Perhaps, bFGF, when added to actively growing chondrocytes, prevents at each cell cycle the dedifferentiation of a given percentage of the population into an irreversible fibroblastic stage, resulting in an increase in proteoglycan synthesis in the confluent stage.[20,21]

Although production of an abundant proteoglycan matrix is a marker of chondrocytes, they lose proteoglycan-synthetic activity after terminal differentiation. bFGF prevented the terminal differentiation-associated decrease in proteoglycan synthesis. Accordingly, FGF treatment resulted in a 10-fold increase in [^{35}S]sulfate incorporation into proteoglycans in the hypertrophic stage in pelleted cultures.[22]

Chondrocytes produce two proteoglycan species: (1) large, chondroitin sulfate proteoglycan that forms aggregates in the presence of hyaluronic acid in cartilage matrix, and (2) small proteoglycans that are ubiquitously found in both chondrogenic and nonchondrogenic tissues. The aggregated form of large proteoglycan, which is associated with a collagen network, confers cartilage resiliency. Although bFGF decreased [^{35}S]sulfate incorporation into the large proteoglycan in the mitotic stage, bFGF increased [^{35}S]sulfate incorporation into the large proteoglycan in the matrix-forming stage and the hypertrophying stage.[22] bFGF did not alter the 6S:4S ratio of chondroitin sulfate glycosaminoglycans.[22] No glycosaminoglycans atypical of a cartilage matrix were detected in bFGF-maintained cultures. bFGF had little effect on [^{35}S]sulfate incorpora-

TABLE 2
Effects of bFGF and TGF-ß1 on Expressions of Mineralization-Related Phenotypes by Chondrocytes

Addition	ALPase (units)[a]	1,25(OH)$_2$D binding (fmol/mg protein)	^{45}Ca uptake (dpm × 10^{-3}/μg DNA)
None	0.58	28	5.9
bFGF	0.03	6	1.1
TGF-ß1	0.09	9	1.2

Note: Rabbit growth plate chondrocytes were seeded (8 × 10^4 cells per culture) as pelleted mass and maintained in medium with 10% serum in centrifuge tubes. bFGF (0.4 ng/ml) or TGF-ß1 (3 ng/ml), at the optimal dose, was added 15 d after cell seeding. Alkaline phosphatase (ALPase) activity, 1,25(OH)$_2$vitamin D$_3$ binding and ^{45}Ca incorporation into insoluble material were determined on day 20 or 21. Values are averages for three to four cultures.

[a]   μmol pNP hydrolysis/30 min/μg DNA.

tion into small proteoglycans throughout all stages of growth and cytodifferentiation.

bFGF decreased type II collagen synthesis by 80% in cultures of proliferating chondrocytes.[7] However, bFGF treatment during cell growth resulted in a marked increase in type II collagen synthesis in confluent cultures.[20] The postmitotic exposure to bFGF had little effect on type II collagen synthesis.

Under certain culture conditions, rabbit chondrocytes produced a short-chain collagen in the hypertrophic stage.[23] Under reducing and nonreducing conditions, the pepsin-resistant and collagenase-digestible macromolecule migrated in SDS polyacrylamide gels at a position corresponding to 64 to 70 kDa. FGF consistently abolished the synthesis of the short-chain collagen (type X).

G. FGF IS AN INHIBITOR OF TERMINAL DIFFERENTIATION

In growth plates and at sites of bone fracture, chondrocytes undergo terminal differentiation and induce matrix calcification. Since only cartilage that is calcified can be replaced by bone, the calcification is essential for endochondral bone formation. However, little is known about regulation of cartilage calcification by hormones and growth factors because no good experimental model has been available. Recently, we found that rabbit growth plate chondrocytes maintained as a pelleted mass in a centrifuge tube produce alkaline phosphatase and 1,25(OH)$_2$vitamin D$_3$ receptor, markers of hypertrophic chondrocytes, at as high levels as those in growth plates *in vivo.*[22-26] Furthermore, in these cultures, the extracellular matrix was calcified in the absence of added ß-glycerophosphate or inorganic phosphate. This *in vitro* calcification is not merely a physicochemical phenomenon, but coupled with terminal differentiation of the chondrocytes. The development of this well-differentiated chondrocyte system allowed us to examine the effects of FGF on the expression of the mineralization-related phenotype by chondrocytes.

bFGF abolished the increases in alkaline phosphatase activity, ^{45}Ca deposition, and the calcium content in pelleted chondrocyte cultures (Table 2).[22] bFGF also suppressed the synthesis of $1,25(OH)_2$vitamin D_3 receptor in the hypertrophic stage (Table 2). These effects were dose-dependent, reversible, and observed in the presence of cytosine arabinoside, an inhibitor of DNA synthesis. The inhibitory effects could be observed only when chondrocytes were exposed to bFGF in a transition period (day 12 to day 18) between the matrix-forming and hypertrophic stages. As chondrocytes differentiated into hypertrophic cells, bFGF became less effective in inhibiting the expressions of mineralization-related phenotypes. aFGF had a similar effect on the induction of alkaline phosphatase and calcification. The physiological significance of the inhibition of calcification by FGF is not known. However, precocious hypertrophy and calcification have to be limited during rapid enlargement of cartilage because terminal differentiation causes decreases in DNA- and proteoglycan-synthetic activities of chondrocytes.

H. FGF RECEPTORS

Reduction in responsiveness to bFGF in the hypertrophic stage may be due to loss of bFGF receptor.[27] In cultures of growth plate chondrocytes, substantial binding of ^{125}I-bFGF to 140-kDa receptors was observed during the periods of cell proliferation and matrix formation, but decreased to a very low level as the cells became hypertrophic. Scatchard plot analysis showed that the decrease in binding of bFGF was due to a decrease in receptor number, not in the affinity of the receptor. Rapid decrease in bFGF binding was not observed with articular chondrocytes or bFGF-exposed, growth plate chondrocytes in culture, perhaps because they rarely underwent terminal differentiation. The terminal differentiation-associated decrease in bFGF binding was also observed *in situ* with sequential slices of growth plates. Five sequential slices (0.2 mm) were removed under a surgical microscope from rib growth plates and three slices from the permanent cartilage region. The hypertrophic zone was identified by the presence of alkaline phosphatase activity. No differences in ^{125}I-bFGF binding were observed among slices from the permanent cartilage region. However, growth plate cartilage showed decreased binding of ^{125}I-bFGF as it became hypertrophic.

The physiological significance of loss of bFGF receptor is known. It is tempting to speculate that the stage-specific loss of receptor for FGF, a potent inhibitor of terminal differentiation, is necessary for full expression of terminal differentiation phenotypes by hypertrophied chondrocytes, while maintaining higher levels of the responsiveness of proliferating and maturing chondrocytes to this factor. It remains unknown what turns off the expression of bFGF receptor.

The specificity of the decrease in bFGF binding in the hypertrophic stage was indicated by the three following observations:[27] (1) bindings of EGF and TGF-ß were constant throughout any stage of chondrocyte cytodifferentiation; (2) binding of $1,25(OH)_2$vitamin D_3 to its receptor increased 40-fold in the hypertrophic stage; and (3) binding of parathyroid hormone or IGF-I in the matrix-forming stage was two- to fourfold higher than that in the mitotic and hypertrophic stages. Thus, there are four patterns of changes in receptor levels during cytodifferentiation. Since all of the above-mentioned hormones and factors affect chondrocyte phenotypic expression, differential expressions of these receptors must determine the characters of chondrocytes at each stage even though the biological responses of chondrocytes to growth factors may also be modulated at postreceptor levels.

Recent studies have shown that liver cells express various isoforms of FGF receptor.[28] The FGF receptor consists of three structural domains: (1) the extracellular aminoterminal domain, (2) the juxtamembrane domain, and (3) the intracellular carboxylterminal domain. The nucleotide sequence of complementary DNA predicts that 3 aminoterminal domain motifs, 2 juxtamembrane motifs, and 2 intracellular carboxylterminal domain motifs combine to form at least 6, and possibly 12, homologous polypeptides. One aminoterminal variant does not have signal sequence and may be an intracellular form. The two intracellular juxtamembrane motifs differ in a potential serine-threonine kinase phosphorylation site. One carboxylterminal motif is a putative tyrosine kinase, and the second carboxylterminal motif is probably not a tyrosine kinase. The three distinct structural domains that combine to form FGF receptor isoforms are likely to affect ligand binding, oligomerization, cellular location, metabolism, and signal transduction.

I. BONE FRACTURE AND ARTHRITIS

Intraarticular administration of bFGF induced chondrocyte proliferation and extracellular matrix synthesis.[29] Furthermore, injection of aFGF (1 µg/d) at the site of bone fracture promoted formation of cartilage and bone.[30] However, the aFGF-injected fracture calluses were weaker in tensile strength than controls. No endochondral bone formation was observed in soft callus during the treatment with FGF, perhaps because FGF inhibited terminal differentiation of chondrocytes.[30]

Sano et al.[31] have shown that aFGF is synthesized in synovium, cartilage, bone, bone marrow, and ligamentous and tendinous structures in arthritic joints. The synthesis of aFGF was correlated with the extent and intensity of synovial mononuclear cell infiltration. Production of bFGF by macrophages may be enhanced in rheumatoid and osteoarthritic tissue.[32] In articular chondrocyte cultures, bFGF increased interleukin-1 (IL-1) binding capacity of the cells and enhanced IL-1-mediated protease release.[33,34] Furthermore, IL-1 suppressed DNA synthesis in overtly differentiated chondrocytes in the presence of FGF, whereas it stimulated DNA synthesis in poorly differentiated chondrocytes and fibroblastic cells in the presence or absence of FGF.[35] IL-1ß also abolished colony formation by chondrocytes in soft agar in response to FGF.[35] However, interferon-α interferon-γ, and tumor necrosis factor-α (TNF-α) had little effect on colony formation by chondrocytes in the presence or absence of FGF.[7] These observations suggest that FGF modulates IL-1 actions in arthritic cartilage. It is of interest to note that a weak homology (25 to 27%) exists between FGF and IL-1,

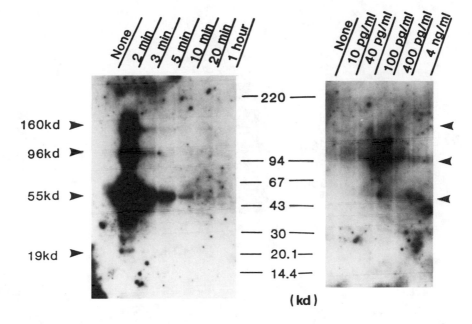

FIGURE 2. Tyrosine phosphorylation of chondrocyte proteins in response to bFGF. Rabbit chondrocytes were exposed to 5 min (right panel) or the indicated times (left panel) to 0.1 ng/ml (left panel) or increasing concentrations (right panel) of bFGF. Western-blot analysis of phosphotyrosyl proteins in chondrocytes was performed using ^{125}I-labeled rabbit antiphosphotyrosine antibodies.

and that precursor structures of IL-1 and FGF have no signal sequence. Three-dimensional structures of aFGF and bFGF are also similar to those of IL-1ß and IL-1α.[36]

J. MECHANISMS

bFGF stimulated release of IGF-I and TGF-ß by chondrocytes in culture.[37,38] Although IGF-I and TGF-ß potentiate the mitogenic effects of FGF in chondrocyte cultures, the physiological relevance of their production is unclear.

One class of bFGF receptor has a typical tyrosine kinase domain.[28,39] Accordingly, bFGF enhanced phosphorylation on tyrosine residues of 160, 90 to 95, and 55-kDa proteins in chondrocytes within 2 min.[7] This stimulation of tyrosine phosphorylation by bFGF was detected at 10 to 40 pg/ml and maximal at 100 pg/ml (Figures 2 and 3). Tyrosine phosphorylation of 90-kDa protein was also induced by bFGF in fibroblasts.[40]

Tyrosine phosphorylation appears to be involved in the control of proteoglycan synthesis. Vanadate, an inhibitor of phosphotyrosyl protein phosphatase, at low concentrations (1 to 6 μM), increased [35S]sulfate incorporation into proteoglycans in chondrocyte cultures, when it increased the level of tyrosine phosphorylation 1.5- to 8-fold.[41,42] In addition, vanadate (1 to 6 μM) increased [3H]thymidine incorporation into DNA in chondrocytes and bone cells.[41,43] It also enhanced the mitogenic action of FGF in bone cells.[43] Furthermore, vanadate, at high con-

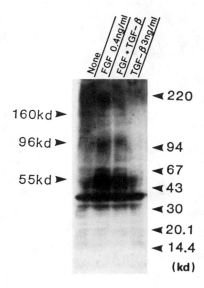

FIGURE 3. Effects of bFGF and TGF-ß on tyrosine phosphorylation in chondrocytes. Rabbit chondrocytes were exposed to bFGF. TGF-ß, or both for 5 min. Western-blot analysis of phosphotyrosyl proteins in chondrocytes was performed using antiphosphotyrosine monoclonal antibody (Amersham) and ^{125}I-protein A.

centrations (20 to 60 μM), induced transformation of chondrocytes while increasing the level of tyrosine phosphorylation 30-fold.[41,42] However, Imamura et al.[44] showed that a mutant aFGF lacking a putative nuclear translocation sequence (amino acid sequence 21 to 27) did not induce DNA synthesis, even though it induced receptor-mediated tyrosine phosphorylation and *c-fos* expression.

bFGF has been shown to stimulate diacylglycerol formation, protein kinase C activation, and Ca^{2+} mobilization in fibroblasts.[45] Protein kinase C may be involved in mediating FGF actions in endothelial cells because the mitogenic action of bFGF was not observed in protein kinase C-deficient endothelial cells (TPA-pretreated cells) or in cells treated with the protein kinase C-inhibitor H-7; whereas, the plasminogen activator-inducing activity of bFGF was unaffected by down-regulation of protein kinase C or by treatment with H-7.[46] However, pretreatment of chondrocytes with TPA did not affect the mitogenic action of bFGF.[7] The cell context appears to affect signaling pathways in response to FGF.

III. TGF-ß

A. SYNTHESIS AND ACTIVATION OF TGF-ß

TGF-ß is a 25,000-mol wt homodimeric peptide and has a broad range of cellular activities such as control of proliferation and differentiation. More than ten TGF-ß-related polypeptides have been reported. Among those, TGF-ß1 to 5 have highly homologous amino acid sequences.

The presence of TGF-ß1 has been shown in precartilaginous tissue, mature cartilage, and mineralizing cartilage.[47-49] Northern blot analyses revealed that high levels of TGF-ß mRNA are associated with growth plates and developing bone.[50] In a model of bone fracture, TGF-ß1 was localized in proliferating, maturing, and hypertrophying cartilaginous callus, and the mesenchymal soft callus.[51] Cartilage also contains TGF-ß2.[49]

bFGF stimulated production of TGF-ß by growth plate chondrocytes in culture, whereas IGF-I, IL-1, PDGF, parathyroid hormone, EGF, and vitamin D metabolites had no effect on the synthesis of TGF-ß.[38]

The majority of cells secrete TGF-ß in an inactive form that cannot interact with TGF-ß receptors.[52] Activation can be achieved by extremes of pH or by treatment with SDS or urea *in vitro*. Various proteases and glucosidases activate TGF-ß. Sato and Rifkin[53] and Lyons et al. [54] suggested that plasmin is a physiological activator of latent TGF-ß.

B. TGF-ß INDUCES CHONDROGENIC DIFFERENTIATION *IN VITRO*

TGF-ß stimulates proliferation and differentiation of chondrogenic cells. Seyedin et al.[55] have shown that TGF-ß1 and 2 stimulate syntheses of type II collagen and proteoglycan core protein in cultures of rat mesenchymal cells in agarose gels. TGF-ß1 also increased [^{3}H]thymidine incorporation into DNA three- to fivefold in fetal or neonatal mouse condyles in organ culture.[56] The majority of labeled cells in TGF-ß-exposed condyles were confined to the precartilage progenitor zone.

C. MITOGENIC EFFECTS

In rabbit chondrocyte cultures, TGF-ß increased [^{3}H]thymidine incorporation into DNA in the presence of FGF, EGF, and serum.[57] However, TGF-ß had little effect on DNA synthesis in the absence of growth factors or serum. TGF-ß stimulated DNA synthesis in chick chondrocytes in 10% serum, but not in 10% dialyzed serum (M_r 12 to 14 kDa cut-off), suggesting involvement of low molecular weight macromolecules in enhancement of mitogenic actions of TGF-ß by serum.[58] Studies with cytofluorometry showed that chondrocytes exposed to TGF-ß in low serum are arrested in late S-phase.[59] The physiological significance of the inhibition of DNA synthesis is not known.

Although TGF-ß1 or 2 alone had a marginal effect on chondrocyte proliferation in soft agar, they markedly stimulated proliferation of chondrocytes in the presence of FGF (Figure 1).[61] The stimulation of chondrocyte proliferation in soft agar was specific for the combination of TGF-ß and FGF.[61] EGF, PDGF, or insulin had little effect on colony formation by chondrocytes even in the presence of TGF-ß or FGF.[61]

D. STIMULATION OF EXTRACELLULAR MATRIX SYNTHESIS

TGF-ß1 and 2 increased incorporations of [^{35}S]sulfate and [^{3}H]glucosamine

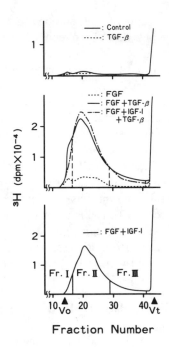

FIGURE 4. Sepharose CL-2B chromatography of proteoglycans located in the cell layers of growth plate chondrocytes in culture exposed to various combinations of bFGF (0.4 ng/ml), TGF-ß1 (1 ng/ml), and IGF-I (60 ng/ml). Cells were seeded at 5×10^4 cells per dish in the presence of 10% serum. After 2 d in culture, they were transferred into 2 ml of a 1:1 mixture of Dulbecco's modified Eagle's medium and Ham's F-12 medium supplemented with 1% bovine serum albumin, 50 µg/ml of ascorbic acid, 2×10^{-7} M hydrocortisone, and 0.5% fetal bovine serum. Growth factors were also added to the low serum medium. Some 8 d after seeding, chondrocytes were exposed to ^{3}H-glucosamine for 10 h.[57] Proteoglycans and glycosaminoglycans were analyzed by gel exclusion chromatography on a Sepharose CL-2B column under dissociative solvent conditions. Fraction I contained hyaluronic acid; fraction II, large chondroitin sulfate proteoglycan (cartilage-characteristic proteoglycan); and fraction III, small proteoglycans.

into proteoglycans synthesized by rabbit chondrocytes in monolayer cultures and in bovine cartilage explants.[57,60,62,63] The increase in incorporation of the radioactive precursors into proteoglycans is not due to changes in pool sizes: TGF-ß caused twofold increase in the content of macromolecules containing uronic acid (proteoglycans) in the cell-matrix layer.[60] TGF-ß selectively increased the synthesis of large, chondroitin sulfate proteoglycan (Figure 4).[60] TGF-ß had little effect on the degree and position (4 or 6) of sulfation, although IGF-I slightly increased the chondroitin 6 sulfate/4 sulfate ratio.[60] Iwamoto et al.[61] and Redini et al.[64] showed that TGF-ß increased collagen synthesis by rabbit chondrocytes in monolayer and soft agar cultures, and did not induce type I collagen synthesis in rabbit and chick chondrocytes.[61] However, Rosen et al.[65] showed that in rat chondrocyte cultures TGF-ß suppressed syntheses of proteoglycan and type II collagen, while stimulating syntheses of type I collagen and fibronectin. It remains unknown why previous studies produced conflicting results about chondrocyte phenotypic expression in the presence of TGF-ß in different culture systems. However, there was a close relationship between the extent of increase in fibronectin synthesis and loss of chondrocyte phenotypic expression in the presence of TGF-ß.[65] Excess synthesis of fibronectin in the presence of TGF-ß may promote dedifferentiation of chondrocyte *in vitro*.

In fracture callus explants, TGF-ß decreased type II collagen transcript levels in the stage of proliferation, but not in the postproliferative stage.[51] The decrease in type II collagen synthesis in the stage of proliferation may be secondary to rapid cell division.

TGF-ß inhibits production of plasminogen activator, collagenase, and stromelysin, while stimulating production of plasminogen activator inhibitor and TIMP in several cell types.[66,67] Thus, TGF-ß promotes the accumulation of extracellular matrices in various tissues by inhibition of matrix degradation. In cartilage explants, TGF-ß inhibited breakdown of proteoglycan, perhaps because it suppresses release of proteoglycan-degrading enzymes such as stromelysin.[63]

E. INHIBITION OF TERMINAL DIFFERENTIATION

In pelleted chondrocyte cultures, TGF-ß1 and 2 suppressed the increases in alkaline phosphatase activity, the number of $1,25(OH)_2$vitamin D_3 receptor, ^{45}Ca incorporation into insoluble material (Table 2), and the calcium content.[23] The effects of TGF-ß on the expression of the terminal differentiation phenotype were dose-dependent and reversible. TGF-ß1 also suppressed the induction of alkaline phosphatase and ^{45}Ca uptake in mouse condyle explants.[56]

F. TGF-ß RECEPTORS

TGF-ß binds to three types of receptors; type I (65 kDa), type II (85 to 110 kDa), and type III (280 to 330 kDa) on various cells. Type III receptor (betaglycan) is a proteoglycan composed of heparan sulfate and chondroitin sulfate, and does not seem to be involved in signal transduction. The three types of TGF-ß receptors were expressed in chondrocyte cultures.[27] Binding of ^{125}I-TGF-ß to I, II, and III receptors in rabbit chondrocytes was constant throughout all stages of growth and cytodifferentiation.[27]

G. MECHANISMS FOR TGF-ß ACTIONS

TGF-ß induced *c-sis* mRNA and a PDGF-like protein in AKR-2B cells and various lymphoid cells, and stimulated release of IL-1 from monocytes.[68,69] On the other hand, TGF-ß suppressed the synthesis of IGF-I by chondrocytes in the presence or absence of FGF.[36] Battegay et al.[70] have shown that TGF-ß induces the synthesis of PDGF-AA, and that anti-PDGF IgG blocks TGF-ß stimulation of DNA synthesis in connective tissue cells. They also suggested that high concentrations of TGF-ß limit the mitogenic response of connective tissue cells including chondrocytes to autocrinely secreted PDGF-AA by decreased expression of the PDGF receptor alpha subunit.[70]

bFGF increased the efficiency of chondrocyte colony formation in soft agar in a dose-dependent manner with an ED_{50} of 0.001 ng/ml in the presence of 3 ng/ml of

TGF-ß compared to ED_{50} at 0.04 ng/ml in its absence (Figure 1).[61] Thus, the concentrations of bFGF needed for the induction of chondrocyte growth in the absence of TGF-ß was 40 times those in its presence. These findings suggest that TGF-ß increases the sensitivity of chondrocytes to FGF. However, TGF-ß treatment had little effect on the affinity and number of FGF receptor and tyrosine phosphorylation on chondrocyte proteins (Figure 3).[7] On the other hand, the synergism between TGF-ß and FGF on DNA synthesis was abolished in chondrocytes whose protein kinase C was down-regulated by TPA.[7] Since TGF-ß stimulated phosphoinositol metabolism and increased intracellular calcium levels in rat fibroblasts (but not in Swiss 3T3 or NRK cells), protein kinase C may be involved in mediating the mitogenic action of TGF-ß in chondrocytes.[71]

Addition of fibronectin mimicked the mitogenic action of TGF-ß in NRK cells in soft agar. Furthermore, RGD peptides suppressed the mitogenic effect of TGF-ß.[72] Thus, the mitogenic action of TGF-ß appears to be mediated, at least in part, by increases of extracellular macromolecules and cell adhesion protein receptors.[72,73]

TGF-ß inhibits DNA synthesis in various cell types such as epithelial, endothelial, and neuronal cells. Addition of TGF-ß1 to lung epithelial cells in late G1 prevented phosphorylation of the retinoblastoma gene product (RB).[74] RB is a nuclear protein of 110 kDa that has growth suppressive activity. RB is expressed ubiquitously in normal mammalian cells and phosphorylated in a transition period between late G1 and S. The inhibition of RB phosphorylation may be linked with the growth-inhibitory activity of TGF-ß.

H. TGF-ß PROMOTES CHONDROGENESIS *IN VIVO*

Daily injections of TGF-ß1 or 2 into the subperiosteal region of newborn rat femurs resulted in intramembranous bone formation and chondrogenesis.[75] TGF-ß2 was more active than TGF-ß1. Interestingly, the cartilage did not calcify during the treatment with TGF-ß. After cessation of TGF-ß injection, the cartilage promptly started to undergo endochondral bone formation. Injections of TGF-ß2 stimulated synthesis of TGF-ß1 in chondrocytes and osteoblasts, suggesting positive autoregulation of TGF-ß.

TGF-ß is present in the hematoma at a site of bone fracture.[51] In a rat fracture model, the hepatoma and soft tissues surrounding the fracture are invaded by mononuclear inflammatory cells and become organized into granulation tissue by day 4. Chondrocytes appear within the granulation tissue near the fracture gap by day 7. Chondrocytes proliferate, produce a cartilaginous matrix, and then become hypertrophic by day 9 to 12. Thereafter, the matrix surrounding these hypertrophic cells calcifies and is replaced by new bone. Anticoagulation treatment impaired the repair of experimental fracture.[76] This suggests that platelet-derived TGF-ß is involved in chondrogenesis at fracture sites.

TGF-ß implanted at subcutaneous or intramuscular sites of the rat did not induce cartilage or bone, but TGF-ß plus a bone-derived osteoinductive factor induced ectopic bone formation at intramuscular sites.[77] Thus, TGF-ß may cooperate with bone morphogenetic proteins (BMPs) to initiate chondrogenesis.

I. BMPs

Several proteins in demineralized bone matrix have been shown to induce ectopic formation of cartilage *in vivo,* and some of them [e.g., such as BMP-2A (BMP-2), BMP-2B (BMP-4), BMP-3, and OP-1 (BMP-7)] are members of a TGF-ß superfamily.[78-81] BMPs have direct effects on both chondrocytes, osteoprogenitor cells, and osteoblasts. BMP-3 (osteogenin) increased syntheses of DNA and proteoglycan in rat chondrocytes in micromass culture.[81] BMP-2A, BMP-2B, and OP-1 increased proteoglycan synthesis by rabbit chondrocytes in monolayer culture.[82] TGF-ß-related proteins may be involved in the control of cytodifferentiation at different stages: BMP-2A mRNA and Vgr-1 mRNA were localized in the proliferating and hypertrophic zones of growth plate, respectively.[83]

IV. CONCLUSIONS

Several members of FGF and TGF-ß families are synthesized by chondrocytes, bone cells, and inflammatory cells, and are stored in extracellular matrices or in the cells. When these peptides are released and activated, they elicit various effects in chondrocytes depending upon cytodifferentiation. Chondrocytes grown in bFGF, aFGF, TGF-ß1 or TGF-ß2 produce cartilage matrix at higher levels after cessation of cell division. Furthermore, FGF and TGF-ß suppress precocious hypertrophy and calcification. In early stages of skeletal development and fracture repair, bFGF, aFGF, TGF-ß1, TGF-ß2, and parathyroid hormone support rapid growth of cartilage with minimum rates of endochondral bone formation. On the other hand, thyroid hormone and calcitonin promote endochondral bone formation by stimulation of chondrocyte hypertrophy and calcification. IGF-1 has little effect on hypertrophy and calcification, while stimulating matrix formation. The three groups of hormones and growth factors appear to have different roles in development and remodeling of cartilage. In addition, a trauma or inflammation could release or activate FGF, TGF-ß, and TGF-ß-related peptides (BMPs) to initiate repair processes. Since FGF- and TGF-ß-like peptides have potent effects on chondrocyte growth, matrix synthesis, and terminal differentiation, they are promising as therapeutic drugs. Further studies are needed to prove involvement of FGF and TGF-ß in cartilage growth, repair, and remodeling *in vivo.*

ACKNOWLEDGMENTS

I am grateful to Dr. M. Iwamoto, Dr. K. Nakashima, Dr. A. Shimazu, Dr. W. Yan, Dr. K. Sato, Dr. T. Koike, Dr. A. Jikko, and Dr. F. Suzuki (Osaka University) for their collaboration in studies on FGF and TGF-ß, and Dr. D. K. Fujii (Sterling Research Group, PA) for his advice in the preparation of this manuscript.

REFERENCES

1. **Neufeld, G. and Gospodarowicz, D.,** Basic and acidic fibroblast growth factors interact with the same cell surface receptors, *J. Biol. Chem.,* 261, 5631, 1986.
2. **Mansukhani, A., Moscatelli, D., Talarico, D., Levytska, V., and Basilico, C.,** A murine fibroblast growth factor (FGF) receptor expressed in CHO cells is activated by basic FGF and Kaposi FGF, *Proc. Natl. Acad. Sci. U.S.A.,* 87, 4378, 1990.
3. **Hebert, J. M., Basilico, C., Goldfarb, M., Haub, O., and Martin, G. R.,** Isolation of cDNAs encoding four mouse FGF family members and characterization of their expression patterns during embryogenesis, *Dev. Biol.,* 138, 454, 1990.
4. **Wilkinson, D. G., Bhatt, S., and McMahon, A. P.,** Expression pattern of the FGF-related proto-oncogene *int-*2 suggests multiple roles in fetal development, *Development,* 105, 131, 1989.
5. **Lobb, R., Sasse, J., Sullivan, R., Shing, Y., D'Amore, P., Jacobs, J., and Klagsbrun, M.,** Purification and characterization of heparin-binding endothelial cell growth factors, *J. Biol. Chem.,* 261, 1924, 1986.
6. **Gonzalez, A.-M., Buscaglia, M., Ong, M., and Baird, A.,** Distribution of basic fibroblast growth factor in the 18-day rat fetus: localization in the basement membranes of diverse tissues, *J. Cell Biol.,* 110, 753, 1990.
7. **Kato, Y., Nakashima, K., and Iwamoto, M.,** unpublished data.
8. **Renko, M., Quarto, N., Morimoto, T., and Rifkin, D. B.,** Nuclear and cytoplasmic localization of different basic fibroblast growth factor species, *J. Cell. Physiol.,* 144, 108, 1990.
9. **Vlodavsky, I., Folkman, J., Sullivan, R., Fridman, R., Ishai-Michaeli, R., Sasse, J., and Klagsbrun, M.,** Endothelial cell-derived basic fibroblast growth factor: synthesis and deposition into subendothelial extracellular matrix, *Proc. Natl. Acad. Sci. U.S.A.,* 84, 2292, 1987.
10. **Saksela, O. and Rifkin, D. B.,** Release of basic fibroblast growth factor-heparan sulfate complexes from endothelial cells by plasminogen activator-mediated proteolytic activity, *J. Cell Biol.,* 110, 767, 1990.
11. **Gospodarowicz, D.,** Fibroblast growth factor and its involvement in developmental processes, *Curr. Top. Dev. Biol.,* 24, 57, 1990.
12. **Kato, Y., Watanabe, R., Hiraki, Y., Suzuki, F., Canalis, E. Raisz, L. G., Nishikawa, K., and Adachi, K.,** Selective stimulation of sulfated glycosaminoglycan synthesis by multiplication-stimulating activity, cartilage-derived factor and bone-derived growth factor: comparison of their actions on cultured chondrocytes with those of fibroblast growth factor and Rhodamine fibrosarcoma-derived growth factor, *Biochim. Biophys. Acta,* 716, 232, 1982.
13. **Kato, Y., Hiraki, Y., Inoue, H., Kinoshita, M., Yutani, Y., and Suzuki, F.,** Differential and synergistic actions of somatomedin-like growth factors, fibroblast growth factor and epidermal growth factor in rabbit costal chondrocytes, *Eur. J. Biochem.,* 12, 680, 1983.
14. **Gospodarowicz, D. and Mescher, A. L.,** A comparison of the responses of cultured myoblasts and chondrocytes to fibroblast and epidermal growth factors, *J. Cell. Physiol.,* 93, 117, 1977.
15. **Hiraki, Y., Kato, Y., Inoue, H., and Suzuki, F.,** Stimulation of DNA synthesis in quiescent rabbit chondrocytes in culture by limited exposure to somatomedin-like growth factors, *Eur. J. Biochem.,* 158, 333, 1986.
16. **Hiraki, Y., Inoue, H., Kato, Y., Fukuya, M., and Suzuki, F.,** Combined effects of somatomedin-like growth factors with fibroblast growth factor or epidermal growth factor in DNA synthesis in rabbit chondrocytes, *Mol. Cell. Biochem.,* 76, 185, 1987.
17. **Kato, Y. and Gospodarowicz, D.,** Growth requirements of low density rabbit costal chondrocyte cultures maintained in serum-free medium, *J. Cell. Physiol.,* 120, 354, 1984.
18. **Kato, Y., Iwamoto, M., and Koike, T.,** Fibroblast growth factor stimulates colony formation of differentiated chondrocytes in soft agar, *J. Cell. Physiol.,* 133, 491, 1987.
19. **Kato, Y. and Gospodarowicz, D.,** Effect of exogenous extracellular matrices on proteoglycan synthesis by cultured rabbit costal chondrocytes, *J. Cell Biol.,* 100, 486, 1985.

20. **Kato, Y. and Gospodarowicz, D.,** Sulfated proteoglycan synthesis by confluent cultures of rabbit costal chondrocytes grown in the presence of fibroblast growth factor, *J. Cell Biol.*, 100, 477. 1985.

21. **Kato, Y.,** Effects of exogenous extracellular matrices on chondrocyte proliferation and phenotypic expression, in *Cell Proliferation and Glomerulonephritis*, Shimizu, F., Kihara, I., and Oite, T., Eds., Nishimura, Niigata, Japan, 1986.

22. **Kato, Y. and Iwamoto, M.,** Fibroblast growth factor is an inhibitor of chondrocyte terminal differentiation, *J. Biol. Chem.*, 265, 5903, 1990.

23. **Kato, Y., Iwamoto, M., Koike, T., Suzuki, F., and Takano, Y.,** Terminal differentiation and calcification in rabbit chondrocyte cultures grown in centrifuge tubes: regulation by transforming growth factor beta and serum factors, *Proc. Natl. Acad. Sci. U.S.A.*, 85, 9552, 1988.

24. **Kato, Y., Shimazu, A., Nakashima, K., Suzuki, F., Jikko, A., and Iwamoto, M.,** Effects of parathyroid hormone and calcitonin on alkaline phosphatase activity and matrix calcification in rabbit growth-plate chondrocyte cultures, *Endocrinology*, 127, 114, 1990.

25. **Kato, Y., Shimazu, A., Iwamoto, M., Nakashima, K., Koike, T., Suzuki, F., Nishii, Y., and Sato, K.,** Role of 1,25-dihydroxycholecalciferol in growth-plate cartilage: inhibition of terminal differentiation of chondrocytes in vitro, *Proc. Natl. Acad. Sci. U.S.A.*, 87, 6522, 1990.

26. **Iwamoto, M., Sato, K., Nakashima, K., Shimazu, A., and Kato, Y.,** Hypertrophy and calcification of rabbit permanent chondrocytes in pelleted cultures: synthesis of alkaline phosphatase and 1,25-dihydroxycholecalciferol receptor, *Dev. Biol.*, 136, 500, 1989.

27. **Iwamoto, M., Shimazu, A., Nakashima, K., Suzuki, F., and Kato, Y.,** Reduction in basic fibroblast growth factor receptor is coupled with terminal differentiation of chondrocytes, *J. Biol. Chem.*, 266, 461, 1991.

28. **Hou, J., Kan, M., McKeehan, K., McBride, G., Adams, P., and McKeehan, W. L.,** Fibroblast growth factor receptors from liver vary in three structural domains, *Science*, 251, 665, 1991.

29. **Cuevas, P., Burgos, J., and Baird, A.,** Basic fibroblast growth factor (FGF) promotes cartilage repair in vivo, *Biochem. Biophys. Res. Commun.*, 31, 611, 1988.

30. **Jingushi, S., Heydemann, A., Kana, S. K., Macey, L. R., and Bolander, M. E.,** Acidic fibroblast growth factor (aFGF) injection stimulates cartilage enlargement and inhibits cartilage gene expression in rat fracture healing, *J. Orthop. Res.*, 8, 364, 1990.

31. **Sano, H., Forough, R., Maier, J. A. M., Case, J. P., Jackson, A., Engleka, K., Maciag, T., and Wilder, R. L.,** Detection of high levels of heparin binding growth factor-1 (acidic fibroblast growth factor) in inflammatory arthritic joints, *J. Cell Biol.*, 110, 1417, 1990.

32. **Baird, A., Mormede, P., Ying, S.-Y., Wehrenberg, W. B., Ueno, N., Ling, N., and Guillemin, R.,** Immunoreactive fibroblast growth factor in cells of peritoneal exudates suggests its identity with macrophage-derived factor, *Biochem. Biophys. Res. Commun.*, 126, 358, 1985.

33. **Chandrasekhar, S. and Harvey, A. K.,** Induction of interleukin-1 receptors on chondrocytes by fibroblast growth factor: a possible mechanism for modulation of interleukin-1 activity, *J. Cell. Physiol.*, 138, 236, 1989.

34. **Phadke, K.,** Fibroblast growth factor enhances the interleukin 1-mediated chondrocytic protease release, *Biochem. Biophys. Res. Commun.*, 142, 448, 1987.

35. **Iwamoto, M., Koike, T., Nakashima, K., Sato, K., and Kato, Y.,** Interleukin 1: a regulator of chondroctye proliferation, *Immunol. Lett.*, 21, 153, 1989.

36. **Zhu, X., Komiya, H., Chirino, A., Faham, S., Fox, G. M., Arakawa, T., Hsu, B. T., and Rees, D. C.,** Three-dimensional structures of acidic and basic fibroblast growth factors, *Science*, 251, 90, 1991.

37. **Elford, R. and Lamberts, S. W. J.,** Contrasting modulation by transforming growth factor-beta of insulin-like growth factor-1 production in osteoblasts and chondrocytes, *Endocrinology*, 127, 1635, 1990.

38. **Gelb, D. E., Rosier, R. N., and Puzas, J. E.,** The production of transforming growth factor-beta by chick growth plate chondrocytes in short term monolayer culture, *Endocrinology*, 127, 1941, 1990.

39. **Lee, P. L., Johnson, D. E., Cousens, L. S., Fried, V. A., and Williams, L. T.,** Purification and complementary DNA cloning of a receptor for basic fibroblast growth factor, *Science,* 245, 57, 1989.

40. **Coughlin, S. R., Barr, P. J., Cousens, L. S., Fretto, L. J., and Williams, L. T.,** Acidic and basic fibroblast growth factors stimulate tyrosine kinase activity in vivo, *J. Biol. Chem.,* 263, 988, 1988.

41. **Kato, Y., Iwamoto, M., Koike, T., and Suzuki, F.,** Effect of vanadate on cartilage-matrix proteoglycan synthesis in rabbit costal chondrocyte cultures, *J. Cell Biol.,* 104, 311, 1987.

42. **Ohwada, M. K., Iwamoto, M., Koike, T., and Kato, Y.,** Effects of vanadate on tyrosine phosphorylation and the pattern of glycosaminoglycan synthesis in rabbit chondrocytes in culture, *J. Cell. Physiol.,* 138, 484, 1989.

43. **Lau, K. W., Tanimoto, H., and Baylink, D. J.,** Vanadate stimulates bone cell proliferation and bone collagen synthesis in vitro, *Endocrinology,* 123, 2858, 1988.

44. **Imamura, T., Engleka, K., Zhan, XI., Tokita, Y., Forough, R., Roeder, D., Jackson, A., Maier, J. A. M., Hia, T., and Maciag, T.,** Recovery of mitogenic activity of a growth factor mutant with a nuclear translocation sequence, *Science,* 249, 1567, 1990.

45. **Kaibuchi, K., Tsuda, T., Kikuchi, A., Tanimoto, T., Yamashita, T., and Takai, Y.,** Possible involvement of protein kinase C and calcium ions in growth factor induced expression of *c-myc* gene in Swiss 3T3 cells, *J. Biol. Chem.,* 261, 1187, 1986.

46. **Presta, M., Maier, J. A. M., and Ragnotti, G.,** The mitogenic signaling pathway but not the plasminogen activator-inducing pathway of basic fibroblast growth factor is mediated through protein kinase C in fetal bovine aortic endothelial cells, *J. Cell Biol.,* 109, 1877, 1989.

47. **Heine, U. I., Munoz, E. F., Flanders, K. C., Ellingsworth, L. R., Lam. H.-Y. P., Thompson, N. L., Roberts, A. B., and Sporn, M. B.,** Role of transforming growth factor-beta in the development of the mouse embryo, *J. Cell Biol.,* 105, 2861, 1987.

48. **Ellingsworth, L. P., Brennan, J. E., Fok, K., Rosen, D. M., Bentz, H., Piez, K. A., and Seyedin, S. M.,** Antibodies to the N-terminal portion of cartilage-inducing factor A and transforming growth factor beta, *J. Biol. Chem.,* 261, 12362, 1986.

49. **Carrington, J. L., Roberts, A. B., Flanders, K. C., Roche, N. S., and Reddi, A. H.,** Accumulation, localization, and compartmentation of transforming growth factor beta during endochondral bone development, *J. Cell Biol.,* 107, 1969, 1988.

50. **Sandberg, M., Vuorio, T., Hirvonen, H., Alitalo, K., and Vuorio, E.,** Enhanced expression of TGF-beta and *c-fos* mRNAs in the growth plates of developing human long bones, *Development,* 102, 461, 1988.

51. **Joyce, M. E., Terek, R. M., Jingushi, S., and Bolander, M. E.,** Role of transforming growth factor-beta in fracture repair, *Ann. NY Acad. Sci.,* 593, 107, 1990.

52. **Lawrence, D. A., Pircher, R., Kryceve-Martinerie, C., and Jullien, P.,** Normal embryo fibroblasts release transforming growth factors in a latent form, *J. Cell. Physiol.,* 121, 184, 1984.

53. **Sato, Y. and Rifkin, D. B.,** Inhibition of endothelial cell movement by pericytes and smooth muscle cells: activation of a latent TGF-beta 1-like molecule by plasmin during coculture, *J. Cell Biol.,* 109, 309, 1989.

54. **Lyons, R. M., Gentry, L. E., Purchio, A. F., and Moses, H. L.,** Mechanism of activation of latent recombinant transforming growth factor beta 1 by plasmin, *J. Cell Biol.,* 110, 1361, 1990.

55. **Seyedin, S. M., Thomas, T. C., Thompson, A. Y., Rosen, D. N., and Piez, K. A.,** Purification and characterization of two cartilage-inducing factors from bovine demineralized bone, *Proc. Natl. Acad. Sci. U.S.A.,* 82, 2267, 1985.

56. **Silbermann, M., Iwamoto, M., Kato, Y., and Suzuki, F.,** TGF beta stimulates DNA synthesis in precartilage cells but depresses alkaline phosphatase activity in chondrocytes and the calcification of the cartilage extracellular matrix in vitro, *Calcif. Tissue Int.,* 42 (Suppl.), A25, 1988.

57. **Hiraki, Y., Inoue, H., Hirai, R., Kato, Y., and Suzuki, F.,** Effect of transforming growth factor beta on cell proliferation and glycosaminoglycan synthesis by rabbit growth plate chondrocytes in culture, *Biochim. Biophys. Acta,* 969, 91, 1988.

58. **O'Keefe, R. J., Puzas, J. E., Brand, J. S., and Rosier, R. N.,** Effect of transforming growth factor-beta on DNA synthesis by growth plate chondrocytes: modulation by factors present in serum, *Calcif. Tissue Int.,* 43, 352, 1988.

59. **Vivien, D., Galera, P., Lebrun, E., Loyau, G., and Pujol, J.-P.,** Differential effects of transforming growth factor-beta and epidermal growth factor on the cell cycle of cultured rabbit articular chondrocytes, *J. Cell. Physiol.,* 143, 534, 1990.

60. **Inoue, H., Kato, Y., Iwamoto, M., Hiraki, Y., Sakuda, M., and Suzuki, F.,** Stimulation of cartilage-matrix proteoglycan synthesis by morphologically transformed chondrocytes grown in the presence of fibroblast growth factor and transforming growth factor-beta, *J. Cell. Physiol.,* 138, 329, 1989.

61. **Iwamoto, M., Sato, K., Nakashima, K., Fuchihata, H., Suzuki, F., and Kato, Y.,** Regulation of colony formation of differentiated chondrocytes in soft agar by transforming growth factor-beta, *Biochem. Biophys. Res. Commun.,* 159, 1006, 1989.

62. **O'Keefe, R. J., Puzas, J. E., Brand, J. S., and Rosier, R. N.,** Effects of transforming growth factor-beta on matrix synthesis by chick growth plate chondrocytes, *Endocrinology,* 122, 2953, 1988.

63. **Morales, T. I. and Roberts, A. B.,** Transforming growth factor beta regulates the metabolism of proteoglycans in bovine cartilage organ cultures, *J. Biol. Chem.,* 263, 12828, 1988.

64. **Redini, F., Galera, P., Mauviel, A., Loyau, G., and Pujol, J.-P.,** Transforming growth factor beta stimulates collagen and glycosaminoglycan biosynthesis in cultured rabbit articular chondrocytes, *FEBS Lett.,* 234, 172, 1988.

65. **Rosen, D. M., Stempien, S. A., Thompson, A. Y., and Seyedin, S. M.,** Transforming growth factor-beta modulates the expression of osteoblast and chondroblast phenotypes in vitro, *J. Cell. Physiol.,* 134, 337, 1988.

66. **Saksela, O., Moscatelli, D., and Rifkin, D. B.,** The opposing effects of basic fibroblast growth factor and transforming growth factor beta on the regulation of plasminogen activator activity in capillary endothelial cells, *J. Cell Biol.,* 105, 957, 1987.

67. **Edwards, D. R., Murphy, G., Reynolds, J. J., Whitham, S. E., Docherty, A. J. P., Angel, P., and Health, J. K.,** Transforming growth factor beta modulates the expression of collagenase and metallopoteinase inhibition, *EMBO. J.,* 6, 1899, 1987.

68. **Leof, E. B., Proper, J. A., Goustin, A. S., Shipley, G. D., Di Corleto, P. E., and Moses, H. L.,** Induction of *c-sis* mRNA and activity similar to platelet-derived growth factor by transforming growth factor beta: a proposed model for indirect mitogenesis involving autocrine activity, *Proc. Natl. Acad. Sci. U.S.A.,* 83, 2453, 1986.

69. **Wahl, S. M., Hunt, D. A., Wakefield, L. M., McCartney-Francis, N., Wahl, L. M., Roberts, A. B., and Sporn, M. B.,** Transforming growth factor beta induces monocyte chemotaxis and growth factor production, *Proc. Natl. Acad. Sci. U.S.A.,* 84, 5788, 1987.

70. **Battegay, E. J., Raines, E. W., Seifert, R. A., Bowen-Pope, D. F., and Ross, R.,** TGF-beta induces bimodal proliferation of connective tissue cells via complex control of an autocrine PDGF loop, *Cell,* 63, 515, 1990.

71. **Muldoon, L. L., Rodland, K. D., and Magun, B. E.,** Transforming growth factor beta modulates epidermal growth factor-induced phosphoinositide metabolism and intracellular calcium levels, *J. Biol. Chem.,* 263, 5030, 1988.

72. **Ignotz, R. A. and Massague, J.,** Transforming growth factor-beta stimulates the expression of fibronectin and collagen and their incorporation into the extracellular matrix, *J. Biol. Chem.,* 261, 4337, 1986.

73. **Roberts, C. J., Birkenmeier, T. M., McQuillan, J. J., Akiyama, S. K., Yamada, S. S., Chen, W., Yamada, K. M., and McDonald, J. A.,** Transforming growth factor beta stimulates the expression of fibronectin and of both subunits of the human fibronectin receptor by cultured human lung fibroblasts, *J. Biol. Chem.,* 263, 4586, 1988.

74. **Laiho, M., DeCaprio, J. A., Ludlow, J. W., Livingston, D. M., and Massague, J.,** Growth inhibition by TGF-beta linked to suppression of retinoblastoma protein phosphorylation, *Cell,* 62, 175, 1990.

75. **Joyce, M. E., Roberts, A. B., Sporn, M. B., and Bolander, M. E.,** Transforming growth factor-beta and the initiation of chondrogenesis and osteogenesis in the rat femur, *J. Cell Biol.,* 110, 2195, 1990.

76. **Rokkanen, P. and Slatis, P.,** Repair of experimental fracture during long term anticoagulation treatment, *Acta Orthop. Scand.,* 35, 21, 1964.

77. **Bentz, H., Nathan, R. M., Rosen, D. M., Armstrong, R. M., Thompson, A. Y., Segarini, P. R., Mathews, M. C., Dasch, J. R., Piez, K. A., and Seyedin, S. M.,** Purification and characterization of a unique osteoinductive factor from bovine bone, *J. Biol. Chem.,* 264, 20805, 1989.

78. **Wozney, J. M., Rosen, V., Celeste, A. J., Mitsock, L. M., Whitters, M. J., Kriz, R. W., Hewick, R. M., and Wang, E. A.,** Novel regulators of bone formation: molecular clones and activities, *Science,* 242, 1528, 1988.

79. **Samphath, T. K., Coughlin, J. E., Whetstone, R. M., Banach, D., Corbett, C., Ridge, J. R., Ozkaynak, E., Oppermann, H., and Rueger, D. C.,** Bovine osteogenic protein is composed of dimers of OP-1 and BMP-2A, two members of the transforming growth factor-beta superfamily, *J. Biol. Chem.,* 265, 13198, 1990.

80. **Celeste, A., Iannazzi, J. A., Taylor, R. C., Hewick, R. M., Rosen, V., Wang, E. A., and Wozney, J. M.,** Identification of transforming growth factor beta family members present in bone-inductive protein purified from bovine bone, *Proc. Natl. Acad. Sci. U.S.A.,* 87, 9843, 1990.

81. **Vukicevic, S., Luyten, F. P., and Reddi, A. H.,** Stimulation of the expression of osteogenic and chondrogenic phenotypes in vitro by osteogenin, *Proc. Natl. Acad. Sci. U.S.A.,* 86, 8793, 1989.

82. **Kato, Y., Shimazu, A. and Ueno, N.,** unpublished data.

83. **Lyona, K. M., Pelton, R. W., and Hogan, B. L. M.,** Patterns of expression of murine Vgr-1 and BMP-2a RNA suggest that transforming growth factor-beta-like genes coordinately regulate aspects of embryonic development, *Genes Dev.,* 3, 1657, 1989.

Chapter 6

ROLE OF INSULIN-LIKE GROWTH FACTORS IN THE REGULATION OF CHONDROCYTES

Stephen B. Trippel

TABLE OF CONTENTS

I. INTRODUCTION

The insulin-like growth factors, or somatomedins, are a family of low molecular weight peptides that are anabolic for multiple cell types, including chondrocytes.[1,2] Two principal classes of insulin-like growth factor (IGF) have been demonstrated by amino acid and DNA sequence analysis: insulin-like growth factor I (IGF-I), or somatomedin C (Sm-C),[3-7] and insulin-like growth factor II (IGF-II) (Figure 1).[8-10] IGF-I is functionally distinct from IGF-II in being more highly human growth-hormone dependent and possessing greater growth-promoting activity.[11] IGF-II may be the principal somatomedin in the embryo and fetus.[12-14] As for many peptide growth factors, the nomenclature of the IGFs has been confounded by the independent identification in various laboratories of factors that ultimately proved to be the same peptide. A few of the principal historical designations for the somatomedins are listed in Table 1. Currently, the generic terms "insulin-like growth factor" and "somatomedin" are used interchangeably, while the terms "IGF-I" and "IGF-II" are in predominant use for the specific peptides.

II. HISTORICAL PERSPECTIVE

That the somatomedins play a role in the regulation of chondrocytes is a matter of definition. Classically defined, a somatomedin is a factor that fulfills the three criteria of growth hormone dependence, insulin-like effect on extraskeletal tissues, and the ability to promote incorporation of sulfate into proteoglycans of cartilage.[15,16] Indeed, the IGFs were first identified on the basis of their ability to stimulate sulfate uptake by cartilage in culture. In 1957, Salmon and Daughaday[17] reported the now classic experiments examining the mechanism by which growth hormone stimulates skeletal growth. They observed that cultured costal cartilage from hypophysectomized (growth hormone-deficient) rats incorporated less sulfate than cultured costal cartilage from normal rats. Similarly, while serum from normal rats potently increased cartilage sulfate incorporation, growth hormone-deficient serum from hypophysectomized rats had little effect. Surprisingly, the addition of growth hormone to the cultured cartilage failed to restore stimulation of sulfate uptake. However, when serum from hypophysectomized rats that had been treated with growth hormone was added to the cartilage in culture, the ability to enhance sulfate incorporation was restored. This growth hormone-dependent activity in the serum was termed "sulfation factor".

Following the identification of sulfation factor by Salmon and Daughaday, a growth hormone-dependent factor was found that enhanced DNA synthesis by cultured cells.[18] The term "thymidine factor" was used to distinguish this from sulfation factor.[19] Efforts to separate thymidine factor activity from sulfation factor activity were unsuccessful[20] and ultimately revealed that the two factors were the same. Due to their positive role as mediators of growth hormone action, these activities were designated "somatomedin".[21] With a more detailed char-

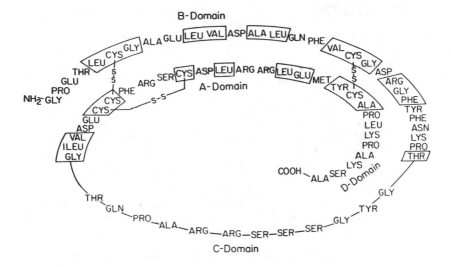

FIGURE 1. Primary structure of somatomedin-C/insulin-like growth factor I. Residues enclosed in boxes in A and B domains of IGF-I are identical to amino acids in corresponding positions in the human proinsulin molecule. The D domain is an eight-residue extension at the carboxy terminus that does not exist in proinsulin. B and A domains of IGF-II are approximately 70% identical to similar regions in IGF-I. The C domain of IGF-II is composed of 8 amino acid residues that are not homologous with the 12 in IGF-I. The D domain of IGF-II is also two residues shorter and not homologous with the D domain of IGF-I. (From Underwood, L., E. and Van Wyk, J. J., *Textbook of Endocrinology,* 7th ed., Wilson, J. D. and Foster, D. W., Eds., W. B. Saunders, Philadelphia, 1985, 155. With permission.)

acterization of the serum fraction containing somatomedin activity, two separate classes of peptides were distinguished and nomenclature for these has evolved to the designations IGF-I and IGF-II.

During the early investigation of serum somatomedin activity, it became apparent that the responsible substance(s) induced cells to engage in more than just sulfation and DNA synthesis. Using partially purified IGF preparations, investigators demonstrated increased incorporation of uridine into RNA[22] leucine into protein[22] and conversion of proline into hydroxyproline, a step in the synthesis of collagen.[23] Thus, by 1970, the yet-to-be-named somatomedins had already been shown to stimulate a broad range of cartilage cellular functions, including synthesis of DNA, RNA, sulfated proteoglycans, collagen and noncollagen protein. This scope of activities distinguished the IGFs early in their history as potentially important anabolic regulators of cartilage.

Considerable effort was expended in the early 1970s to develop accurate and simplified bioassays for somatomedin. Since many of these studies employed cartilage as the responding tissue, they contributed extensively to defining the effects of somatomedin on various forms of cartilage. Somatomedin, at least in an impure form, was found capable of regulating DNA synthesis and sulfate incorporation into matrix by embryonic,[24-26] costal,[17,22,27] articular,[27-29] and epiphyseal cartilage.[28-30]

TABLE 1
Partial List of Insulin-Like Growth Factors/Somatomedins

Insulin-like growth factor-I	IGF-I
Insulin-like growth factor-II	IGF-II
Somatomedin-A	Sm-A
Somatomedin-B	Sm-B
Somatomedin-C	Sm-C
Multiplication stimulating activity	MSA, rIGF-II
Insulin-like activity	ILA
Nonsuppressible insulin-like activity	NSILA

Note: An apparent reduction in the number of somatomedins has occurred with the demonstration that many of the substances independently identified in different laboratories appear to be identical or homologous. The somatomedins IGF-I and Sm-C are identical. A second class of somatomedin includes IGF-II and ILA (human) and MSA (rat IGF-II). NSILA denotes a mixture of IGF-I and IGF-II that antedates their separation. Sm-A, though incompletely characterized, appears to consist principally of IGF-I. Sm-B, while somewhat growth hormone dependent, does not stimulate cartilage and has been excluded as somatomedin.[16]

The specific biochemical steps in proteoglycan metabolism that are regulated by the IGFs have been the subject of several studies. Salmon and Duvall[22] showed that when cartilage from hypophysectomized rats was incubated with somatomedin, the resulting stimulation of labeled sulfate incorporation (sulfation) followed that of labeled leucine incorporation (protein synthesis). This suggested that somatomedin does not simply increase the sulfation of a preexisting proteoglycan pool, but may be acting on synthesis of the protein core. Subsequently, Kilgore et al.[31] demonstrated in chick embryo cartilage that somatomedin enhanced the activity of xylosyltransferase, the enzyme responsible for initiating chondroitin sulfate chain formation on core protein.[32] Interestingly, somatomedin proved not to be acting on the enzyme itself, but to be increasing endogenous acceptor protein levels. These data suggested that the mechanism of somatomedin action on cartilage sulfation is, at least in part, at the level of core protein synthesis. Kemp and Hintz[33] have provided further support for this view by demonstrating that cyclohexamide (an inhibition of protein synthesis) virtually eliminated the stimulatory effect of somatomedin on sulfate incorporation. In contrast, when beta D xyloside (a substrate for the initiation of chondroitin sulfate chains that bypasses the need for a protein core) was added to the cartilage cultures, incorporation of [^{35}S]sulfate was restored, but the stimulatory effect of somatomedin was not.[34] These data are consistent with the hypothesis that somatomedin stimulates proteoglycan core protein synthesis and that, in the absence of alternative substrates for sulfation (e.g., xyloside), this core protein synthesis is the rate-limiting step in the somatomedin-stimulated sulfation process.

The effect of IGF-I on the structure of the final proteoglycan product is somewhat controversial. When glycosaminoglycans (GAGs) produced by so-

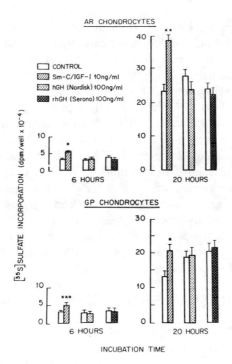

FIGURE 2. Effect of Sm-C/IGF-I, hGH, or recombinant methionyl hGH (rhGH) on [^{35}S]sulfate incorporation into proteoglycans obtained from rabbit articular (upper panel) and rabbit growth plate (lower panel) chondrocyte primary cultures. Both types of chondrocytes were used at confluency after 20-h deprivation of fetal bovine serum. [^{35}S]sulfate incorporation was performed during a period of 6 or 20 h, as indicated. Each value represents the mean + SEM of six wells. Each treated group was compared to its control. *, $p < 0.02$, **, $p < 0.01$, ***, $p < 0.001$. (From Trippel, S. B., Corvol, M. T., Dumontier, M. F., Rappaport, R., Hung, H. H., and Mankin, H. J., *Pediatr. Res.,* 25, 76, 1989. With permission.)

matomedin-stimulated and unstimulated control species cartilage were fraction-ated by Sephadex 200 chromatography, the growth factor was observed to shift the molecular weight distribution of the GAGs to a larger range, suggesting a somatomedin effect on GAG chain length.[33] More recently, others have detected no change in the hydrodynamic size of GAGs when comparing IGF-I-treated and control bovine articular cartilage explants[35] or rat rib chondrocytes.[36]

Although early studies defining the biological effects of the somatomedins on cartilage were undertaken with only partially purified somatomedin prepara-tions, subsequent studies employing highly purified native IGF-I or bacterially produced recombinant IGF-I have generally confirmed early results and have served to expand upon them. Highly purified human IGF-I[37] and recombinant human IGF-I[35,38] stimulate proteoglycan synthesis in bovine articular cartilage explant cultures, and in growth plate[39] and articular[39] isolated chondrocyte cultures (Figure 2).

The previously observed somatomedin stimulation of collagen synthesis by chondrocytes[23] has recently been shown to be specific for type II collagen.[40]

While the precise level of regulation of collagen synthesis remains to be established, Sandell and Dudek have demonstrated that in chondrocytes isolated from adult bovine articular cartilage, IGF-I stimulates expression of collagen type II mRNA, consistent with regulation of the type II collagen gene.[40]

Following the separation of IGF-I from IGF-II, it became possible to compare their biological effects. IGF-I has generally been found to be more active than IGF-II in its anabolic effects on cartilage. In studies of rat costal cartilage, IGF-I was more active than IGF-II at low concentrations (2.5 ng/ml),[41] although at maximal concentrations (25 ng/ml), IGF-I nd IGF-II stimulated sulfation to a similar extent. In cultured chick embryo cartilage, both IGF-I and IGF-II manifested similar potencies as promotors of 3H-uridine incorporation into RNA and [^{35}S]sulfate incorporation into glycosaminoglycans.[2] A comparison of recombinant human IGF-I and highly purified rat IGF-II (rIGF-II) found rIGF-II to be substantially less potent than IGF-I in increasing proteoglycan synthesis and in diminishing proteoglycan catabolism: 100 ng/ml of IGF-II was approximately 60% as effective as 5 ng/ml of IGF-I.[38]

III. EPIPHYSEAL GROWTH PLATE CARTILAGE

A. SOMATOMEDIN HYPOTHESIS

The early work of Salmon and Daughaday[17] and of others[15,16] led to the postulate that the stimulatory effects of growth hormone (GH) on skeletal growth are mediated by somatomedin. This has come to be known as the somatomedin hypothesis. Several lines of evidence support this postulate. Serum somatomedin concentrations are GH dependent,[42] administration of insulin-like growth factor I stimulates longitudinal skeletal growth in hypophysectomized rats,[43,44] IGF-I exerts negative feedback on GH at both the hypothalmus and pituitary,[45] and most,[16-18,39,46-50] but not all,[51-54] studies of IGF-I and GH in cultured chondrocytes have found that IGF-I stimulates DNA and matrix synthesis while GH fails to do so. Also consistent within this hypothesis is the intriguing observation recently reported by Eigenmann et al that differences in the body size of poodles from toy to miniature to standard is paralleled by differences in serum insulin-like growth factor I levels, but not by growth hormone capacity.[55]

According to the classical view of the somatomedin hypothesis, GH produced in the pituitary is released into the circulation which transports it to the liver[56] and other tissues.[57] There, it stimulates production of IGF-I which is, in turn, released into the circulation to act in an endocrine fashion on its target tissues, including the growth plate.

Seemingly inconsistent with this hypothesis is data suggesting that the action of GH on chondrocytes is a direct one, not involving mediation by IGF-I. Madsen et al.[51] reported that GH enhanced proliferation of rabbit ear and rib chondrocytes in culture and also stimulated proteoglycan synthesis in cultured rat chondrocytes.[52] Eden et al.[58] have demonstrated specific binding of GH to rabbit ear and epiphyseal plate chondrocytes, indicating that these cells possess receptors for GH. *In vivo* studies have also been interpreted as revealing a direct

GH effect on cartilage. GH injected into the growth plates of hypophysectomized rats elicited an increased growth rate at that physis,[59,60] suggesting a direct action on longitudinal bone growth. Similarly, when GH is administered intra-arterially to one limb of a hypophysectomized rat, the growth rate is increased in that limb, but not the contralateral limb.[61] GH administered systemically to hypophysectomized rats is a much more effective stimulus of skeletal growth than IGF-I, even when GH is administered at 50-fold lower doses.[62]

This somewhat confusing discrepancy in the literature over the validity of the somatomedin hypothesis is perhaps more apparent than real. Reports addressing the ability of GH to stimulate cartilage may be divided into two categories according to their use of *in vivo* or *in vitro* conditions. So divided, such studies also tend to distribute according to the effect of GH. Reports based on *in vivo* experiments are strikingly uniform in finding stimulation of cartilage by GH.[59-63] In contrast, results of *in vitro* studies generally demonstrate lack of direct stimulation by GH.[39,46-50] This *in vivo-in vitro* dichotomy is consistent with the hypothesis that the effects of GH observed *in vivo* are dependent on mediating substances, such as IGF-I that are available *in vivo* but not available under most *in vitro* conditions.

Classically, the IGFs have been viewed as endocrine factors, circulating with their binding proteins and acting on a target cell at a distance from their source. While IGF-I may indeed act in this fashion, more recent data suggest that IGF-I also acts in an autocrine or paracrine fashion at the growth plate. In support of this view, Schlechter et al.[63] have shown that administration of an anti-IGF-I antibody blocks the stimulatory effect of intraarterial GH on growth plate stimulation (Figure 3). These data suggest that the observed GH effect on the growth plate is mediated, at least in part, by locally produced IGF-I. Indeed, recent reports suggest that GH stimulates IGF-I synthesis by growth plate chondrocytes *in vitro*,[64] and increases expression of IGF-I mRNA by growth plate chondrocytes *in vivo*.[65,66]

Currently, the somatomedin hypothesis appears to be best viewed from both an endocrine and an autocrine perspective. According to this model, GH stimulates production of IGF-I by multiple tissues.[57] This IGF-I then either enters the circulation to act in an endocrine fashion or is retained locally to act in an autocrine or paracrine fashion (Figure 4). These two mechanisms are not mutually exclusive. The relative contribution of each to the regulation of cartilage at different sites and different stages of development remains to be established.

The role of IGF-I and GH in regulating growth plate cartilage may also involve control of cell differentiation. Zezulak and Green[67] noted that GH, but not IGF-I, could promote the differentation of cultured preadipocytes to adipocytes and that these newly differentiated cells were more sensitive to the mitogenic effect of IGF-I than the precursor cells. Such data led to the formulation of the "dual effector" theory of GH action[68] which postulates that IGF-I produces clonal expansion of cell populations previously induced to differentiate by GH. The growth plate is an attractive location to apply this model. Reserve zone cells

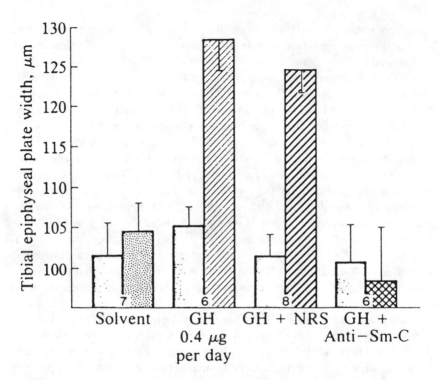

FIGURE 3. Effects on tibial epiphyseal plate width of chronic infusion via minipump of solvent or rat GH (0.4 µg/d) alone or with normal rabbit serum (NRS; 80% of pump volume) or rabbit antiserum to human Sm-C (Anti-Sm-C; 80% of pump volume) into the arterial supply of one hindlimb of male hypophysectomized rats. The number of animals per group is given at the bottom of the bars. Solid bars represent the width of the noninfused epiphyseal plates of that group. (From Schlecter, N. L., Russell, S. N., Spencer, E. M., and Nicoll, C., *Proc. Natl. Acad. Sci. U.S.A.,* 83, 7932, 1986. With permission.)

represent appropriate putative precursors which, in response to GH, could become susceptable to clonal expansion as proliferative zone cells under the influence of IGF-I. To date, this theory has, at least in part, received empirical support,[39,46,54] and will be an important subject for future investigation.

GH-initiated mechanisms may not be the sole means by which IGF-I participates in regulation of cartilage growth. Burch et al.[69] have recently shown that embryonic chick cartilage releases somatomedin-like peptide into GH-free defined medium and that an anti-IGF antibody inhibits the cartilage growth that normally occurs in this medium. These data indicate that embryonic cartilage constitutively produces a somatomedin-like peptide and responds to it in an autocrine fashion. However, because the released IGF detected by IGF-I RIA was produced at very low concentrations (0.003 ng/ml) and because the immunoinhibition was achieved with an antibody that effectively binds both IGF-I and IGF-II, it is possible that IGF-II rather than IGF-I was the somatomedin responsible for the observed cartilage growth.

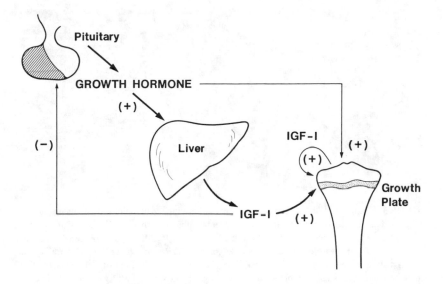

FIGURE 4. Somatomedin hypothesis. The classical model of the somatomedin hypothesis postulates that pituitary GH stimulates production by the liver of IGF-I which is carried by the circulation in an endocrine fashion to the growth plate. Recent data are not consistent with this model, but support the depicted modification in which GH also stimulates local production in the growth plate of IGF-I which acts locally in an autocrine or paracrine fashion.

B. GROWTH PLATE CHONDROCYTE POPULATIONS

The growth plate is composed of several distinct, zonally distributed populations of cells differing considerably in morphology,[70-73] biochemistry,[74-77] and function.[75,78-80] Given the heterogeneity of growth plate chondrocytes, it is plausible that the role of cellular regulators such as the IGFs may be heterogeneous among these different cell populations. Available data seem to support this view. In studies localizing radiolabeled IGF-I binding in the bovine growth plate,[81] IGF-I specifically bound to cells in all zones of the growth plate, but proliferative zone cells bound significantly more IGF-I than cells in the reserve, maturation, or hypertrophic zones (Figure 5). The greatest IGF-I binding was localized to the upper (immature) portion of the proliferative zone with twofold greater binding than to cells of the lower (more mature) proliferative zone and three- to fourfold greater binding than to the other zones. These differences indicate that growth plate chondrocytes vary in their capacity to bind IGF-I according to their stage of differentiation.

In the above studies, it was also observed that cellular incorporation of [³H]thymidine was localized principally to the upper proliferative zone, the same cell population that manifested the greatest IGF-I binding.[81] Although cells in the maturation and hypertrophic zones demonstrated no thymidine incorporation (indicating no mitotic activity), these cells did specifically bind IGF-I and have previously been shown to respond to IGF-I by increasing synthesis of

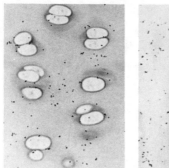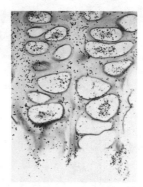

FIGURE 5. Binding of ^{125}I-IGF-I to different zones of bovine distal radial growth plates. Sections (160 μm) of neonatal calf growth plate cartilage were incubated with ^{125}I-IGF-I at 4°C for 18 h and then at 37°C for 10 min to permit internalization of bound ^{125}I-IGF-I. Fixed cartilage was sectioned at 4 μm and ^{125}I-IGF-I was localized by autoradiography. (A) Reserve zone; (B) proliferative zone; and (C) maturation and hypertrophic zones. Specific cellular localization of ^{125}I-IGF-I, as determined by grain counts on multiple sections, was significantly ($p < 0.05$) greater to cells of the proliferative zone than to cells of the other zones. (From Trippel, S. B., Van Wyk, J. J., and Mankin, H. J., *J. Bone Joint Surg.*, 68A, 897, 1986. With permission.)

proteoglycan but not of DNA.[30] These data suggest that the role of IGF-I in regulating growth plate chondrocytes depends, in part, on the maturational stage of the cell. The data further suggest an uncoupling of the proliferative response to IGF-I as the cells progress from the proliferative to the maturation zones. In this transition, the cells appear to lose the ability to divide in response to IGF-I, but retain the ability to respond with respect to differentiated functions such as matrix synthesis.

Additional data suggest that cells of the reserve and proliferative zones differ in their responsiveness to IGF-I. In a recent study, the proliferative and reserve zones of bovine distal ulnar growth plates were separated by microdissection and the isolated chondrocytes from each zone were exposed to highly purified IGF-I. The proliferative zone cells responded with a more pronounced increase in both [^{3}H]thymidine incorporation and [^{35}S]sulfate incorporation than their reserve zone counterparts. In addition, lower concentrations of IGF-I were required to elicit a mitogenic response from proliferative zone cells than were required to elicit this response from reserve zone cells.[39]

The *in vitro* observation that proliferative zone cells possess the highest IGF-I binding capacity[81] suggests that *in vivo* localization of IGF-I within the growth plate would be principally to the proliferative zone cells. Indeed, Nilsson et al.[82] have reported that, in the rat growth plate, immunohistochemical localization of IGF-I is primarily to the proliferative zone. These authors further demonstrated that labeling was increased by GH. This increased proliferative zone IGF-I staining after GH treatment may simply reflect an increase in IGF-I bound to the cells with the greatest IGF-I binding capacity, but may also represent a differential production of IGF-I by the different populations of growth plate chondrocytes.

Support for this latter possibility has recently come from the ultrastructural localization of IGF-I in rat costal growth plates. Using electron microscopic immunohistochemistry, Wroblewski et al.[83] showed that IGF-I, though present over all chondrocytes, occurred particularly over the proliferative zone cells. Thus, available data suggest that growth plate chondrocytes differ with respect to IGF-I binding, IGF-I synthesis, and IGF-I responsiveness, and that these differences are, at least in part, a function of the stage of cellular differentiation.

C. TRANSGENIC STUDIES

Further insight into the role of IGF-I in regulating skeletal growth at the growth plate has been provided by the recent development of transgenic mice that express high levels of IGF-I. Mice with an IGF-I gene under the control of the metallothionein I gene promoter have serum IGF-I levels about 1.5 times higher, and weigh approximately 1.3 times more, than their nontransgenic littermates.[84] Interestingly, they do not appear to undergo an increase in linear skeletal growth,[84] as estimated by radiographs of long bones and by tibial epiphyseal growth over a 10-d period.[85]

Transgenic mice have also been developed that overexpress GH. Compared to the IGF-I transgenic mice, the GH transgenic animals expressed more IGF-I and grew to a larger size.[85] Inasmuch as the IGF-I transgenic animals also had normal GH-producing capacity, and because GH transgenic animals exhibited both elevated GH and elevated IGF-I, these data are not particularly helpful in clarifying the relationship between IGF-I, GH, and skeletal growth. To help distinguish the roles of IGF-I and GH, mice were made transgenic for both the IGF-I gene and for ablation of GH-expressing cells. The mice carrying both transgenes (IGF-I and absence of GH) grew larger than their GH-deficient transgenic littermates and exhibited weight and linear growth indistinguishable from that of their normal nontransgenic siblings.[85]

Taken together, these transgenic studies indicate that expression of IGF-I alone (in the absence of GH) can generate normal growth, a finding of potential clinical significance for treatment of patients with syndromes of GH resistance. They also reveal that, in the presence of normal GH secretion, additional chronic expression of IGF-I may not elicit supranormal skeletal growth. These data emphasize the high degree of complexity in GH/IGF-I relationships in regulating skeletal development.

D. IGF-II AS A FETAL SOMATOMEDIN

IGF-II, when administered *in vivo,* has been less potent than IGF-I in stimulating indices of skeletal growth. Schoenle et al.[86] delivered highly purified IGF-II, IGF-I, or saline to hypophysectomized rats for 6 d and measured tibial epiphyseal width and costal cartilage thymidine incorporating activity. In comparison to saline, IGF-II at a dose of 131 µg/d significantly stimulated both these growth parameters, but failed to achieve the level of stimulation achieved by IGF-I at either 43 µg/d or 103 µg/d. These results are consistent with a limited but potentially significant role for IGF-II in skeletal growth in young rats. Since

IGF-II may be relatively more important in prenatal than postnatal development[12,13] (and indeed may be more important than GH-dependent factors such as IGF-I in prenatal tissues),[12] such studies of IGF-II in older animals may not reflect the true importance of IGF-II in early skeletal development.

Further evidence for a primarily prenatal role of IGF-II in skeletal growth has recently been provided by the development of mice carrying a disrupted IGF-II gene. Chimeric mice possessing one normal and one inactivated IGF-II allele were proportionately smaller than their wild-type littermates, but otherwise appeared normal.[87] This mutation seemed to exert its influence only during the embryonic period since postnatal growth of the heterozygous mice was the same as that of the wild-type animals. That the disruption of the IGF-II allele appears to retard prenatal but not postnatal skeletal growth in this model lends further support to the hypothesis that IGF-II is the principal IGF regulating cartilage in the fetus.

IV. IGF RECEPTORS

A. LIGAND-RECEPTOR INTERACTION

Evidence for a specific cell membrane receptor for the somatomedins first emerged from studies of the insulin receptor. Using adipocytes and liver membrane preparations, Hintz et al.[88] found that somatomedin competed with ^{125}I-labeled insulin for insulin-binding sites, indicating an ability of somatomedin to bind to the insulin receptor. Using fetal chick chondrocytes, these authors made the unexpected observation that somatomedin was much more effective than even high concentrations of insulin in displacing radiolabeled insulin from chondrocyte membrane binding sites. They interpreted this to mean that, in contrast to the insulin receptor on adipocytes and liver membranes (which bound insulin more effectively than somatomedin), the receptor on chondrocytes is more highly keyed for somatomedin than for insulin.[88] When ^{125}I-labeled IGF-I became available, this group discovered that the labeled IGF-I was completely displaced from chick embryo chondrocyte membranes by low levels (2 U/ml) of unlabeled IGF-I, but was poorly displaced by even high concentrations (20 µg/ml) of insulin. Similarly, ^{125}I-insulin was fully displaced by 20 Ug/ml unlabeled insulin and much less affected by 2 µ/ml IGF-I.[1] These results offered telling evidence that the two hormones bind to different receptor sites and that there exists on chondrocytes an IGF-I receptor distinct from the insulin receptor.[1] Stuart et al.[89] confirmed the presence of distinct insulin and somatomedin receptors on embryonic chick cartilage membranes and observed that the molar potency of insulin to compete for binding was less than 1% that of IGF-I.

With the eventual availability of pure IGFs and the development of techniques for the isolation of intact chondrocytes suitable for receptor studies,[90] it became possible to more fully characterize the IGF receptors on these cells. Bovine growth plate chondrocytes were found to possess high-affinity (K_a 4–5 $\times 10^8$ M^{-1}) IGF-I receptors with a concentration of approximately 2×10^4 receptors per cell. Competition studies comparing the ability of IGF-I, rat IGF-

II, and insulin to displace ^{125}I-IGF-I revealed that the IGF-I receptor binds all three of these peptides, but that rIGF-II is one order of magnitude and insulin is three orders of magnitude less potent than IGF-I in competing for the receptor[90] (Figure 7). These binding properties are similar to those of the IGF-I receptor in other tissues.[91] Other peptides, including fibroblast growth factor, epidermal growth factor, and GH failed to cross-react with the IGF-I receptor.[90] These ligand-receptor interactions suggest an IGF-I effector system with a strong preference for IGF-I, but with the capacity to be activated or otherwise modulated by related peptides. Interestingly, chondrocytes from bovine articular cartilage, which have a lower cellular proliferative requirement than their counterparts in the growth plate, revealed a similar affinity constant for IGF-I binding, but four times fewer IGF-I receptors per cell than growth plate chondrocytes.[90] These data are consistent with the hypothesis that IGF-I receptor expression is one mechanism of regulating chondrocyte responsiveness to this growth factor.

The difference observed in the foregoing studies between growth plate and articular chondrocyte IGF-I receptors may not apply to all IGF receptors. Early studies with partially purified insulin-like activity (ILA), a peptide related or identical to IGF-II, revealed receptor binding on both articular and growth plate rabbit chondrocytes in culture.[92] In contrast to IGF-I, binding of ILA was greater to articular than growth plate chondrocytes. Whether this difference between cell types in ILA binding reflects a difference in receptor affinity, in receptor number per cell, or in other factors is not known. It has been suggested that the greater amount of matrix observed around the cultured growth plate cells compared to the articular cells may have resulted in poorer accessibility of the growth factor to its receptor sites.[92]

Receptors for IGF-I have also been characterized on cultured rat articular chondrocyte membrane preparations,[93] passaged rat costal chondrocytes,[94] and isolated bovine articular chondrocytes.[95] In all cases, both IGF-II and insulin cross-reacted with the IGF-I receptor, but with decreasing competition potency from IGF-I to IGF-II to insulin. It is evident from these studies that the apparent affinity constant (K_a) for IGF-I binding to its receptor varies with experimental conditions, ranging from $11 \times 10^9\ M^{-1}$ for articular chondrocyte membranes to $5.7 \times 10^7\ M^{-1}$ for passaged rat costal chondrocytes.[93]

The fate of IGF-I after binding to its receptor has been investigated in at least two chondrocyte models using complementary methods. In one study, when ^{125}I-IGF-I was bound to bovine growth plate chondrocytes at 4°C (a subphysiologic temperature that would inhibit such cellular activities as receptor internalization), most bound ^{125}I-IGF-I could be dissociated from the cell, indicating its presence at the cell surface. In contrast, at the physiologic temperature of 37°C, only a minority of bound ^{125}I-IGF-I was still dissociable,[90] suggesting that at physiologic temperatures, most of receptor-bound IGF-I is internalized by the cell. To determine whether IGF-I is subsequently degraded by growth plate chondrocytes, ^{125}I-IGF-I integrity was assayed with a specific anti-IGF-I antibody. At 37°C, only 50% of the radioactivity recovered in the incubation medium remained

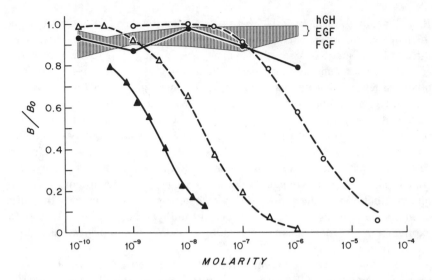

FIGURE 6. Specificity of ^{125}I-IGF-I binding to isolated bovine growth plate chondrocytes. The effects of graded concentrations of unlabeled IGF-I (▲ - ▲), rIGF-II (△ - △), nerve growth factor (● - ●) and insulin (○ - ○) on ^{125}I-IGF-I binding are expressed as competition curves. The entire range of the bound to free ratio (B/B$_0$) for EGF, hGH, and FGF is indicated by brackets. Nonspecific binding averaged 0.9%. (From Trippel, S. B., Van Wyk, J. J., Foster, M. B., and Svoboda, M. E., *Endocrinology,* 112, 2128, 1983. With permission.)

immunoreactive at 6 h, while ^{125}I-IGF-I remaining on the cell surface retained 85% of its immunoreactivity. At 15°C, virtually all of the ^{125}I-IGF-I in the medium and on the cell surface remained immunoreactive.[90] The preservation of immunoreactivity of the membrane-bound, but not internalized, ^{125}I-IGF-I indicates that the process of binding per se does not cause IGF-I degradation and suggests that exposure to the medium is not sufficient to degrade the bound peptide. These data are consistent with the hypothesis that IGF-I is internalized and degraded by lysosomal proteases, and that the fragments are then extruded from the cell. However, at least some of the diminished immunoreactivity observed in these studies might be due to proteases released into the medium and acting on free IGF-I. Clarification of this issue has recently been obtained from passaged rat costal chondrocytes. Using the lysosomotropic agents chloroquine and ammonium chloride, Schalch et al.[94] demonstrated a greater increase in ^{125}I-IGF-I in cells incubated with these two agents than in controls, suggesting that IGF-I is internalized and degraded through a lysosomal proteolytic pathway. Transduction of the IGF-I stimulus by the IGF-I receptor in nonchondroid cells involves activation of receptor tyrosine kinase activity and does not appear to require internalization.[96,97] Thus, this internalization and degradation pathway may represent a means of turning off the IGF-I signal by down-regulating surface receptors and altering cellular responsiveness to the peptide.

Chondrocytes also possess receptors for IGF-II. In studies of isolated bovine growth plate chondrocytes, the IGF-II receptor was expressed at a concentration of approximately 2.4×10^4 receptors per cell with an affinity constant of approximately $5 \times 10^8 \, M^{-1}$.[98] Just as rIGF-II was able to displace ^{125}I-IGF-I from growth plate chondrocytes, so unlabeled IGF-I competed with ^{125}I-rIGF-II for binding to these cells. Thus, the IGF-II receptor appears to recognize both IGF-II and IGF-I. In contrast to the IGF-I receptor, the IGF-II receptor failed to recognize insulin even at high concentrations, a phenomenon not limited to chondrocytes. Passaged rat costal chondrocytes[99] and rabbit articular chondrocyte membranes[93,100] also possess IGF-II receptors that are characterized by inability to recognize insulin and exhibit a similar affinity for the ligand ($K_a = 3.8 \times 10^8$ M^{-1} and $16 \times 10^8 \, M^{-1}$ respectively).[93,99] Interestingly, unlike bovine growth plate chondrocytes[90] or rat articular chondrocyte membranes,[93] the IGF-II receptor appears to be the predominant IGF receptor on the passaged rat chondrocytes.[94]

Matrix effects on IGF-receptor interactions remain a relatively unexplored aspect of IGF regulation of cartilage. The role of the cartilage matrix is likely to be significant, at least in part, because both IGF-I and IGF-II *in vivo* are generally bound to a complex of high molecular weight (150 kDa)[101] binding proteins. These binding proteins may be expected to influence diffusivity through cartilage matrix. Findings consistent with such a phenomenon have recently been reported by Maroudas et al.[102] in human articular cartilage explants. The partition coefficient of IGF-I between cartilage and external medium was decreased by the addition to the IGF-I of 20% fetal calf serum (presumably containing IGF binding proteins). Thus, IGF associated with its binding proteins probably has little access to the cells residing in normal articular cartilage matrix. Whether this access is improved in degenerated articular cartilage remains to be seen, but certainly the known loss of proteoglycan that occurs in degenerative joint disease would argue for enhanced growth factor penetration.

B. IGF RECEPTOR STRUCTURE

Receptors for the IGF-I and IGF-II groups of peptides have been designated type I and type II receptors, respectively.[103] Early structural studies of these receptors on rat liver plasma membranes,[104,105] and several human and rodent tissues[103,106,107] have shown that the two receptors differ in molecular mass and subunit composition. On isolated bovine growth plate chondrocytes, the type I receptor, examined by affinity crosslinking with radiolabeled IGF-I, appears as a complex composed of subunits linked by disulfide bonds. The intact growth plate chondrocyte receptor migrates on SDS polyacrylamide gel electrophoresis with an apparent M_r of greater than 300,000. Under reducing conditions, in which disulfide bonds are broken, the labeled component of the IGF-I receptor on these cells migrates with an apparent M_r of 140,000.[98] The IGF-I receptor on rabbit articular chondrocyte membranes shows a similar pattern on covalent crosslinking.[93]

Affinity labeling of the IGF-II receptor reveals a very different structure. Under nonreducing conditions, it migrates with an apparent M_r of 230,000 to 260,000.[93,98,99] The apparent molecular mass of this receptor increases under reducing conditions,[93,98] indicating the absence of a disulfide-linked subunit structure, and the presence of intramolecular disulfide bonds whose reduction results in unfolding of the receptor molecule.

The recent cloning of both the type I[108] and type II[109] receptors has confirmed and further elucidated the structure suggested by earlier biochemical methods. The mammalian type I receptor is a heterotetrameric complex ($M_r \approx 350,000$) composed of subunits, designated α and ß, that are joined by disulfide bonds (Figure 7). The two α subunits ($M_r \approx 140,000$ each) contain the receptor's ligand binding region and the two ß subunits ($M_r \approx 90,000$ each) traverse the cell membrane to activate postreceptor signal transduction pathways. In contrast to the type I receptor, the type II receptor is a single molecule with a long extracellular domain and a short intracellular region (Figure 8) that may regulate signal transduction via associated G proteins.[110,111]

Affinity labeling data indicate that the binding subunit of the IGF-I receptor on chondrocyte membranes has a higher apparent molecular mass than that of placental membranes.[98] This suggests species or tissue specificity of receptor subunit structure. Similar intertissue differences in apparent molecular mass, observed for the insulin receptor, have been shown to be due to differences in receptor glycosylation.[112]

As noted previously, both type I and type II receptors on chondrocytes are able to bind both IGF-I and IGF-II, though with differing affinities. This receptor-ligand cross-reactivity suggests an interplay of IGF-I and IGF-II biologic action mediated by each others' receptors. Since chondrocytes possess both types of receptors, it has been difficult to distinguish the specific biologic effects mediated by binding of each growth factor with its own receptor. Studies in noncartilagenous tissues have generally found that the biological effects of both IGF-I and IGF-II are mediated through the IGF-I receptor.[113-116] Presently, the biologic role of IGF binding to the type II receptor is controversial. Although most studies of IGF-II biologic effects argue that they are mediated by IGF-II binding to the type I receptor, an increasing number of reports are providing evidence that IGF-II can stimulate cell activity via the type II receptor in some cell types.[110,111,117]

At least one purpose of the IGF-II receptor in some tissues is to regulate lysosomal enzyme transport. This function stems from the unusual dual role of this molecule in serving both as the IGF-II receptor and as the cation-independent mannose-6-phosphate receptor.[109,118] Although rapid progress has been made in characterizing the IGF-I and IGF-II receptors, the distinctive contributions of each to the various metabolic functions of chondrocytes remain to be elucidated.

V. ARTICULAR CARTILAGE

Articular cartilage has been known for decades to be under the influence of GH or GH-dependent factors. Long before the somatomedins were discovered, abnormal proliferative activity in articular cartilage was noted in patients with

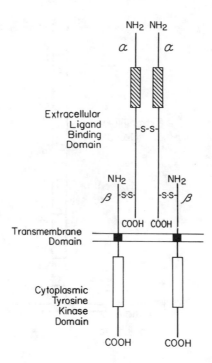

FIGURE 7. Schematic representation of the IGF-I receptor. Cysteine-rich regions associated with the ligand-binding extracellular domains are shown as hatched boxes, transmembrane domains as solid boxes, and tyrosine kinase domains involved in signal transduction as open boxes. (Adapted from Ullrich, A. et al., *EMBO J.*, 5, 2503, 1986.)

acromegaly, a disease characterized by elevated GH levels.[119] During the search for improved bioassays for somatomedins, articular chondrocytes were among the cell types chosen for analysis. Articular chondrocytes were responsive to somatomedin and successful bioassays were developed based on this responsiveness.[19,27,28] Such work led to an early characterization of the effect of partially purified somatomedin preparations on articular cartilage[11,27,29,120] and established that somatomedin stimulates both [³H]thymidine incorporation into DNA and sulfate incorporation into proteoglycans by these cells.

With the availability of highly purified native and recombinant IGF-I, further studies have corroborated and elaborated upon these early results. Several laboratories have demonstrated that IGF-I increases the rate of proteoglycan synthesis by articular cartilage in explant culture.[35,37,38] The importance of IGF-I in articular cartilage proteoglycan homeostasis has been emphasized by the observations of Luyten et al.[38] who showed that this peptide also inhibits proteoglycan catabolism in articular cartilage. Interestingly, the half-maximal catabolic response to IGF-I occurred at an IGF-I concentration of 1.5 ng/ml, while the half-maximal anabolic response was not achieved until 4.5 ng/ml. The finding that the cartilage catabolic response was more sensitive to IGF-I than the

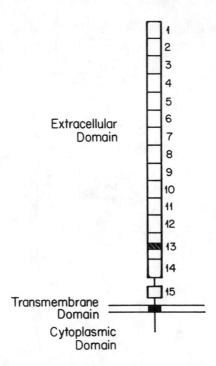

FIGURE 8. Schematic representation of the IGF-II/mannose-6-phosphate receptor. Repeating units in the extracellular domain are shown as open boxes. These repeats are only 20% identical, but contain a highly conserved pattern of cysteine residues. The transmembrane region is shown as a solid box. The thirteenth repeat contains a region of fibronectin homology (hatched box). (Adapted from Morgan, D. O., Edman, J. C., Standring, D. N., Fried, V. A., Smith, M. D., Roth, R. A., and Rutter, W. J., *Nature (London)*, 329, 301, 1987.)

anabolic response suggests that the two components of proteoglycan metabolism can be experimentally uncoupled and may be differentially regulated.[38] By simultaneously stimulating proteoglycan synthesis and inhibiting its catabolism, IGF-I may serve to generate net proteoglycan production by articular cartilage chondrocytes.

It is possible that the differentiated activity (e.g., matrix synthesis) of articular chondrocytes is more sensitive to IGF-I than is mitotic activity (e.g., DNA synthesis). Using adult bovine articular cartilage in explant culture, Osborn et al. found that 15 ng/ml of IGF-I stimulated sulfate incorporation, but not thymidine incorporation.[37] Similar results were obtained by Luyten et al.[38] using young (newborn to 10 months old) bovine articular cartilage. IGF-I at concentrations of either 5 μg/ml or 20 μg/ml enhanced sulfate incorporation, but failed to increase DNA content.

Not surprisingly, IGF-II has been found to be markedly less potent than IGF-I in its effect on both synthesis and catabolism of proteoglycan. In bovine articular cartilage, 100 ng/ml of IGF-II generated only 60% of the response produced by 5 ng/ml of IGF-I.[38] These data are consistent with findings in other cell types that the IGF-II response is mediated by IGF-II binding to the IGF-I receptor, a receptor which has a lower affinity for IGF-II than for IGF-I.

Given the existence of a factor that is anabolic for articular cartilage, the question arises whether this factor is available to articular cartilage *in vivo* and whether it might play an *in vivo* role in articular cartilage metabolism. This question was addressed early by Coates et al.[121] using a porcine costal cartilage bioassay to measure the thymidine incorporation and sulfation activities of synovial fluid from patients with various arthritides. These authors showed that synovial fluid stimulates both DNA synthesis and proteoglycan synthesis by cartilage, consistent with the presence of somatomedin in the synovial fluid. Of considerable interest was their observation that, while synovial fluid from joints with psoriatic, Reiter's, posttraumatic and postoperative effusions possessed stimulatory somatomedin-like activity, fluid from patients with rheumatoid arthritis possessed either an impaired capacity to stimulate DNA synthesis or was inhibitory. These data suggest that while most synovial fluid contains somatomedin activity, in rheumatoid arthritis, this activity is lost or blocked.

To determine how much of the stimulatory effect of synovial fluid is specifically attributable to IGF-I, Schalkwijk et al.[122] incubated articular cartilage in synovial fluid with or without a specific anti-IGF-antibody and measured proteoglycan synthesis. The antibody strikingly reduced the anabolic activity of the synovial fluid. These data argue that the stimulatory effect of synovial fluid on proteoglycan synthesis is generated principally by the IGFs and strengthen the view that the IGFs play a role in preserving the well-being of articular cartilage *in vivo*.

Pursuing the issue of cartilage responsiveness to somatomedin in joint disease, Schalkwijk et al.[123] incubated IGF-I with cartilage from a murine zymosan-induced model of inflammatory arthritis. Articular chondrocytes from the arthritic tissue responded more poorly to IGF-I than did chondrocytes from normal articular cartilage. Although it remains to be seen whether a similar unresponsiveness exists in noninflammatory arthritic cartilage, it is evident that both IGF-I levels and cellular responsiveness to IGF-I play a role in regulating cartilage metabolism in diseased as well as normal joints.

The relationship between IGF and age may also play a role in articular cartilage disease. It is well established that serum IGF levels decrease with advancing years.[121,124] In addition, human articular cartilage manifests a progressive decrease in basal [^{3}H]thymidine and [^{35}S]sulfate incorporation with increasing age.[120] Human articular cartilage responsiveness to plasma growth factors (including IGFs) increases between ages 12 and 17 years and then declines.[120] A similar decreased response to serum stimulation during ageing has been observed for rat cartilage.[125]

These data suggest that, with increasing age, both the amount of available IGF-I and the chondrocyte responsiveness to the factor diminish. This phenomenon may well render ageing cartilage progressively less capable of maintaining its structural and functional integrity and could predispose to the degenerative changes so uncomfortably familiar to the elderly.

VI. NEOPLASTIC CARTILAGE

The demonstration that IGFs are mitogenic for chondrocytes raises the question whether these peptides play a role in the accelerated growth that characterizes cartilage neoplasms. Although the potential role of IGFs in various neoplasias has been entertained for some time,[126-130] information regarding these factors in chondroid tumors is limited. Most of this information is derived from studies of the Swarm rat chondrosarcoma, a transplantable tumor that shares several features with human chondrosarcoma.[131] Though limited, the evidence for IGF involvement in the behavior of this tumor is fairly incriminating. Swarm rat chondrosarcoma chondrocytes possess receptors for both IGF-I and IGF-II[131] and are stimulated by both peptides *in vitro*.[132-134] The tumor also appears to require GH-dependent factors to achieve its rapid level of growth *in vivo*.[135-137]

It is possible that IGF-I also plays a role in human chondrosarcomas. Chondrocytes isolated from human chondrosarcomas have been found to specifically bind IGF-I, presumably reflecting the expression of IGF receptors.[138] The IGF-I binding by chondrosarcoma cells was substantially greater than by cells from giant cell tumors, a nonchondrogenic skeletal neoplasm. This indicates that the ability to bind IGF-I is not simply a characteristic of neoplasms in general or skeletal neoplasms in particular. Interestingly, the tissue content of IGF-I, determined by radioimmunoassay, was also higher in human chondrosarcomas than in giant cell tumors.[138] These observations suggest that human chondrosarcomas possess both the IGF-I needed to generate a response and the receptors necessary to mediate it. Thus, available data support the hypothesis that chondrosarcoma is a somatomedin-regulated tumor. Further studies will be required to confirm or refute this hypothesis. If the hypothesis is confirmed, it will be of considerable interest to learn whether inhibition of IGF may be used to inhibit tumor growth.

VII. IGF INTERACTIONS WITH OTHER GROWTH FACTORS

The most significant functions of the IGFs in regulating cartilage may ultimately prove to be in their interactions with other growth factors. The enormous complexity of successfully operating the cellular machinery of a chondrocyte almost certainly requires multiple levels of control and multiple controlling factors. It is probable that many of the major noninsulin-like peptide growth factors, including fibroblast growth factor (FGF), epidermal growth factor (EGF), transforming growth factor-ß (TGF-ß), and platelet-derived growth factor (PDGF), interact with the IGF's to influence cartilage. Each of these factors, acting alone, has been reported to stimulate some aspect of cartilage metabolism.[139-143] Since it is unlikely that any one of these factors will have access to cartilage in complete isolation from the others, an understanding of growth factor regulation of cartilage must address the ways in which these factors interact with each other.

Although data derived from the investigation of individual factors may appear to have defined their role in cartilage regulation, there is now strong evidence to support the concept that, acting in concert with other factors, a peptide may assume a reduced or enhanced role in comparison to its effect in isolation. When adult bovine articular cartilage in organ culture is exposed to IGF-I, it responds, as we have seen, with an increase in sulfate incorporation.[35,37,38] When insulin, at concentrations high enough to cross-react with the IGF-I receptor, was substituted for IGF-I, there was also an increase in sulfate incorporation.[37] When cartilage was stimulated with maximally effective doses of insulin, the addition of IGF-I to the insulin generated no further sulfate incorporation. [37] The lack of additive effect between these factors suggests that both insulin and IGF-I activate a common receptor pathway such that once that pathway is maximally stimulated, additional peptide exerts no further effect.

Growth plate chondrocytes also seem to be susceptible to interaction between IGF-I and certain other growth factors. IGF-I and basic FGF each have been shown to stimulate [^{3}H]thymidine incorporation into DNA by growth plate chondrocytes in culture.[36,144] The combination of these factors was additive or synergistic.[144] These findings are consistent with the fact that the effector pathways of IGF-I, FGF, and PDGF are initiated by binding to different cell surface receptors and presumably activate different postreceptor signal transduction mechanisms.

IGF-II, much like IGF-I, modulates the effects of other growth factors on cartilage. Early studies by Kato et al.[145] demonstrated that rat IGF-II and FGF each stimulated thymidine incorporation by rabbit costal chondrocytes. In combination, these factors became additive or synergistic.[145] EGF also enhanced thymidine incorporation by these cells and generated additive or synergistic effects in combination with rIGF-II.[145]

Cartilage neoplasms may be susceptible to interactive regulation by growth factors as well. Matsumura et al.[146] have shown that cultured Swarm rat chondrosarcoma chondrocytes respond in a dose-dependent fashion to both IGF-I and TGF-ß. The effect of TGF-ß was biphasic, increasing thymidine incorporation at concentrations up to 10 µg/ml and decreasing this incorporation between 10 and 30 ng/ml. Interestingly, IGF-I and TGF-ß exerted additive or synergistic effects on thymidine incorporation by these cells up to a point, but their combined effect was also biphasic: the initial dose-dependent increase in thymidine incorporation generated by these combined factors was followed by a decrease at higher concentrations.[146]

A recent addition to our understanding of the relationship between these factors is a study by O'Keefe et al.[147] of cultured 3- to 5-week-old chick growth plate chondrocytes. IGF-I enhanced, in a dose-dependent fashion, the mitogenic effects of TGF-ß and of FGF. Measuring changes in alkaline phosphatase, these authors found that IGF-I also additively increased the effect of TGF-ß or FGF on this index of cell differentiated activity. In contrast, IGF-I did not modify the effects of FGF or TGF-ß on protein synthesis or on proteoglycan synthesis.[147] Thus, interactive growth factor effects may apply to certain cell functions and not to others.

VIII. CONCLUSIONS

The somatomedins, or insulin-like growth factors, are important regulators of cartilage growth and differentiation. In this capacity, they participate in the control of skeletal growth and development, articular cartilage homeostasis, and possibly the abnormal behavior of cartilage neoplasms. For this reason, the IGFs may play a role in the skeletal deformities and growth disturbances that result from disorders of cartilage growth and differentiation, in the debilitating joint diseases stemming from loss of articular cartilage homeostasis, and in the morbidity and mortality imposed by cartilage tumors. While substantial progress has been made in our understanding of IGF actions on cartilage, we know little of the detailed mechanisms by which the IGFs regulate normal cartilage and even less regarding their role in the pathogenesis of cartilage disease. If we are to hope for some system of effective therapeutic intervention in such disease, it is essential that the contributions of growth factors such as the IGFs be further analyzed and understood. As this understanding is achieved, it is probable that these factors will find therapeutic application.

ACKNOWLEDGMENTS

The author thanks Ms. Mildred Hudson for expert assistance in preparing the manuscript. Supported in part by USPHS Grant AR 31068.

REFERENCES

1. **Van Wyk, J. J., Underwood, L. E., Hintz, R. L., Clemmons, D. R., Voina, S. J., and Weaver, R. P.,** The somatomedins: a family of insulin-like hormones under growth hormone control, *Rec. Prog. Horm. Res.,* 30, 259, 1974.
2. **Zapf, J., Rinderknecht, E., Humbel, R. E., and Froesch, D. R.,** Nonsuppressible insulin-like growth activity (NSILA) from human serum: recent accomplishments and their physiologic implications, *Metabolism,* 27, 1803, 1978.
3. **Van Wyk, J. J., Svoboda, M. E., and Underwood, L. E.,** Evidence from radioligand assays that somatomedin-C and insulin-like growth factor-I are similar to each other and different from other somatomedins, *J. Clin. Endocrinol. Metab.,* 50, 206, 1980.
4. **Rinderknecht, E. and Humbel, R. E.,** The amino acid sequence of human insulin-like growth factor I and its structural homology with proinsulin, *J. Biol. Chem.,* 253, 2769, 1978.
5. **Svoboda, M. E. and Van Wyk, J. J.,** Purification of somatomedin-C/insulin-like growth factor-I, *Methods Enzymol.,* 109, 798, 1985.
6. **Jansen, M., Van Schaik, F. M. A., Ricker, A. T., Bullock, B., Woods, D. E., Gabbay, K. H., Sussenbach, J. S., and Van den Brande, J. L.,** Sequence of a DNA encoding human insulin-like growth factor I precursor, *Nature (London),* 306, 609, 1983.
7. **Underwood, L. E. and Van Wyk, J. J.,** Normal and aberrant growth, in *Textbook of Endocrinology,* 7th ed., Wilson, J. D. and Foster, D. W., Eds., W. B. Saunders, Philadelphia, 1985, 155.

8. Marquardt, H., Todaro, G. J., Henderson, L. E., and Oraszlan, S., Purification and primary structure of a polypeptide with multiplication stimulating activity from rat liver cell cultures, *J. Biol. Chem.*, 256, 6859, 1981.

9. Rinderknecht, E. and Humbel, R. E., Primary structure of human insulin-like growth factor II, *FEBS Lett.*, 1978, 89, 283.

10. Dull, T. J., Gray, A., Hayflick, J. S., and Ullrich, A., Insulin-like growth factor II precursor gene organization in relation to insulin gene family, *Nature (London)*, 310, 777, 1984.

11. Underwood, L. E., D'Ercole, A. J., Clemmons, D. R., and Van Wyk, J. J., Paracrine functions of somatomedins, *Clin. Endocrinol. Metab.*, 15, 59, 1986.

12. Adams, S. O., Nissley, S. P., Handwerger, S., and Rechler, M. M., Development patterns of insulin-like growth factor I and II synthesis and regulation of rat fibroblasts, *Nature (London)*, 302, 150, 1983.

13. Lund, P. K., Moats-Staats, B. M., Hynes, M. A., Simmons, J. G., Jansen, M., D'Ercole, A. J., and Van Wyk, J. J., Somatomedin-C/insulin-like growth factor I and insulin-like growth factor II mRNA's in rat fetal and adult tissues, *J. Biol. Chem.*, 261, 14539, 1986.

14. Nissley, S. P., Short, P. A., Rechler, M. M., White, R. M., Knight, A. B., and Higa, O. Z., Increased levels of multiplication stimulating activity, an insulin-like growth factor, in fetal serum, *Proc. Natl. Acad. Sci. U.S.A.*, 77, 3649, 1980.

15. Van Wyk, J. J. and Underwood, L. E., The somatomedins and their actions, in *Biochemical Actions of Hormones*, Vol V, Litwack, G., Ed., Academic Press, New York, 1978, 101.

16. Phillips, L. S. and Vasilopoulou-Sellin, R., Somatomedins, *N. Engl. J. Med.*, 302, 371, 1980.

17. Salmon, W. D., Jr. and Daughaday, W. H., A hormonally controlled serum factor which stimulates sulfate incorporation by cartilage in vitro, *J. Lab. Clin. Invest.*, 49, 825, 1957.

18. Daughaday, W. J. and Reeder, C., Synchronous activation of DNA synthesis in hypophysectomized rat cartilage by growth hormone, *J. Lab. Clin. Med.*, 68, 357, 1966.

19. Van den Brande, J. L., Van Wyk, J. J., Weaver, R. P., and Mayberry H. E., Partial characterization of sulphation and thymidine factors in acromegalic plasma, *Acta Endocrinol.*, 66, 65, 1971.

20. Van Wyk, J. J., Hall, K., Van den Brande, J. L., and Weaver, R. P., Further purification and characterization of sulfation factor and thymidine factor from acromegalic plasma, *J. Clin. Endocrinol. Metab.*, 32, 389, 1971.

21. Daughaday, W. H., Hall, K., Raben, M. S., Salman, W. D., Jr., Van den Brande, J. L., and Van Wyk, J. J., Somatomedin: a proposed designation for sulfation factor, *Nature (London)*, 235, 107, 1971.

22. Salmon, W. D. and DuVall, M. R., A Serum fraction with "sulfation factor activity" stimulates in vitro incorporation of leucine and sulfate into protein-polysaccharide complexes, uridine into RNA, and thymidine into DNA of costal cartilage from hypophysectomized rats, *Endocrinology*, 86, 721, 1970.

23. Daughaday, W. J. and Mariz, I. K., Conversion of proline-U-C^{14} to labelled hydroxyproline by rat cartilage in vitro: effects of hypophysectomy, growth hormone, and cortisol, *J. Lab. Clin. Med.*, 5, 741, 1962.

24. Garland, J. T., Jennings, J., Levitsky, L. L., and Buchanan, F., Stimulation of DNA synthesis in isolated chondrocytes by somatomedin-II. Validation of the assay for clinical use and comparison with the stimulation of protein synthesis, *J. Clin. Endocrinol. Metab.*, 43, 847, 1976.

25. Hill, D. J., Holder, A. T., Seid, J., Preece, M. A., Tomlinson, S., and Milner, R. D. G., Increased thymidine incorporation into fetal rat cartilage in vitro in the presence of human somatomedin, epidermal growth factor and other growth factors, *J. Endocrinol.*, 96, 489, 1983.

26. Ashton, I. K. and Francis, M. J. O., Response of chondrocytes isolated from human fetal cartilage to plasma somatomedin activity, *J. Endocrinol.*, 76, 473, 1978.

27. Ashton, I. K. and Francis, M. J. O., An assay for plasma somatomedin: [^{35}H]thymidine incorporation by isolated rabbit chondrocytes, *J. Endocrinol.*, 14, 205, 1977.

28. **Ash, P. and Francis, M. J. O.,** Response of isolated rabbit articular and epiphyseal chondrocytes to rat liver somatomedin, *J. Endocrinol.,* 66, 71, 1975.
29. **Corvol, M. T., Dumontier, M. F., Rappaport, R., Guyda, H., and Posner, B. I.,** The effect of a slightly acidic somatomedin (ILAs) on the sulphation of proteoglycans from articular and growth plate chondrocytes in culture, *Acta Endocrinol.,* 89, 263, 1978.
30. **Hill, D. J.,** Stimulation of cartilage zones in the calf costochondral growth plate in vitro by growth hormone dependent plasma somatomedin activity, *J. Endocrinol.,* 83, 219, 1979.
31. **Kilgore, B. S., McNutt, M. L., Meador, S., Lee, J. A., Hughes, E. R., and Elders, M. J.,** Alteration of cartilage glycosaminoglycan protein acceptor by somatomedin and cortisol, *Pediatr. Res.,* 13, 96, 1979.
32. **Lash, J. W. and Vasan, N. S.,** Glycosaminoglycans of cartilage, in *Cartilage,* Vol. 1, Hall, B. K., Ed., Academic Press, New York, 1983, 215.
33. **Kemp, S. F. and Hintz, R. L.,** The action of somatomedin on glycosaminoglycan synthesis in cultured chick chondrocytes, *Endocrinology,* 106, 744, 1980.
34. **Schwartz, N. B., Galligani, L., Ho, P.-L., and Dorfman, A.,** Stimulation of synthesis of free chondroitin sulfate chains by beta-D-xylosides in cultured cells, *Proc. Natl. Acad. Sci. U.S.A.,* 71, 4047, 1974.
35. **McQuillan, D. S., Handley, C. J., Campbell, M. A., Bolis, S., Milway, V. E., and Herington, A. C.,** Stimulation of proteoglycan biosynthesis by serum and insulin-like growth factor I in cultured bovine articular cartilage, *Biochem. J.,* 240, 423, 1986.
36. **Makower, A.-M, Wroblewski, J., and Pawlowski, A.,** Effects of IGF-I, EGF, and FGF on proteoglycans synthesized by fractionated chondrocytes of rat rib growth plate, *Exp. Cell Res.,* 179, 498, 1988.
37. **Osborn, K. D., Trippel, S. B., and Mankin, H. J.,** Growth factor stimulation of adult articular cartilage, *J. Orthop. Res.,* 7, 35, 1989.
38. **Luyten, F. P., Hascall, V. C., Nissley, S. P., Morales T. I., and Reddi, A. H.,** Insulin-like growth factors maintain steady state metabolism of proteoglycans in bovine articular cartilage explants, *Arch. Biochem. Biophys.,* 267, 416, 1988.
39. **Trippel, S. B., Corvol, M. T., Dumontier, M. F., Rappaport, R., Hung, H. H., and Mankin, H. J.,** Effect of somatomedin-C/insulin-like growth factor I and growth hormone on cultured growth plate and articular chondrocytes, *Pediatr. Res.,* 25, 76, 1989.
40. **Sandell, L. J. and Dudek, E. J.,** Insulin-like growth factor I stimulates type II collagen gene expression in cultured chondrocytes, *Orthop. Trans.,* 12, 377, 1988.
41. **Zapf, J., Schoenle, E., and Froesch, E. R.,** Insulin-like growth factors I and II: some biological actions and receptor binding characteristics of two purified constituents of nonsuppressible insulin-like activity of human serum, *Eur. J. Biochem.,* 87, 285, 1978.
42. **Copeland, K. C., Underwood, L. E., and Van Wyk, J. J.,** Induction of immunoreactive somatomedin-C in human serum by growth hormone: dose response relationships and effect on chromatographic profiles, *J. Clin. Endocrinol. Metab.,* 50, 690, 1980.
43. **Schoenle, E., Zapf, J., Humbel, R. E., and Froesch, E. R.,** Insulin-like growth factor I stimulates growth in hypophysectomized rats, *Nature (London),* 296, 252, 1982.
44. **Froesch, E. R., Schmid, C., Schwander, J., and Zapf, J.,** Actions of insulin-like-growth factors, *Ann. Rev. Physiol.,* 47, 443, 1985.
45. **Berelowitz, M., Szabo, M., Frohman, L. A., Firestone, S., and Chu, L.,** Somatomedin-C mediates growth hormone negative feedback by effects on both the hypothalmus and the pituitary, *Science,* 212, 1279, 1981.
46. **Vetter, U., Zapf, J., Heit, W., Helbing, G., Heinze, E., Froesch, E. F., and Teller, W.,** Human fetal and adult chondrocytes. Effect of insulin-like growth factors I and II, insulin and growth hormone on clonal growth, *J. Clin. Invest.,* 77, 1903, 1986.
47. **Balk, S. D., Morisi, A., Gunther, H. S., Svoboda, M. E., Van Wyk, J. J., Nissley, S. P., and Scanes, C. G.,** Somatomedins (insulin-like growth factors), but not growth hormone, are mitogenic for chicken heart mesenchymal cells and act synergistically with epidermal growth factor and brain fibroblast growth factor, *Life Sci.,* 35, 335, 1984.

185

48. **Burch, W., Corda, G., and Leung, F. C.,** Homologous and heterologous growth hormone fail to stimulate avian cartilage growth in vitro, *J. Clin. Endocrinol. Metab.,* 60, 747, 1985.
49. **Makower, A.-M., Wroblewski, J., and Pawlowski, A.,** Effects of IGF-I, rGH, FGF, EGF and NCS on DNA synthesis, cell proliferation and morphology of chondrocytes isolated from rat rib growth cartilage, *Cell. Biol. Int. Rep.,* 13, 259, 1989.
50. **Jones, K. L., Villela, J. F., and Lewis, V. J.,** The growth of cultured rabbit articular chondrocytes is stimulated by pituitary growth factors, but not by purified human growth hormone or ovine prolactin, *Endocrinology,* 118, 2588, 1986.
51. **Madsen, K., Friberg, U., Ross, P., Eden, S., and Isaksson, O.,** Growth hormone stimulates the proliferation of cultured chondrocytes from rabbit ear and rat rib growth cartilage, *Nature (London),* 304, 545, 1983.
52. **Madsen K., Makower, A.-M., Friberg, U., Eden, S., and Isaksson, O.,** Effect of human growth hormone on proteoglycan synthesis in cultured rat chondrocytes, *Acta Endocrinol.,* 108, 338, 1985.
53. **Lindahl, A., Isgaard, J., Nilsson, A., and Isaksson, O. G. P.,** Growth homrone potentiates colony formation of epiphyseal chondrocytes in suspension culture, *Endocrinology,* 118, 1843, 1986.
54. **Lindahl, A., Isgaard, J., Carlsson, L., and Isaksson, O. G. P.,** Differential effects of growth hormone and insulin-like growth factor I on colony formation of epiphyseal chondrocytes in suspension culture in rats of different ages, *Endocrinology,* 121, 1061, 1987.
55. **Eigenmann, J. E., Patterson, D. F., and Froesch, E. F.,** Body size parallels insulin-like growth factor I levels but not growth hormone secretory capacity, *Acta Endocrinol.,* 106, 448, 1984.
56. **McConaghy, P. and Sledge, C. B.,** Production of sulphation factor by the perfused liver, *Nature (London),* 225, 1249, 1970.
57. **D'Ercole, A. J., Stiles, A. D., and Underwood, L. E.,** Tissue concentrations of somatomedin-C: further evidence for multiple sites of synthesis and paracrine or autocrine mechanisms of action, *Proc. Natl. Acad. Sci. U.S.A.,* 81, 935, 1984.
58. **Eden, S., Isaksson, O. G. P., Madsen, K., and Friberg, U.,** Specific binding of growth hormone to isolated chondrocytes from rabbit ear and epiphyseal plate, *Endocrinology,* 112, 1127, 1983.
59. **Isaksson, O. G. P., Jansson, J.-O., and Gause, I. A. M.,** Growth hormone stimulates longitudinal bone growth directly, *Science,* 216, 1237, 1982.
60. **Russell, S. M. and Spencer, E. M.,** Local injections of human or rat growth hormone or of purified human somatomedin-C stimulate unilateral tibial epiphyseal growth in hypophysectomized rats, *Endocrinology,* 116, 2563, 1985.
61. **Schlechter, N. L., Russell, S. M., Greenberg, S., Spencer, E. M., and Nicoll, C. S.,** A direct growth effect of growth hormone in rat hindlimb shown by arterial infusion, *Am. J. Physiol.,* 250, E231, 1986.
62. **Skottner, A., Clark, R. G., Robinson, I. C. A. F., and Fryklund, L.,** Recombinant human insulin-like growth factor: testing the somatomedin hypothesis in hypophysectomized rats, *J. Endocrinol.,* 112, 123, 1987.
63. **Schlechter, N. L., Russell, S. M., Spencer, E. M., and Nicoll, C. S.,** Evidence suggesting that the direct growth-promoting effect of growth hormone on cartilage in vivo is mediated by local production of somatomedin, *Proc. Natl. Acad. Sci. U.S.A.,* 83, 7932, 1986.
64. **Trippel, S. B., Hung, H. H., and Mankin, H. J.,** Synthesis of somatomedin-C by growth plate chondrocytes, *Orthop. Trans.,* 11, 422, 1987.
65. **Isgaard, J., Muller, C., Isaksson, O. G. P., Nilsson, A., Matthews, L. S., and Norstedt, G.,** Regulation of insulin-like growth factor messenger ribonucleic acid in rat growth plate by growth hormone, *Endocrinology,* 122, 1515, 1988.
66. **Isgaard, J., Carlsson, L., Isaksson, O. G. P., and Jansson, J.-O.,** Pulsatile intravenous growth hormone (GH) infusion to hypophysectomized rats increases insulin-like growth factor I messenger ribonucleic acid in skeletal tissues more effectively than continuous GH infusion, *Endocrinology,* 123, 2605, 1988.

67. **Zezulak, K. M. and Green, H.,** The generation of insulin-like growth factor-I-sensitive cells by growth hormone action, *Science,* 233, 551, 1986.
68. **Green, H., Morikawa, M., and Nixon, T.,** A dual effector theory of growth-hormone action, *Differentiation,* 29, 195, 1985.
69. **Burch W. M., Weir, S., and Van Wyk, J. J.,** Embryonic chick cartilage produces its own somatomedin-like peptide to stimulate cartilage growth in vitro, *Endocrinology,* 119, 1370, 1986.
70. **Brighton, C. T., Sugioka, Y., and Hunt, R. M.,** Cytoplasmic structures of epiphyseal plate chondrocytes. Quantitative evaluation using electron micrographs of rat costochondral junctions with special reference to the fate of hypertrophic cells, *J. Bone Joint Surg.,* 55A, 771, 1983.
71. **Ham, A. W.,** *Histology,* 7th ed., J. B. Lippincott, Philadelphia, 1971, 388.
72. **Holtrop, M. E.,** The ultrastructure of the epiphyseal plate, I. The flattened chondrocyte, *Calcif. Tissue Res.,* 9, 131, 1972.
73. **Holtrop, M. E.,** The ultrastructure of the epiphyseal plate. II. The hypertrophic chondrocyte, *Calcif. Tissue Res.,* 9, 140, 1972.
74. **Brighton, C. T. and Hunt, R. M.,** The role of mitochondria in growth plate calcification as demonstrated in a rachitic model, *J. Bone Joint Surg.,* 60A, 630, 1978.
75. **Greer, R. B., Jacicke, G. H., and Mankin, H. J.,** Protein-polysaccharide synthesis at three levels of the normal growth plate, *Calcif. Tissue Res.,* 2, 157, 1968.
76. **Rosenberg, L., Johnson, T., Lanzo, D. A., Liber, L., Macy, N., Poole, A. R., and Wolff, A.,** Identification of a 14K matrix protein present in high concentration in the hypertrophic zone of cartilage growth plate, *Trans. Orthop. Res.,* 9, 190, 1984.
77. **Wuthier, R. E.,** A zonal analysis of inorganic and organic constituents of the epiphysis during endochondral ossification, *Calcif. Tissue Res.,* 4, 20, 1969.
78. **Brighton, C. T.,** Structure and function of the growth plate, *Clin. Orthop.,* 136, 22, 1978.
79. **Kember, N. F.,** Cell population kinetics of bone growth: the first ten years of autoradiographic studies with tritiated thymidine, *Clin. Orthop.,* 76, 213, 1971.
80. **Kember, N. F.,** Cell kinetics and the control of growth in long bones, *Cell Tissue Kinet.,* 11, 477, 1978.
81. **Trippel, S. B., Van Wyk, J. J., and Mankin, H. J.,** Localization of somatomedin-C binding to bovine growth plate chondrocytes in situ, *J. Bone Joint Surg.,* 68A, 897, 1986.
82. **Nilsson, A., Isgaard, J., Lindahl, A., Dahlstrom, A., Skottner, A., and Isaksson, O.,** Regulation by growth hormone of number of chondrocytes containing IGF-I in rat growth plate, *Science,* 233, 571, 1986.
83. **Wroblewski, J., Engstrom, M., Skottner, A., Madsen, K., and Friberg, U.,** Subcellular location of IGF-I in chondrocytes from rat rib growth plate, *Acta Endocrinol.,* 115, 37, 1987.
84. **Quaife, C. J., Mathews, L. S., Pinkert, C. A., Hammer, R. E., Brinster, R. L., and Palmiter, R. D.,** Histopathology associated with elevated levels of growth hormone and insulin-like growth factor I in transgenic mice, *Endocrinology,* 124, 40, 1989.
85. **Behringer, R. R., Lewin, T. M., Quaife, C. J., Palmiter, R. D., Brinster R. L., and D'Ercole, A. J.,** Expression of insulin-like growth factor I stimulates normal somatic growth in growth hormone-deficient transgenic mice, *Endocrinology,* 127, 1033, 1990.
86. **Schoenle, E., Zapf, J., Hauri, C., Steiner, T., and Froesch, E. R.,** Comparison of in vivo effects of insulin-like growth factors (IGF) I and II and of growth hormone (GH) in hypophysectomized rats, *Acta Endocrinol.,* 108, 167, 1985.
87. **DeChiara, T. M., Efstratiadis, A., and Robertson, E. J.,** A growth-deficiency phenotype in heterozygous mice carrying an insulin-like growth factor II gene disrupted by targeting, *Nature (London),* 345, 78, 1990.
88. **Hintz, R. L., Clemmons, D. R., Underwood, L. E., and Van Wyk, J. J.,** Competitive binding of somatomedins to the insulin receptor of adipocytes, chondrocytes, and liver membranes, *Proc. Natl. Acad. Sci. U.S.A.,* 69, 2351, 1972.
89. **Stuart, C. A., Furlanetto, R. W., and Lebovitz, H. E.,** The insulin receptor of embryonic chicken cartilage, *Endocrinology,* 105, 1293, 1979.

90. **Trippel, S. B., Van Wyk, J. J., Foster, M. B., and Svoboda, M. E.,** Characterization of a specific somatomedin-C receptor on isolated bovine growth plate chondrocytes, *Endocrinology,* 112, 2128, 1983.

91. **Clemmons, D. R. and Van Wyk, J. J.,** Somatomedin: physiological control and effects on cell proliferation, in *Tissue Growth Factors, Handbook of Experimental Pharmacology,* Baserga, R. Ed., Vol. 57, Springer-Verlag, New York, 1981, 161.

92. **Postel-Vinay, M. C., Corvol, M. T., Lang, F., Fraud, F., Guyda, H., and Posner, B.,** Receptors for insulin-like growth factors in rabbit articular and growth plate chondrocytes in culture, *Exp. Cell Res.,* 148, 105, 1983.

93. **Jansen, J., van Buul-Offers, S. C., Hoogerbrugge, C. M., de Poorter, T. L., Corvol, M. G., and Van den Brande, J. L.,** Characterization of specific insulin-like growth factor (IGF)-I and IGF-II receptors on cultured rabbit articular chondrocyte membranes, *J. Endocrinol.,* 120, 245, 1989.

94. **Schalch, D. S., Sessions, C. M., Farley, A. C., Masakawa, A., Ember, C. A., and Dills, D. G.,** Interaction of insulin-like growth factor I/somatomedin-C with cultured rat chondrocytes: receptor binding and internalization, *Endocrinology,* 118, 1590, 1986.

95. **Watanabe, N., Rosenfeld, R. G., Hintz, R. L., Dollar, L. A., and Smith, R. L.,** Characterization of a specific insulin-like growth factor-I/somatomedin-C receptor on high density, primary monolayer cultures of bovine articular chondrocytes: regulation of receptor concentration by somatomedin, insulin and growth hormone, *J. Endocrinol.,* 107, 275, 1985.

96. **Jacobs, S., Kull, F., C., Earp, S. C., Svoboda, M. E., Van Wyk, J. J., and Cuatrecasas, P.,** Somatomedin-C stimulates the phosphorylation of the ß-subunit of its own receptor, *J. Biol. Chem.,* 258, 9581, 1983.

97. **Rubin, J. B., Shia, M. G., and Pilch, P. F.,** Stimulation of tyrosine-specific phosphorylation in vitro by insulin-like growth factor I, *Nature (London),* 305, 438, 1983.

98. **Trippel, S. B., Chernausek, S. D., Van Wyk, J. J., Moses, A. C., and Mankin, H. J.,** Demonstration of type I and type II somatomedin receptors on bovine growth plate chondrocytes, *J. Orthop. Res.,* 6, 817, 1988.

99. **Sessions, C. M., Ember, C. A., and Schalch, D. S.,** Interaction of insulin-like growth factor II with rat chondrocytes: receptor binding, internalization, and degradation, *Endocrinology,* 120, 2108, 1987.

100. **Ashton, I. K. and Soul, J. H.,** Receptors for multiplication stimulating activity (MSA) on rabbit chondrocytes, *Calcif. Tissue Int.,* 36, 576, 1984.

101. **Drop, S. L. S. and Hintz, R. O.,** *Insulin-Like Growth Factor Binding Proteins,* Excerpta Medica, Amsterdam, 1989.

102. **Maroudas, A., Popper, O., and Grushko, G.,** Partition coefficients of IGF-I between cartilage and external medium in the presence and absence of FCS, presented at 37th Annual Meeting, Orthop. Res. Soc., Anaheim, March 4–7, 1991, 398.

103. **Massaque, J. and Czech, M. P.,** The subunit structures of two distinct receptors for insulin-like growth factors I and II and their relationship to the insulin receptor, *J. Biol. Chem.,* 257, 5038, 1982.

104. **Kasuga, N., Van Obberghen, E., Nissley, S. P., and Rechler, M. M.,** Demonstration of two subtypes of insulin-like growth factor receptors by affinity cross-linking, *J. Biol. Chem.,* 256, 5305, 1981.

105. **Nissley, S. P. and Rechler, M. M.,** Multiplication stimulating activity (MSA): a somatomedin-like polypeptide from cultured rat liver cells, *Natl. Cancer Inst. Monogr.,* 48, 167, 1978.

106. **Bhaumick, B., Bala, R. M., and Hollenberg, M. D.,** Somatomedin receptor of human placenta: solubilization, photolabelling, partial purification, and comparison with insulin receptor, *Proc. Natl. Acad. Sci. U.S.A.,* 78, 4279, 1981.

107. **Chernausek, S. D., Jacobs, S., and Van Wyk, J. J.,** Structural similarities between human receptors for somatomedin-C and insulin: analysis by affinity labeling, *Biochemistry,* 20, 7345, 1981.

108. **Ullrich, A., Gray, A., Tam, A. W., Yang-Feng, T., Tsubokawa, M., Collins, C., Henzel, W., Le Bon, T., Kathuria, S., Chen, E., Jacobs, S., Francke, V., Ramachandran, J., and Fugita-Yamaguchi, Y.,** Insulin-like growth factor I receptor primary structure: comparison with insulin suggests structural determinants that define functional specificity, *EMBO J.,* 5, 2503, 1986.

109. **Morgan, D. O., Edman, J. C., Standring, D. N., Fried, V. A., Smith, M. D., Roth, R. A., and Rutter, W. J.,** Insulin-like growth factor II receptor as a multifunctional binding protein, *Nature (London),* 329, 301, 1987.

110. **Nishimoto, I., Hata, Y., Ogata, E., and Kojima, I.,** Insulin-like growth factor II stimulates calcium influx in competent BALB/c3T3 cells primed with epidermal growth factor, *J. Biol. Chem.,* 262, 12120, 1987.

111. **Nishimoto, I., Murayama, Y., and Okamoto, T.,** Signal transduction mechanism of IGF-II/man-6-p receptor, in *Modern Concepts of Insulin-Like Growth Factors,* Spencer, E. M., Ed., Elsevier, New York, 1991, 517.

112. **Heidenreich, K. A., Zahniser, N. R., Berhanu, P., Brandenburg, D., and Olefsky, J. M.,** Structural differences between insulin receptors in the brain and peripheral target tissues, *J. Biol. Chem.,* 258, 8527, 1983.

113. **Mottola, C. and Czech, M. P.,** The type II insulin-like growth factor receptor does not mediate increased DNA synthesis in H-35 hepatoma cells, *J. Biol. Chem.,* 259, 2705, 1984.

114. **Shimizu, M., Webster, C., Morgan, D. O., Blau, H. M., and Roth, R. A.,** Insulin and insulin-like growth factor receptors and responses in cultured human muscle cells, *Am. J. Physiol.,* 251, E611, 1986.

115. **Conovor, C. A., Misra, P., Hintz, R. L., and Rosenfeld, R. G.,** Effect of an anti-insulin-like growth factor I receptor antibody on insulin-like growth factor II stimulation of DNA synthesis in human fibroblasts, *Biochem. Biophys. Res. Commun.,* 139, 501, 1986.

116. **Ewton, D. Z., Falen, S. L., and Florini, J. R.,** The type II insulin-like growth factor (IGF) receptor has low affinity for IGF-I analogues: pleiotypic actions of IGFs on myoblasts are apparently mediated by the type I receptor, *Endocrinology,* 120, 115, 1987.

117. **Hari, J., Pierce, S. B., Morgan, D. O., Sara V., Smith, M. C., and Roth, R. A.,** The receptor for insulin-like growth factor II mediates an insulin-like response, *EMBO J.,* 6, 3367, 1987.

118. **Keiss, W., Thomas, C. L., Greenstein, L. A., Lee, L., Sklar, M. M., Rechler, M. M., Sahagian, G. G., and Nissley, S. P.,** Insulin-like growth factor II (IGF-II) inhibits both the cellular uptake of ß-galactosidase and the binding of ß-galactosidase to purified IGF-II/mannose 6-phosphate receptor, *J. Biol. Chem.,* 264, 4710, 1989.

119. **Wayne H., Bennett, G. A., and Bauer, W.,** Joint disease associated with acromegaly, *Am. J. Med. Sci.,* 209, 671, 1945.

120. **Ashton, I. K. and Matheson, J. A.,** Change in response with age of human articular cartilage to plasma somatomedin activity, *Calcif. Tissue Int.,* 29, 89, 1979.

121. **Coates, C. J., Burwell, R. G., Lloyd-Jones, K., Swannell, A. J., Walker, G., and Selby, C.,** Somatomedin activity in synovial fluid from patients with joint diseases, *Ann. Rheum. Dis.,* 37, 303, 1978.

122. **Schalkwijk, J., Joosten, L. A. B., van den Berg, W. B., Van Wyk, J. J., and van de Putte, L. B. A.,** Insulin-like growth factor stimulation of chondrocyte proteoglycan synthesis by human synovial fluid, *Arthritis Rheum.,* 32, 66, 1989.

123. **Schalkwijk, J., Joosten, L. A. B., von den Berg, W. B., and van de Putte, L. B. A.,** Chondrocyte nonresponsiveness to insulin-like growth factor I in experimental arthritis, *Arthritis Rheum.,* 32, 894, 1989.

124. **Florini, J. R. and Roberts, S. B.,** Effect of rat age on blood levels of somatomedin-like growth factors, *J. Gerontol.,* 35, 23, 1980.

125. **Kumar, N.,** Unresponsiveness of cartilage to serum somatomedins—further observations, *Ind. J. Exp. Biol.,* 17, 571, 1979.

126. **Scranton, P. E., McMaster, J. H., Kenny, F. M., Foley, T. P., and Taylor, F. M.,** Investigation of carbohydrate metabolism and somatomedin in osteosarcoma patients, *J. Surg. Oncol.,* 7, 403, 1978.

127. **Knauer, D. J., Iyer, A. P., Bannerjee, M. R., and Smith, G. L.,** Identification of so-matomedin-like polypeptides produced by mammary tumors of Balb/c mice, *Cancer Res.,* 40, 4368, 1980.

128. **Vassilopoulou-Sellin, R., Wallis, C. J., and Samaan, N. A.,** Hormonal evaluation in pa-tients with osteosarcoma, *J. Surg. Oncol.,* 28, 209, 1985.

129. **Huff, K. K., Kaufman, D., Gabbay, K. H., Spencer, E. M., Lippman, M. E., and Dickson, R. B.,** Human breast cancer cells secret an insulin-like growth factor-I related polypeptide, *Cancer Res.,* 46, 4613, 1986.

130. **Van Wyk, J. J.,** The somatomedins: biological actions and physiologic control mechanisms, in *Hormonal Proteins and Peptides,* Vol. 12, Li, C. H., Ed., Academic Press, Orlando, FL, 1984, 81.

131. **Foley, T. P., Jr., Nissley, S. P., Stevens, R. L., King, G. L., Hascall, V. C., Humbel, R. E., Short, P. A., and Rechler, M. M.,** Demonstration of receptors for insulin-like growth factors on Swarm rat chondrosarcoma chondrocytes, *J. Biol. Chem.,* 257, 663, 1982.

132. **Stevens, R. L., Nissley, S. P., Kimura, J. H., Rechler, M. M., Caplan, A. I., and Hascall, V. C.,** Effects of insulin and multiplication stimulation activity on proteoglycan biosynthesis in chondrocytes from the Swarm rat chondrosarcoma, *J. Biol. Chem.,* 256, 2045, 1981.

133. **Hascall, V. C., Humbel, R. E., Short, P.A., and Rechler, M. M.,** Demonstration of re-ceptors for insulin and insulin-like growth factors on Swarm rat chondrosarcoma chondrocytes. Evidence that insulin stimulates proteoglycan synthesis through the insulin receptor, *J. Biol. Chem.,* 257, 663, 1982.

134. **Seong, S. C., Trippel, S. B., and Mankin, H. J.,** Effects of somatomedin-C on Swarm rat chondrosarcoma chondrocytes in culture, *Orthop. Trans.,* 11, 306, 1987.

135. **Salomon, D. S., Paglia, L. M., and Verbruggen, L.,** Hormone-dependent growth of a rat chondrosarcoma in vivo, *Cancer Res.,* 39, 4387, 1979.

136. **McCumbee, W. D. and Lebovitz, H. E.,** Hormone responsiveness of a transplantable rat chondrosarcoma. I. In vitro effects of growth-hormone dependent serum factors and insulin, *Endocrinology,* 106, 905, 1979.

137. **McCumbee, W. D., McCarty, Jr., K. S., and Lobovitz, H. E.,** Hormone responsiveness of a transplantable rat chondrosarcoma. II. Evidence for in vivo hormone dependence, *Endo-crinology,* 106, 1930, 1980.

138. **Trippel, S. B., Hung, H. H., Van Wyk, J. J., and Mankin, H. J.,** Somatomedin-C content and binding in human chondrosarcoma and giant cell tumors, *Orthop. Trans.,* 10, 383, 1986.

139. **Prins, A. P. A., Lipman, J. M., McDevitt, C. A., and Sokoloff, L.,** Effect of purified growth factors on rabbit articular chondrocytes in monolayer culture. II. Sulfated proteoglycan synthesis, *Arthritis Rheum.,* 25, 1228, 1982.

140. **Prins, A. P. A., Lipman, J. M., and Sokoloff, L.,** Effect of purified growth factors on rabbit articular chondrocytes in monolayer culture. I. DNA synthesis, *Arthritis Rheum.,* 25, 1217, 1982.

141. **Sachs, B. L., Goldberg, V. M., Moskowitz, R. W., and Malemud, C. J.,** Response of articular chondrocytes to pituitary fibroblast growth factor (FGF), *J. Cell. Physiol.,* 112, 51, 1981.

142. **Kato, Y. and Gospodarowicz, D.,** Growth requirements of low density rabbit costal chondrocyte cultures maintained in serum-free medium, *J. Cell. Physiol.,* 120, 354, 1984.

143. **Seyedin, S. M., Thomas, T. C., Thompson, A. Y., Rosen, D. M., and Piez, K. A.,** Purification and characterization of two cartilage-inducing factors from bovine demineral-ized bone, *Proc. Natl. Acad. Sci. U.S.A.,* 82, 2267, 1985.

144. **Trippel, S. B., Doctrow, S., Whelan, M. C., and Mankin, H. J.,** Modulation of growth plate chondrocyte response to Sm-C/IGF-I and basic FGF by growth factor effect interaction, *Orthop. Trans.,* 13, 335, 1989.

145. **Kato, Y., Hiraki, Y., Inoue, H., Kinoshita, M., Yutani, Y., and Suzuki, F.,** Differential and synergistic actions of somatomedin-like growth factors, fibroblast growth factor and epider-mal growth factor in rabbit costal cartilage, *Eur. J. Biochem.,* 192, 885, 1983.

146. **Matsumura, T., Whelan, M., Mankin, H. J., and Trippel, S. B.,** Interaction of IGF-I and TGF-ß in regulating Swarm rat chondrosarcoma chondrocytes, presented at 37th Annual Meeting Orthop. Res. Soc., Anaheim, March 4–7, 1991, 122.

147. **O'Keefe, R. J., Crabb, I. D., Puzas, J. E., and Rosier, R. N.,** Synergistic effects of IGF-I with TGF-ß and FGF are specific for DNA synthesis in growth plate chondrocytes, *Orthop. Trans.,* 13, 336, 1989.

Chapter 7

REGULATION OF CHONDROCYTES BY CYTOKINES

Jeremy Saklatvala

TABLE OF CONTENTS

I. INTRODUCTION

Articular chondrocytes maintain a matrix consisting largely of aggregating proteoglycans, collagen, and water. The proper composition and architecture of this matrix is essential to the mechanical functions of articular cartilage which are to provide a frictionless bearing, to distribute load, and to absorb shock. The normal state is maintained by a delicate balance between the synthesis and catabolism of the matrix macromolecules. In common diseases of joints, such as rheumatoid arthritis and osteoarthritis, this balance is disturbed; the cartilage becomes abnormal, damaged, and is eventually lost. Changes in synthetic or degradative activity of the cells result in a matrix of altered composition which is functionally inadequate and render the cartilage prone to mechanical injury. The combination of metabolic disturbance and mechanical damage finally result in a state which is beyond the limited capacity of the chondrocytes to repair.

The ability of each chondrocyte to make a correct matrix is inherent in the cell lineage, but is also affected by extrinsic influences such as mechanical pressure, hormones, supply of nutrients, and so on. Over the last 10 years, the ability of certain cytokines to profoundly influence cartilage metabolism has been discovered. Cytokines are now thought to be particularly important in causing damage to cartilage in inflammatory joint disease such as rheumatoid arthritis, and possibly also in osteoarthritis, which may have an inflammatory component.

Cytokines are small proteins that are released by activated leucocytes (and other cells) and regulate inflammatory and immune responses. They include interleukins, tumor necrosis factors (TNFs), interferons, colony stimulating factors, and certain growth factors. They have paracine, autocrine, and endocrine actions, and many of them regulate the behavior of a range of cell types. The two cytokines which have the most important effects on cartilage matrix are interleukin 1 (IL-1) and tumor necrosis factor (TNF). These two quite different molecules are strikingly similar in their overall biological actions. This review deals with them in some detail, while touching on the roles of other cytokines, particularly with regard to modulation of the actions of IL-1 and TNF.

II. PROPERTIES OF IL-1 AND TNF

IL-1[1,2] is the name given to a pair of proteins which have a broad range of inflammatory effects. The two proteins, called α and β, are produced mainly by activated macrophages and monocytes. They are made as 35-kDa cytoplasmic precursors which are translocated through the cell membrane and cleaved to the mature 17-kDa forms by removal of their N-terminal portions. The sequence of these events is not yet clear, but the process is thought to be an active one, rather than simply a passive leakage from dying or damaged cells. (Dead and dying cells at an inflammatory site may nevertheless represent an important source of IL-1.)

The reason for the existence of two forms of IL-1 is unknown: they are 28%

homologous. Their production may be differently regulated, but they are indistinguishable in their biological effects.

There are also two tumor necrosis factors:[3,4] α, which arises mainly from activated monocytes and macrophages, and ß, which is produced by activated lymphocytes. They are 17-kDa and 25-kDa proteins, respectively, and have 30% homology in amino acid sequence. Unlike the IL-1s, they are produced via the normal secretory path. Some TNF, however, remains associated with the cell surface.

When they are produced at an inflammatory site, both IL-1 and TNF induce expression of molecules on the vascular endothelium to which leucocytes adhere. They also stimulate production of chemotactic factors such as IL-8 and monocyte chemotactic protein. These changes result in migration of leucocytes out of the vascular compartment and into the tissues. This is a prime physiological function of both cytokines. They also have local actions on connective tissues (including cartilage) and, in concert with other cytokines, they stimulate lymphocyte functions. On entering the systemic circulation, they cause fever (by their action on the hypothalamus) and they stimulate the pituitary adrenal axis. They also induce neutrophilia and contribute to the acute phase response. When produced in large amounts in chronic disease, they cause cachexia; and if they are made in overwhelming amounts, as in septic shock, they precipitate circulatory failure and death.

III. ACTIONS OF IL-1 AND TNF ON CHONDROCYTES

Stimulation of chondrocytes in articular cartilage by IL-1 or TNF causes degradation of the proteoglycan matrix and inhibition of its resynthesis.

Originally, Fell and Jubb[5] and others[6] showed that cultured synovial tissue made a factor which induced chondrocytes in explants of cartilage to degrade the proteoglycan. Previously, only vitamin A had been known to have such an effect. Two cartilage-resorbing factors were eventually purified from porcine leucocyte cultures and were identified as porcine forms of IL-1.[7-9] Other species of IL-1 had similar effects on cartilage.[10-12] The actions of IL-1 on cartilaginous tissue were investigated in detail in organ culture of cartilage of bovine nasal septum[7-9,13] and porcine articular cartilage.[14] IL-1 caused a dose-dependent release of proteoglycan in the range 10^{-11} to 10^{-9} M: complete release occurred in 2 to 3 d from nasal septal cartilage and about 6 to 8 d in the case of porcine articular cartilage, depending the dose of IL-1. TNF[15] had a similar effect, but had to be used in higher dose and caused a slower release. Combinations of the cytokines in low doses were additive. These early experiments strongly suggested that the chondrocytes were proteolytically cleaving the proteoglycan, with the result that large fragments were released from the aggregates into the culture medium.[14,15] Culture with cytokines had little early effect on the collagenous matrix of the cartilage, but some collagen breakdown was observed on prolonged culture with IL-1.

IL-1[15,16-18] and TNF[15,19] also inhibited production of proteoglycan by cartilage. This action has been less intensively investigated than the degradation,
although it is likely to be biologically as important in terms of impairing cartilage
function in chronic disease. The proteoglycans made by IL-1-treated cartilage
explants were normal with regard to glycosylation and sulphation, and the
reduction in their amount was probably due to inhibition of production of the
protein core.[16,20] To what extent these changes were transcriptional, posttranscriptional, or posttranslational is unknown. The TNF-induced inhibition of
proteoglycan synthesis is likely to be mediated in the same way.

IL-1 also inhibited collagen synthesis in cartilage explants[21] and in isolated
chondrocytes.[22] Reduction in protein was generally paralleled by reduction in
steady-state levels of the pro-collagen mRNA. All types of collagen investigated
have been reduced, II, IX, XI,[23] as well as I and III in isolated chondrocytes.[22]
In contrast, divergent findings on the effects of IL-1 on collagen synthesis in
fibroblasts have been reported: some have found stimulation, especially if
prostaglandin synthesis inhibitors are included,[24] and others inhibition.[25] It remains to be established to what extent collagen synthesis in connective tissues
in vivo is strongly regulated by IL-1.

The degradation of proteoglycan in cartilage occured also when IL-1 was
injected into joints of living animals.[26-29] Single injections into knees of rabbits
caused a 20% depletion of proteoglycan after 24 h. By 72 h, the proteoglycan
content had been restored to normal.[26] The effect was a direct one on cartilage
and was not secondary to the massive influx of leucocytes into the synovial
cavity which the cytokine elicited.[27] TNF has not been shown to cause proteoglycan
degradation *in vivo* — this may be because of its lower potency coupled with the
short half-life of a small protein injected into the joint. At a site of inflammation,
both cytokines will generally be found and will of course act in concert.

These experiments *in vivo* were important because they confirmed that the
results obtained *in vitro* were not an artifact of tissue culture. They showed the
sensitivity of cartilage to IL-1 and they showed that chondrocytes could repair
the initial damage.

The effects of the cytokines on intact cartilage prompted much investigation
of their action on isolated chondrocytes. Many of the actions were indistinguishable from those found in fibroblasts. IL-1 or TNF stimulate cultured chondrocytes
to secrete metalloproteinases such as specific collagenase[8,10,30-32] and the general proteolytic enzyme stromelysin.[10,31,33,34] These metalloproteinases, which
are closely related structurally, are released from cells as inactive proenzymes.
The physiological mechanism of their activation is uncertain, but they can be
activated by proteolytic cleavage by plasmin.[35] IL-1 (and TNF) will stimulate
production of plasminogen activators[36-39] (of both urokinase and tissue types),
so a potential mechanism exists whereby chondrocytes, given a source of
plasminogen, could activate the matrix metalloproteinases.

The cytokines stimulate chondrocytes to make inflammatory mediators. One
well-known effect of IL-1[1,2] and TNF[3,4] on fibroblasts is to induce prostaglandin

synthesis. Isolated chondrocytes also make large amounts of PGE_2 in response to IL-1[8,36,40,41] or TNF.[34,43] There is activation[42,43] and induction of synthesis of phospholipase A_2[44,45] which generates arachidonic acid from membrane phospholipids. This provides substrate for the prostaglandin-synthesizing enzyme, cyclo-oxygenase, which is located in the endoplasmic reticulum. IL-1 induces synthesis of cyclo-oxygenase,[46] although this has not actually been shown in chondrocytes. Whether prostaglandins have any important local effects on cartilage metabolism is not known. PGE is an inflammatory mediator which causes vasodilation and augments vascular permeability and perception of pain.

Besides prostaglandins, connective tissue cells stimulated by IL-1 or TNF also produce other cytokines such as IL-6, IL-8, colony stimulating factors, interferon, and even PDGF.[1-4] Chondrocytes have been reported to make IL-6 upon stimulation.[47] It has been suggested (see later) that the depression of proteoglycan synthesis induced by IL-1 or TNF is secondary to production of IL-6:[48] but whether IL-6 has important effects on chondrocyte function remains to be established. It does not generally have IL-1-like actions on connective tissue cells such as fibroblasts.[49] IL-6 is a stimulator of B cell functions and is an important inducer of the hepatic acute phase response. Chondrocytes can express IL-1 mRNA, and IL-1 activity may be made by cultured cartilage.[50] Whether chondrocytes *in vivo* ever produce significant quantities of mature IL-1 protein remains to be established.

The important effects of IL-1 (and TNF) in stimulating glucose transport[51] and glycolysis[52] in fibroblasts probably occur also in chondrocytes, but have not been specifically shown. The cytokines are probably not directly involved in regulation of chondrocyte growth.[17,19,40]

IV. THE MECHANISM OF PROTEOGLYCAN DEGRADATION BY CHONDROCYTES

The chondroitin sulphate-rich proteoglycan fragments released rapidly from cartilage during stimulation with IL-1 or TNF are monomers of proteoglycan that have undergone limited proteolysis and lost their hyaluronate binding region.[14,15] The link protein, and fragments containing the hyaluronate binding region, together with hyaluronate, come out of the explants more slowly.[53] The fragments released from stimulated cartilage are indistinguishable from those released by untreated tissue which suggests the cytokines may simply accelerate an existing process. Recently published analyses of the released fragments based either on their immunological reactivity,[54] or their content of certain marker peptides,[55] show that there is a cleavage of the proteoglycan between the G1 and G2 domains. Amino acid sequencing of the large fragments released from bovine articular cartilage suggests that the cleavage occurs between glutamic acid and alanine in the consensus sequence NXTEXE*ARGXVILTXK.[56]

The enzyme (or enzymes) responsible for the proteolytic degradation of proteoglycan in cartilage has eluded identification for a long time. Early candidate enzymes were plasmin[57] and lysosomal enzymes such as cathepsins D and B.[58] More recently, it has become quite generally accepted that the matrix metalloproteinases are involved and that stromelysin is probably responsible for the degradation.[10,39,59] It is well known that IL-1 stimulates isolated chondrocytes to release procollagenase and prostromelysin. However, it has been difficult to show that chondrocytes in cartilage make stromelysin in response to stimulation with IL-1. There is one report of identification of stromelysin in chondrocytes in cartilage in joints of rats with experimental arthritis induced by type II collagen.[60] Stromelysin will cleave cartilage proteoglycan between the G1 and G2 domains, but this occurs at a site some way N-terminal to that found in proteoglycan released from cartilage stimulated in culture.[61]

One approach to try to identify the proteinase (or proteinases) causing the degradation has been to use enzyme inhibitors. There is one report[62] on the successful use of metalloproteinase inhibitors to block proteoglycan degradation in IL-1-stimulated cartilage. These inhibitors were a thiol tripeptide, a carboxyalkyl peptide, and four hydroxamic acid peptides. They had IC_{50} values of about 10^{-8} to 10^{-7} M on stromelysin or collagenase and inhibited proteoglycan breakdown induced in explants of rabbit articular cartilage by IL-1, retinoic acid, or bacterial lipopolysaccharide. They were effective at 10^{-4} M, but much less so, at 10^{-5} M. The compounds were not thought to interfere with the chondrocyte response by a toxic or nonspecific mechanism, although this is very hard to exclude. In another study, metalloproteinase inhibitors, including TIMP (tissue inhibitor of metalloproteinases), failed to inhibit the IL-1-induced degradation in bovine nasal septal cartilage, as did a range of other inhibitors of serine, cysteine, and aspartate proteinases.[63]

More recently, it has been found that lipophilic inhibitors of cysteine proteinases will inhibit IL-1-induced degradation in bovine nasal septum cartilage in organ culture.[64] The epoxide Ep475 is a hydrophilic inhibitor of cysteine proteinases, but has no effect on IL-1-stimulated proteoglycan degradation when present in cartilage cultures at 2 mM. The lipid-soluble ethyl ester of this compound, Ep453, while being relatively inactive as a proteinase inhibitor, will pass through cell membranes and is readily cleaved by esterases to the more active compound Ep475. Addition of Ep453 to cartilage cultures significantly inhibited IL-1-induced degradation at 10^{-6} M, and was highly effective at 10^{-4} M. One interpretation of these findings is that degradation of proteoglycan is due to the action of lysosomal cysteine proteinases (cathepsins B, H, L, and S). Prior blockade of the enzymes in the cell inhibits degradation, whereas introduction of an inhibitor to the extracellular milieu, even in high concentration, is ineffective. Perhaps, enzyme and substrate concentrations are so high when the two meet that the inhibitor is ineffective; or perhaps enzyme and substrate interact in a privileged compartment from which inhibitor has been excluded: such a compartment might occur in clefts at the cell surface or in endocytic vacuoles.

Other interpretations of the inhibition are that a cysteine proteinase might be needed to activate some other degradative enzyme, or that the epoxide is simply interfering nonspecifically with a cellular process needed for the response. Ep453 apparently did not inhibit protein synthesis, lactate production, or other responses to IL-1.

In conclusion, the identity of the physiological proteoglycan-degrading enzyme in cartilage is unknown. It seems unlikely to be stromelysin, but inhibitor studies suggest that cysteine proteinases and metalloproteinases may be involved.

V. MODULATION OF ACTION OF IL-1 OR TNF BY OTHER AGENTS

The degradative effects of IL-1 or TNF on explants of cartilage and the inhibition of proteoglycan synthesis can be partly reversed by the presence of IGF-I. Antagonism is overridden by high doses of cytokine.[54,65]

The proteoglycan degradation induced in cartilage by IL-1 has also been found to be partly inhibited by transforming growth factor (TGF)-ß.[66,67] TGF-ß on its own promotes proteoglycan synthesis, while slowing catabolism,[68] so it might be expected to function as a physiological antagonist of IL-1 and TNF.

The degradation of proteoglycan induced by IL-1[69] or TNF[34] is reduced in the presence of γ-interferon. This latter finding is surprising because γ-interferon generally enhances the actions of TNF.[3,4] γ-Interferon also inhibits the production of prostaglandin and metalloproteinase by isolated chondrocytes stimulated with IL-1[69,70] or TNF.[34]

While γ-interferon, TGF-ß, and IGF-I are physiologically antagonistic to IL-1 and TNF on cartilage, certain growth factors may potentiate or augment the action of the cytokines. Fibroblast growth factor (FGF)[71] and platelet derived growth factor (PDGF)[72] both strongly potentiated prostaglandin synthesis in IL-1-stimulated chondrocytes. PDGF also augmented the amount of metalloproteinase induced by IL-1. Both growth factors increased IL-1 receptor expression.[71-73] IL-1 increased PDGF receptor expression, but decreased FGF receptor expression. Interactions with growth factors on cartilage itself have not been reported.

The actions of IL-1 on articular cartilage *in vivo*[74] or isolated chondrocytes[75] are antagonized by the IL-1 receptor antagonist protein. This monocyte product is homologous to IL-1 and binds to the IL-1 receptor, apparently without initiating any intracellular response.[1]

VI. MECHANISM OF ACTION OF IL-1 AND TNF

The mechanisms of action of IL-1 and TNF on chondrocytes (and other cells) are largely unknown. Chondrocytes express the type I IL-1 receptor and, in culture, have up to 10,000 sites per cell.[71-73,76] This is an 80-kDa transmembrane glycoprotein which mediates the main cellular effects of IL-1.[77] There is a related type II IL-1 receptor of 67 kDa which occurs on B cells and may be quite widespread; it has a very short cytoplasmic portion and its signaling function is not established.[78] There are also two TNF receptors; they are 55[79] and 75[80] kDa proteins which are unrelated to the IL-1 receptors. The TNF receptors tend to be coexpressed on cells. They have not been extensively studied on chondrocytes. The structures of the cytoplasmic portions of IL-1 and TNF receptors contain no clues as to their signaling mechanisms.

After combining with their receptors, both IL-1 and TNF cause certain rapid and similar increases in protein phosphorylation.[81-84] Therefore, it is likely that the cytokines work by activating protein kinases. One of these kinases strongly phosphorylates the small heat shock protein (hsp 27)[82] in a wide range of cells, including chondrocytes.[83] Interestingly, IL-1 induces expression of hsp 65 in chondrocytes.[85] Another kinase (or kinases) phosphorylates the epidermal growth factor receptor[84] and there are likely to be many more substrates and enzymes to be identified. The kinases activated by the cytokines in fibroblasts mainly phosphorylate serine and possibly threonine, and are distinct from protein kinase C or cAMP-dependent kinase.[83,84] Studies of chondrocytes also suggest that responses to IL-1 are not mediated by PKC.[86,87] Associated with the wave of kinase activity caused by IL-1 or TNF, there is induction of transcription factors such as NF kappa B, AP1, and NFIL6, and eventually there are prolonged characteristic changes in gene expression.[88] Strong circumstantial evidence suggests that the kinases activated by the cytokines somehow, directly or indirectly, activate transcription factors, but the enzymes and the substrates relevant to transcriptional control all remain to be discovered.

VII. ACTION OF INTERFERONS ON CARTILAGE

One other group of cytokines that may have some effects on cartilage are the interferons.[89] The classical leucocyte and fibroblast (α and ß) interferons are antiviral, induce class I MHC antigens, and have a variety of immunostimulatory actions. γ-Interferon, or immune interferon, is a product of activated T cells. Some of its actions are similar to those of the α- or ß-interferons; but unlike them, it is only weakly antiviral and it combines with a distinct receptor. It uniquely induces class II MHC antigens on cells and is an important activator of macrophages and monocytes. It has been found to inhibit collagen synthesis in dermal and synovial fibroblasts[90] and in chondrocytes.[91] It induces class II MHC antigens on chondrocytes, but the significance of this is not known. α- and ß-Interferons also inhibit collagen synthesis, but very high doses are required.[91] γ-Interferon has been found to antagonize the production of prostaglandins and metalloproteinases induced by IL-1 or TNF (see earlier).

VIII. CONCLUSIONS

This review has discussed the role of cytokines in cartilage metabolism in terms of single agents acting on isolated tissue. Physiologically, cartilage would never be exposed to such single stimuli, but it is difficult to predict the long-term effects of complex mixtures of cytokines, growth factors, and other inflammatory mediators. IL-1 and TNF clearly have a detrimental action on the major proteoglycan component of the matrix. Their long-term effect on collagenous components is less clear, but the inhibition of collagen synthesis in culture suggests it is probably adverse. Absence of proteoglycan will of itself expose the collagen to mechanical damage.

It should also be stressed that in inflammatory disease cartilage matrix may be damaged by enzymes arising from other cells in the joint. There may be dying and disintegrating polymorphonuclear leucocytes which are a rich source of proteinases. There may be synovial tissue proliferating as invasive pannus, which itself makes proteinases in response to cytokine stimulation.

Increasing knowledge of the molecular mechanism of action of cytokines and growth factors on chondrocytes should provide new insights into the control of synthesis and degradation of cartilage matrix. Hopefully, such knowledge may lead to new ways of controlling chondrocyte metabolism which could be exploited for therapy of common diseases.

REFERENCES

1. **Dinarello, C. A.,** Interleukin 1 and interleukin 1 antagonism, *Blood,* 77, 1627, 1991.
2. **DiGiovine, F. S. and Duff, G. W.,** Interleukin 1: the first interleukin, *Immunol. Today,* 11, 13, 1990.
3. **Sherry, B. and Cerami, A.,** Cachectin/tumor necrosis factor exerts endocrine, paracine and autocrine control of inflammatory responses, *J. Cell Biol.,* 107, 1269, 1988.
4. **Vilcek, J. and Lee, T. H.,** Tumor necrosis factor. New insights into the molecular mechanisms of its multiple actions, *J. Biol. Chem.,* 266, 7313, 1991.
5. **Fell, H. B. and Jubb, R. W.,** The effect of synovial tissue on the breakdown of articular cartilage in organ culture, *Arthritis Rheum.,* 20, 1359, 1977.
6. **Steinberg, J., Sledge, C. B., Noble, J., and Stirrat, C. R.,** A tissue culture model of cartilage breakdown in rheumatoid arthritis. Quantitative aspects of proteoglycan release, *Biochem. J.,* 180, 403, 1979.
7. **Saklatvala, J., Curry, V. A., and Sarsfield, S. J.,** Purification to homogeneity of pig leucocyte catabolin, a protein that causes cartilage resorption in vitro, *Biochem. J.,* 215, 385, 1983.
8. **Saklatvala, J., Pilsworth, L. M. C., Sarsfield, S. J., Gavrilovic, J., and Heath, J. K.,** Pig catabolin is a form of interleukin 1: cartilage and bone resorb, fibroblasts make prostaglandin and collagenase, and thymocyte proliferation is augmented in response to one protein, *Biochem. J.,* 224, 461, 1984.
9. **Saklatvala, J., Sarsfield, S. J., and Townsend, Y.,** Pig interleukin 1. Purification of two immunologically different leukocyte proteins that cause cartilage resorption, lymphocyte activation and fever, *J. Exp. Med.,* 162, 2163, 1985.
10. **Gowen, M., Wood, D. D. Ihrie, E. J., Meats, J. E., and Russell, R. G. G.,** Stimulation by human interleukin 1 of cartilage breakdown and production of collagenase and proteoglycanase by human chondrocytes but not by human osteoblasts in vitro, *Biochim. Biophys. Acta,* 797, 186, 1984.
11. **Krakauer, T., Oppenheim, J. J., and Jasin, H. E.,** Human interleukin 1 mediates cartilage matrix degradation, *Cell. Immunol.,* 91, 92, 1985.
12. **Hubbard, J. R., Steinberg, J. J., Bednar, M. S., and Sledge, C. B.,** Effect of purified interleukin 1 on cartilage degradation, *J. Orthop. Res.,* 6, 180, 1988.

13. **Saklatvala, J.,** Characterization of catabolin, the major product of pig synovial tissue that induces resorption of cartilage proteoglycan in vitro, *Biochem. J., 199*, 705, 1981.
14. **Tyler, J. A.,** Chondrocyte-mediated depletion of articular cartilage proteoglycans in vitro, *Biochem. J., 225*, 493, 1985.
15. **Saklatvala, J.,** Tumor necrosis factor α stimulates resorption and inhibits synthesis of proteoglycan in cartilage, *Nature (London), 322*, 547, 1986.
16. **Tyler, J. A.,** Articular cartilage cultured with catabolin (pig interleukin 1) synthesizes a decreased number of normal proteoglycan molecules, *Biochem. J., 227*, 869, 1985.
17. **Ikebe, T., Hirata, M., and Koga, T.,** Human recombinant interleukin 1-mediated suppression of glycosaminoglycan synthesis in cultured rat costal chondrocytes, *Biochem. Biophys. Res. Commun., 140*, 386, 1986.
18. **Van den Berg, W. B., van de Loo, F. A. J., Zwarts, W. A., and Otterness, I. G.,** Effects of recombinant IL1 on intact homologous cartilage, *Ann. Rheum. Dis., 47*, 855, 1988.
19. **Ikebe, T., Hirata, M., and Koga, T.,** Effects of human recombinant tumor necrosis factor α and interleukin 1 on the synthesis of glycosaminoglycan and DNA in cultured rat costal chondrocytes, *J. Immunol., 140*, 827, 1988.
20. **Benton, H. P. and Tyler, J. A.,** Inhibition of cartilage proteoglycan synthesis by interleukin 1, *Biochem. Biophys. Res. Commun., 154*, 421, 1988.
21. **Tyler, J. A. and Benton, H. P.,** Synthesis of type II collagen is decreased in cartilage cultured with interleukin 1 while the rate of intracellular degradation remains unchanged, *Collagen Rel. Res., 8*, 393, 1988.
22. **Goldring, M. B. and Krane, S. M.,** Modulation by recombinant interleukin 1 of synthesis of types I and II collagens and associated procollagen mRNA levels in cultured human cells, *J. Biol. Chem., 262*, 16724, 1987.
23. **Tyler, J. A., Bird, J. L. E., and Giller, T.,** IL1 inhibits the production of types II, IX and XI procollagen mRNA in cartilage, *Ann. N.Y. Acad. Sci., 580*, 512, 1990.
24. **Krane, S. M., Dayer, J.-M., Simon, S. S., and Byrne, S.,** Mononuclear cell-conditioned medium containing mononuclear cell factor (MCF) homologous with interleukin 1 stimulates collagen and fibronectin synthesis by adherent rheumatoid synovial cells: effects of prostaglandin E$_2$ and indomethacin, *Collagen Rel. Res., 5*, 66, 1985.
25. **Mauviel, A., Teyton, L., Bhatnagar, R., Penfornis, H., Laurent, M., Hartmann, D., Bonaventure, J., Loyau, G., Saklatvala, J., and Pujol, J.-P.,** Interleukin 1α modulates collagen gene expression in cultured synovial cells, *Biochem. J., 252*, 247, 1988.
26. **Pettipher, E. R., Higgs, G. A., and Henderson, B.,** Interleukin 1 induces leukocyte infiltration and cartilage proteoglycan degradation in the synovial joint, *Proc. Natl. Acad. Sci. U.S.A., 83*, 8749, 1986.
27. **Pettipher, E. R., Henderson, B., Moncada, S., and Higgs, G. A.,** Leukocyte infiltration and cartilage proteoglycan loss in immune arthritis in the rabbit, *Br. J. Pharmacol., 95*, 169, 1988.
28. **Page Thomas, D. P., King, B., Stephens, T., and Dingle, J. T.,** In vivo studies of cartilage regeneration after damage induced by catabolin/IL1, *Ann. Rheum. Dis., 50*, 75, 1991.
29. **O'Byrne, E. M., Blancuzzi, V., Wilson, D. E., Wong, M., and Jeng, A. Y.,** Elevated substance P and accelerated cartilage degradation in rabbit knees injected with interleukin 1 and tumour necrosis factor, *Arthritis Rheum., 33*, 1023, 1990.
30. **Phadke, K.,** Fibroblast growth factor enhances the interleukin-1 mediated chondrocytic protease release, *Biochem. Biophys. Res. Commun., 142*, 448, 1987.
31. **Campbell, I. K., Golds, E. E., Mort, J. S., and Roughley, P. J.,** Human articular cartilage secretes characteristic metal-dependent proteinases upon stimulation by mononuclear cell factor, *J. Rheumatol., 13*, 20, 1986.
32. **Stephenson, M. L., Goldring, M. B., Birkhead, J. R., Krane, S. M., Rahmsdorf, H. J., and Angel, P.,** Stimulation of procollagenase synthesis parallels increases in cellular procollagenase mRNA in human articular chondrocytes exposed to recombinant interleukin 1ß or phorbol ester, *Biochem. Biophys. Res. Commun., 144*, 583, 1987.

33. **Schnyder, J., Payne, T., and Dinarello, C. A.,** Human monocyte or recombinant interleukin 1 are specific for the secretion of a metalloproteinase from chondrocytes, *J. Immunol.,* 138, 496, 1987.

34. **Bunning, R. A. D. and Russell, R. G. G.,** The effect of tumour necrosis factor α and γ-interferon on the resorption of human articular cartilage and the production of prostaglandin E and of caseinase activity by human articular chondrocytes, *Arthritis Rheum.,* 32, 780, 1989.

35. **Werb, Z., Mainardi, C. L., Vater, C. A., and Harris, E. D., Jr.,** Endogenous activation of latent collagenase by rheumatoid synovial cells: evidence for a role of plasminogen activator, *N. Engl. J. Med.,* 296, 1017, 1977.

36. **McGuire-Goldring, M. B., Meats, J. E., Wood, D. D., Ihrie, E. J., Ebsworth, N. M., and Russell, R. G. G.,** In vitro activation of human chondrocytes and synoviocytes by human interleukin-1-like factor, *Arthritis Rheum.,* 27, 654, 1984.

37. **Bunning, R. A. D., Crawford, A., Richardson, H. J., Opdenakker, G., Van Damme, J., and Russell, R. G. G.,** Interleukin 1 preferentially stimulates the production of tissue type plasminogen activator by human articular chondrocytes, *Biochim. Biophys. Acta,* 924, 473, 1987.

38. **Leizer, R., Clarris, B. J., Ash, P. E., van Damme, J., Saklatvala, J., and Hamilton, J. A.,** Interleukin 1ß and interleukin 1α stimulate the plasminogen activator activity and prostaglandin E_2 levels of human synovial cells, *Arthritis Rheum.,* 30, 562, 1987.

39. **Campbell, I. K., Piccoli, D. S., Butler, D. M., Singleton, D. H., and Hamilton, J. A.,** Recombinant human interleukin-1 stimulates human articular cartilage to undergo resorption and human chondrocytes to produce both tissue and urokinase-type plasminogen activator, *Biochem. Biophys. Acta,* 967, 183, 1988.

40. **Chin, J. E. and Lin, Y.,** Effects of recombinant human interleukin 1ß on rabbit articular chondrocytes and stimulation of prostanoid release and inhibition of cell growth, *Arthritis Rheum.,* 31, 1290, 1988.

41. **Campbell, I. K., Piccoli, D. S., and Hamilton, J. A.,** Stimulation of chondrocyte prostaglandin E_2 production by recombinant human interleukin 1 and tumor necrosis factor, *Biochim. Biophys. Acta,* 1051, 310, 1990.

42. **Chang, J., Gilman, S. C., and Lewis, A. J.,** Interleukin 1 activates phospholipase A_2 in rabbit chondrocytes: a possible signal for IL1 action, *J. Immunol.,* 136, 1283, 1986.

43. **Suffys, P., Roy, F. V., and Fiers, W.,** Tumor necrosis factor and interleukin 1 activate phospholipase in rat chondrocytes, *FEBS Lett.,* 232, 24, 1988.

44. **Giordano-Lyons, B., Davis, G. L., Galbraith, W., Pratt, M. A., and Arner, E. C.,** Interleukin 1ß stimulates phospholipase A_2 mRNA synthesis in rabbit articular chondrocytes, *Biochem. Biophys. Res. Commun.,* 164, 488, 1989.

45. **Kerr, J. S., Stevens, J. M., Davis, G. L., McLaughlin, J. A., and Harris, R. R.,** Effects of recombinant interleukin-1ß on phospholipase A_2 mRNA and eicosanoid formation in rabbit chondrocytes, *Biochem. Biophys. Res. Commun.,* 165, 1079, 1989.

46. **Maier, J. A. M., Hla, T., and Maciag, T.,** Cyclooxygenase is an immediate early gene induced by interleukin-1 in human endothelial cells, *J. Biol. Chem.,* 265, 10805, 1990.

47. **Bunning, R. A. D., Russell, R. G. G., and Van Damme, J.,** Independent induction of interleukin 6 and prostaglandin E_2 by interleukin 1 in human articular chondrocytes, *Biochem. Biophys. Res. Commun.,* 166, 1163, 1990.

48. **Nietfeld, J. J., Wilbrink, B., Helle, M., van Roy, L. A. M., den Otter, W., Swaak, A. J. G., and Huber-Brunning, O.,** Interleukin-1-induced interleukin-6 is required for the inhibition of proteoglycan synthesis by interleukin 1 in human articular cartilage, *Arthritis Rheum.,* 33, 1695, 1990.

49. **Seckinger, P., Yaron, I., Meyer, F. A., Yaron, M., and Dayer, J. M.,** Modulation of the effects of interleukin-1 on glycosaminglycan synthesis by the urine-derived interleukin-1 inhibitor but not by interleukin 6, *Arthritis Rheum.,* 33, 1807, 1990.

50. **Ollivierre, F., Gubler, U., Towle, C. A., Laurencin, C., and Treadwell, B. V.,** Expression of IL-1 genes in human and bovine chondrocytes: a mechanism for autocrine control of cartilage matrix, *Biochem. Biophys. Res. Commun.,* 141, 904, 1986.

51. **Bird, T. A., Davies, A., Baldwin, S. A., and Saklatvala, J.,** Interleukin 1 stimulates hexose transport in fibroblasts by increasing the expression of glucose transporters, *J. Biol. Chem.,* 265, 13578, 1990.

52. **Taylor, D. J., Whitehead, R. J., Evanson, J. M., Westmacott, D., Feldmann, M., Bertfield, H., Morris, M. A., and Woolley, D. E.,** Effect of recombinant cytokines on glycolysis and fructose 2,6-bisphosphate in rheumatoid synovial cells in vitro, *Biochem. J.,* 250, 111, 1988.

53. **Ratcliffe, A., Tyler, J. T., and Hardingham, T. E.,** Articular cartilage cultured with interleukin 1. Increased release of link protein, hyaluronate-binding region and other proteoglycan fragments, *Biochem. J.,* 238, 571, 1986.

54. **Fosang, A. J., Tyler, J. A., and Hardingham, T. E.,** Effect of interleukin 1 and insulin-like growth factor 1 on the release of proteoglycan components and hyaluronate from pig articular cartilage in explant culture, *Matrix,* 11, 17, 1991.

55. **Sandy, J. D., Boynton, R. E., and Flannery, C. R.,** Analysis of the catabolism of aggrecan in cartilage explants by quantitation of peptide from the three globular domains, *J. Biol. Chem.,* 266, 8198, 1991.

56. **Sandy, J. T., Neame, P. J., Boynton, R. E., and Flanner, C. R.,** Catabolism of aggrecan in cartilage explants, *J. Biol. Chem.,* 256, 8683, 1991.

57. **Lack, C. H., Anderson, A. J., and Ali, S. Y.,** Action of plasmin on cartilage matrix in vivo, *Nature (London),* 191, 1402, 1961.

58. **Dingle, J. T.,** Recent studies on the control of joint damage: the contribution of the Strangeways Research Laboratory, *Ann. Rheum. Dis.,* 38, 201, 1979.

59. **Campbell, I. K., Roughley, P. J., and Mort, J. S.,** The action of human articular cartilage metalloproteinases on link protein. Similarities between products of in situ and in vitro degradation, *Biochem. J.,* 237, 117, 1986.

60. **Hasty, K. A., Reife, R. A., Kang, A. H., and Stuart, J. M.,** The role of stromelysin in the cartilage destruction that accompanies inflammatory arthritis, *Arthritis Rheum.,* 33, 388, 1990.

61. **Fosang, A. J., Neame, P. J., Hardingham, T. E., Murphy, G., and Hamilton, J. A.,** Cleavage of cartilage proteoglycan between G1 and G2 domains by stromelysin, *J. Biol. Chem.,* 266, 15579, 1991.

62. **Caputo, C. B., Sygowski, L. A., Wolanin, D. J., Patton, S. P., Caccesse, R. G., Shaw, A., Roberts, R. A., and Di Pasquale, G.,** Effect of synthetic metalloprotease inhibitors on cartilage autolysis in vitro, *J. Pharmacol. Exp. Ther.,* 240, 460, 1987.

63. **Saklatvala, J. and Sarsfield, S. J.,** How do interleukin 1 and tumour necrosis factor induce degradation of proteoglycan in cartilage? in *The Control of Tissue Damage,* Glauert, A. M., Ed., Elsevier, Amsterdam, 1988, 97.

64. **Buttle, D. J., Saklatvala, J., Tamai, N., and Barrett, A. J.,** The inhibition of interleukin-1-stimulated cartilage proteoglycan degradation by lipophilic cysteine endopeptidase inactivators, *Biochem. J.,* 281, 175, 1991.

65. **Tyler, J. A.,** IGF 1 can decrease degradation and promote synthesis of proteoglycan in cartilage exposed to cytokines, *Biochem. J.,* 260, 543, 1989.

66. **Andrews, H. J., Edwards, T. A., Cawston, T. E., and Hazleman, B. L.,** Transforming growth factor-ß causes partial inhibition of interleukin 1-stimulated cartilage degradation in vitro, *Biochem. Biophys. Res. Commun.,* 162, 144, 1989.

67. **Chandrasekhar, S. and Harvey, A. K.,** TGFß is a potent inhibitor of IL1-induced protease activity and proteoglycan degradation, *Biochem. Biophys. Res. Commun.,* 157, 1352, 1988.

68. **Morales, T. I. and Roberts, A. B.,** Transforming growth factor ß regulates the metabolism of proteoglycans in bovine cartilage organ cultures, *J. Biol. Chem.,* 263, 12828, 1988.

69. **Andrews, H. J., Bunning, R. A. D., Dinarello, C. A., and Russell, R. G. G.,** Modulation of human chondrocyte metabolism by recombinant human interferon gamma: in vitro effects on basal and IL1-stimulated proteinase production, cartilage degradation and DNA synthesis. *Biochim. Biophys. Acta,* 1012, 128, 1989.

70. **Andrews, H. J., Bunning, R. A. D., Plumpton, T. A., Clark, I. M., Russell, R. G. G., and Cawston, T. E.,** Inhibition of interleukin-1-induced collagenase production in human articular chondrocytes in vitro by recombinant human interferon γ, *Arthritis Rheum.,* 33, 1733, 1990.

71. **Chin, J. E., Hatfield, C. A., Krzesicki, R. F., and Herblin, W. F.,** Interactions between interleukin-1 and basic fibroblast growth factor on articular chondrocytes, *Arthritis Rheum.,* 34, 314, 1991.

72. **Smith, R. J., Justen, J. M., Sam, L. M., Rohlaff, N. A., Ruppel, P. L., Brunden, M. N., and Chen, J. E.,** Platelet-derived growth factor potentiates cellular responses of articular chondrocytes to interleukin 1, *Arthritis Rheum.,* 34, 697, 1991.

73. **Chandrasekhar, S. and Harvey, A. K.,** Induction of IL1 receptors on chondrocytes by fibroblast growth factor, *J. Cell. Physiol.,* 138, 236, 1989.

74. **Henderson, B., Thompson, R. C., Hardingham, T., and Lewthwaite, J.,** Inhibition of interleukin-1-induced synovitis and articular cartilage proteoglycan loss in the rabbit knee by recombinant human interleukin-1 receptor antagonist, *Cytokine,* 3, 246, 1991.

75. **Smith, R. J., Chin, J. E., Sam, L. M., and Justen, J. M.,** Biologic effects of an interleukin-1 receptor antagonist protein on interleukin-1-stimulated cartilage erosion and chondrocyte responsiveness, *Arthritis Rheum.,* 34, 78, 1991.

76. **Saklatvala, J. and Bird, T. A.,** A common class of receptors for the two types of porcine interleukin-1 on articular chondrocytes, *Lymphokine Res.,* 5 (Suppl. 1), 99, 1986.

77. **Sims, J. E., Acres, R. B., Grubin, C. E. McMahan, C. J., Wignall, J. M., March, C. J., and Dower, S. K.,** Cloning the interleukin 1 receptor from human T cells, *Proc. Natl. Acad. Sci. U.S.A.,* 86, 8946, 1989.

78. **McMahan, C. J., Slack, J. L., Mosley, B., Cosman, D., Lupton, S. D., Brunton, L. L., Grubin, C. E., Wignall, J. M., Jenkins, N. A., Brannan, C. I., Copeland, N. G., Huebner, K., Croce, C. M., Cannizzaro, L. A., Benjamin, D., Dower, S. K., Spriggs, M. K., and Sims, J. E.,** A novel IL-1 receptor cloned from B cells by mammalian expression is expressed in many cell types, *EMBO J.,* 10, 2821, 1991.

79. **Loetscher, H., Pan, Y. C. E., Lahm, H.-W., Gentz, R., Brockhaus, M., Tabuchi, H., and Lesslauer, W.,** Molecular cloning and expression of the human 55kD tumor necrosis factor receptor, *Cell,* 61, 351, 1990.

80. **Smith, C. A., Davis, T., Anderson, D., Solam, L., Beckmann, M. P., Jerzy, R., Dower, S. K., Cosman, D., and Goodwin, R. G.,** A receptor for tumor necrosis factor defines an unusual family of cellular and viral proteins, *Science,* 248, 1019, 1990.

81. **O'Neill, L. A. J., Bird, T. A., and Saklatvala, J.,** Interleukin 1 signal transduction, *Immunol. Today,* 11, 392, 1990.

82. **Guesdon, F. and Saklatvala, J.,** Identification of a cytoplasmic protein kinase regulated by interleukin 1 which phosphorylates the small heat shock protein, hsp 27, *J. Immunol.,* 147, 3402, 1991.

83. **Saklatvala, J., Kaur, P., and Guesdon, F.,** Phosphorylation of the small heat shock protein is regulated by interleukin 1, tumor necrosis factor, growth factors, bradykinin and ATP, *Biochem. J.,* 277, 634, 1991.

84. **Bird, T. A. and Saklatvala, J.,** Down modulation of epidermal growth factor receptor affinity in fibroblasts treated with interleukin 1 or tumor necrosis factor is associated with phosphorylation at a site other than threonine 654, *J. Biol. Chem.,* 265, 235, 1990.

85. **Cruz, T. F., Kandel, R. A., and Brown, I. R.,** Interleukin 1 induces the expression of a heat shock gene in chondrocytes, *Biochem. J.,* 277, 327, 1991.

86. **Arner, E. C. and Pratta, M.,** Modulation of interleukin-1-induced alterations in cartilage proteoglycan metabolism by activation of protein kinase C, *Arthritis Rheum.,* 34, 1006, 1991.

87. **Hulkower, K. I., Georgescu, H. I., and Evans, C. H.,** Evidence that responses of articular chondrocytes to interleukin 1 and basic FGF are not mediated by protein kinase C, *Biochem. J.,* 276, 157, 1991.

88. **Muegge, K. and Durum, S. K.,** Cytokines and transcription factors, *Cytokine,* 2, 1, 1990.
89. **DeMaeyer, E. and De Maeyer-Guignard, J.,** *Interferons and Other Regulatory Cytokines,* John Wiley & Sons, New York, 1988, 1.
90. **Jiminez, S. A., Freundlich, B., and Rosenbloom, J.,** Selective inhibition of human diploid fibroblast collagen synthesis by interferons, *J. Clin. Invest.,* 74, 1112, 1984.
91. **Goldring, M. B., Sandell, L. J., Stephenson, M. L., and Krane, S. M.,** Immune interferon suppresses levels of procollagen mRNA and type II collagen synthesis in cultured human articular and costal chondrocytes, *J. Biol. Chem.,* 261, 9049, 1986.

Chapter 8

INTERCELLULAR REGULATION
BY SYNOVIOCYTES

Geethani Bandara and Christopher H. Evans

TABLE OF CONTENTS

I. INTRODUCTION

Evidence of the synovial capacity to modulate the metabolism of chondrocytes was first published by Fell and Jubb.[1] These authors observed that fragments of porcine synovium maintained under *in vitro* conditions produced substances which accelerated the release of proteoglycans and collagen from the matrix of living articular cartilage. Dead cartilage failed to respond to the synovial factors, demonstrating that destruction of the matrix was mediated by living chondrocytes within the cartilage. Later work by the same authors revealed the ability of synovial factors not only to increase catabolism of the cartilaginous matrix, but also to decrease synthesis of new proteoglycan molecules by chondrocytes.[2] These classical findings opened a vigorous new area of research which has now developed into an important field in its own right.

Subsequent research has had several goals. A major one has been to identify the synovial factors responsible for modulating the catabolic and anabolic activities of chondrocytes. Coupled to this have been studies of the mechanisms which regulate the synthesis and biological activities of these factors. Other aims have been to identify the cells within synovium which produce these substances and to determine their molecular modes of action at the level of the chondrocyte. Elucidating these matters should lead to better therapeutic control of cartilage breakdown in arthritis.

This chapter reviews recent progress in understanding how synoviocytes regulate the metabolism of chondrocytes. Most investigators, including ourselves, approach this question from the perspective of the exaggerated breakdown of cartilage that occurs in arthritic joints. For this reason, the present review concentrates on the responses of adult articular cartilage to substances released by the synovium of such joints. However, it is likely that these sorts of intercellular interactions also help maintain the biochemical homeostasis of the normal joint, as well as playing a role in its growth and development.

II. SYNOVIOCYTES

Healthy synovium is a thin membrane, only one to three cell layers thick, which lines all diarthrodial joints.[3] Unlike other membranes which line body cavities, the synovium has no basement membrane. Synovium is underlain by areolar, fibrous, or adipose tissue, beyond which extends a fibrous capsule. One of the problems in studying synoviocytes lies with obtaining uncontaminated synovial membrane. Unless painstaking care is exercised, it is almost impossible not to include underlying connective tissue when dissecting synovium from joints. The results of most, if not all, published *in vitro* studies of synovial metabolism, including our own, need to be interpreted in this light.

Synovium comprises a collagenous matrix penetrated by lymphatic vessels, blood capillaries, and nerve endings, together with at least two types of resident synoviocytes. Although normal synovium contains capillary endothelial cells and nerve cells, this review concentrates on the synovial cells which line the joint. Historically, these have been divided into two categories: (1) the type A synoviocyte morphologically resembles a macrophage and is phagocytic,[4,5] and (2) the type B synoviocyte is fibroblastic.[4,5] Until recently, histological and anatomical criteria were the only defining properties of these cells. However, the development of discriminating antibodies is helping to provide better precision. Like macrophages, the type A synoviocytes have receptors for the F_c part of the IgG molecule and the C_3 component of complement,[6] and are rich in lysozyme. In addition, they are recognized by monoclonal antibodies which specifically bind to macrophages.[7] Identification of the type B synoviocyte remains less definitive, but has been considerably helped by the recent finding that human synovial fibroblasts can be distinguished from fibroblasts of the subsynovial tissues through their recognition by antibodies to laminin and type IV collagen.[8] This is surprising, as these proteins are important components of basement membrane, a structure which synovium lacks. Further definition of the synovial fibroblast may develop from the recent observation that, at least in humans, it can be distinguished from other synovial and subsynovial cells by a monoclonal antibody "Mab 67".[9]

The literature also bears reference to synoviocytes with a morphology that is intermediate between the A and B types, and synoviocytes with a dendritic morphology. The latter are characteristic of inflamed joints and may reflect an altered metabolic state of the synovial fibroblast, rather than a distinct type of synoviocyte. Prostaglandin E_2 (PGE_2), which is present at high local concentrations in inflamed joints, could be responsible for this morphological switch.[10] However, as Goto et al.[11] have cloned dendritic cells from human rheumatoid synovium for extended periods, this phenotype appears to have the stability of a separate cell type. Also unclear is the status of the intermediate type of cell, sometime referred to as the type C synoviocyte.

In arthritic joints, the synovium is thickened as a result of both increased cell number and greater deposition of matrix. During the synovitis of inflammatory arthritides, the synovium becomes infiltrated to various degrees by lymphocytes, additional macrophages, mast cells, and polymorphonuclear leucocytes. Although these cells strongly influence chondrocyte metabolism, space does not permit discussion of their ability to do this. However, it should be noted that their products can influence chondrocyte metabolism both directly and indirectly via their effects upon the resident synoviocytes (Figure 1). The latter is possible because the A and B types of synoviocyte do not constitutively secrete their entire cytokine repertoire. Instead, the production of these factors can be elevated or reduced in response to various stimuli. Their induction is part of the larger phenomenon of synoviocyte activation, which includes a large increase in the synthesis of neutral metalloproteinases (NMPs), plasminogen activator, and several other products (Figure 2). Synoviocytes can be activated in a paracrine fashion by activating factors secreted from other cells in the joint, and in an autocrine fashion by synoviocytes themselves (Figure 1).[12,13] Other activators include the cartilaginous wear particles of osteoarthritic joints,[14,15] various types of crystals[16] in crystal-induced arthropathies, and certain physical stimuli such as heat shock.[17]

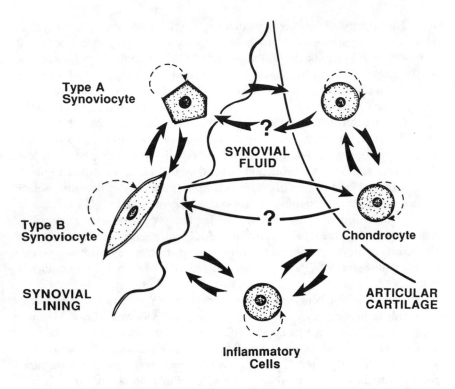

FIGURE 1. Intercellular conversations in the joint. Solid arrows indicate paracrine interactions; dashed arrows indicate autocrine effects. For details, see text.

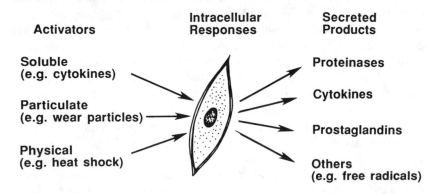

FIGURE 2. Synoviocyte activation. In response to various types of activator, synoviocytes undergo a process of activation as a result of which the synthesis and secretion of several products is greatly enhanced.

III. CHONDROCYTES

Cartilage cells, or chondrocytes, are as unusual as the tissue in which they reside. As cartilage is aneural, avascular, and alymphatic, chondrocytes are not directly linked to the main communication channels of the body. Furthermore, cartilage is a tissue of low cellularity, with a sparse population of cells embedded within an abundant extracellular matrix composed largely of type II collagen and specific proteoglycans. Due to its isolation, apparent low metabolic activity, and largely mechanical role, cartilage was once thought to be metabolically inert. As this chapter will discuss, nothing could be further from the truth.

There are several different types of cartilage, including articular, meniscal, elastic, epiphyseal (growth plate), and nasal. Owing to its relevance to arthritis and proximity to synovium, articular cartilage has been most studied in the present context. However, as the articular cartilage of many species is so thin, nasal cartilage has been widely used to investigate matrix breakdown under the influence of synovial factors. While this provides a convenient experimental system, it raises concerns about possible differences between the properties of nasal chondrocytes and articular chondrocytes, especially as nasal cartilage would not normally be exposed to synovial factors *in vivo*.

Even within one type of cartilage, the chondrocytes show considerable heterogeneity.[18] Subpopulations of articular chondrocytes, for example, can be distinguished on the basis of morphology, buoyant density, and metabolic behavior. It is emerging that different subpopulations of chondrocytes respond differently to the various types of synovial influence discussed in this review.

Most investigations of chondrocyte metabolism employ either fragments of otherwise intact cartilage or cell cultures of chondrocytes which have been enzymically released from their matrix. The former system remains closer to the *in vivo* condition, but is technically limiting. To give but one example, it is proving extremely difficult to purify intact RNA from cartilage.[19] Cell culture obviates many of these problems. Short-term monolayer cultures of articular chondrocytes synthesize type II collagen, cartilage proteoglycans, and display several other characteristics of the differentiated phenotype which typifies these cells.[20] There is also much interest in suspension cultures of chondrocytes in agarose gels.[18] Under these conditions, the cells are rounded and secrete an abundant cartilaginous matrix.

Using such experimental systems, investigators have sought to identify the synovial factors which modulate chondrocyte metabolism and to understand the cellular mechanisms through which they achieve this.

IV. SYNOVIAL INFLUENCE OVER THE DEGRADATION OF THE CARTILAGINOUS MATRIX

A. CATABOLIN AND INTERLEUKIN-1

As described in the introduction, this area of research was initiated by the discovery that synovium secretes factors which increase the breakdown of cartilage by its own chondrocytes. Attempts to purify the active principal from culture medium conditioned by synovium succeeded in identifying a factor, originally called "catabolin".[21] Owing to the limited availability of synovial tissue, the purification to homogeneity of catabolin was accomplished from blood mononuclear cells.[22] Purification and analysis of catabolin showed it to be identical with interleukin-1 (IL-1), a cytokine previously identified as a product of macrophages.[23] At least two different IL-1s exist as the products of separate genes. The acidic (α) form of IL-1 has a pI of about 5 and the neutral (ß) form, a pI of about 7.[24]

Purified recombinant IL-1α and IL-1ß provoke the autolysis of fragments of living porcine and bovine cartilage in the same qualitative manner as catabolin. However, the susceptibility of human articular cartilage to the catabolic effects of IL-1 is a matter of disagreement. Nietfeld et al.[25] failed to influence the release of proteoglycans from slices of human articular cartilage with concentrations of IL-1 up to 100 pg/ml, although proteoglycan synthesis was suppressed. Similar findings have been reported by Bayliss et al.[26] However, several other groups[27-29] have measured enhanced rates of proteoglycan release from human articular cartilage exposed to similar concentrations of IL-1. Animal experiments have confirmed that IL-1 induces cartilage breakdown *in vivo*.[30]

Synovial fibroblasts synthesize IL-1s α and ß *in vitro*.[31,32] However, secretion of IL-1 appears to be under complex regulation[32] and, in contrast to monocytes,[33] involves the preferential release of IL-1α over IL-1ß.[34] *In situ* hybridization confirms the presence of mRNA coding for IL-1ß within the type A synoviocyte.[35] Rheumatoid synovial tissue contains mRNAs coding for IL-1α[36] and ß,[35] and produces large quantities of IL-1α and IL-1ß *in vitro*.[37]

IL-1 is difficult to detect by bioassay of synovial fluid, because inhibitors are present. However, immunological methods, which are not subject to this limitation, have confirmed the presence of biologically active concentrations of IL-1 in such fluids. In one study, for example, synovial fluid aspirated from rheumatoid joints was found to contain 130.3 ± 22 pg/ml IL-1ß and that aspirated from osteoarthritic joints, 27.8 ± 4.5 pg/ml. Fluids from two normal patients contained 20 pg/ml IL-1ß, the lower limit of detection in the assay.[38] When inflammation is low, synovial fluid IL-1 is presumably derived from synoviocytes and, perhaps, chondrocytes (Figure 1).[39] In the inflamed joint, additional macrophages and polymorphonuclear leucocytes[40,41] may also contribute IL-1 to synovial fluid. As discussed in other sections, IL-1 also suppresses the synthesis of the cartilaginous matrix by chondrocytes, and alters several other metabolic parameters.

Studies of bovine articular chondrocytes cultured as suspensions in agarose suggest that cells from the superficial layers of cartilage respond much more vigorously than the deeper cells to IL-1. Age is also an important variable, with chondrocytes of older animals responding more weakly than those of younger individuals.[25]

The mechanisms through which IL-1 brings about the destruction of the cartilaginous matrix are unclear, but it may be linked to the induction of neutral proteinases.[42,43] Articular chondrocytes have the facultative capacity to synthesize and secrete at least four different types of neutral proteinases. Three of them — collagenase, gelatinase, and stromelysin — are metalloenzymes, while the fourth, plasminogen activator, is a serine proteinase. Between them, these enzymes can degrade the entire extracellular macromolecular matrix of cartilage, although the extent of their involvement in either normal turnover or the pathological destruction of articular cartilage remains under discussion.[44] IL-1 induces the synthesis of these NMPs by cultures of articular chondrocytes through a mechanism which involves the cycloheximide sensitive[45] induction of their cognate mRNAs.[46,47] The actions of these proteinases may be increased through reduced production of inhibitors such as the Tissue Inhibitor of Metalloproteinases (TIMP) by chondrocytes exposed to IL-1.[48] This is in contrast to synovial fibroblasts where IL-1 increases the biosynthesis of TIMP.[49]

Monolayer cultures of lapine articular chondrocytes do not secrete lysosomal proteinases into the extracellular medium at higher rates in response to synovial factors.[43] However, concentrations of lysosomal enzymes are modestly higher in medium conditioned by porcine articular cartilage exposed to IL-1.[50]

B. OTHER CATABOLIC CYTOKINES

Although synoviocytes produce IL-1 and IL-1 induces lysis of the cartilaginous matrix, there is evidence that other synovial cytokines can also provoke cartilage breakdown. One of these is tumor necrosis factor-α (TNF-α) which, like IL-1, accelerates the release of proteoglycans from living porcine[51] articular cartilage. As with IL-1, there are conflicting reports concerning the ability of TNF-α to induce the breakdown of human cartilage.[26,52] TNF-α increases the synthesis of NMPs[53] and plasminogen activator[52] by human articular chondrocytes. In all of these activities, TNF-α is a weaker agonist than IL-1.

Lapine chondrocytes respond differently. Addition of human, recombinant TNF-α to monolayer cultures of lapine articular chondrocytes fails to induce neutral metalloproteinases or PGE$_2$.[54] Species mismatching does not account for this discrepancy, as partially purified lapine TNF-α is also ineffective in this regard.[55] However, it does suppress matrix synthesis by lapine chondrocytes.[55] The lack of a catabolic effect is not an *in vitro* artifact, as TNF-α fails to induce the breakdown of articular cartilage in rabbits' knees, although it does promote certain inflammatory changes.[56]

Synovial fibroblasts produce very little TNF-α,[54] suggesting that type A synoviocytes account for most of the synovial production of TNF-α. This conclusion is supported by *in situ* hybridization studies.[35] Although TNF-α is present in synovial fluid,[57] a recent study suggests that normal synovial fluids contain greater concentrations of TNF-α than fluids obtained from arthritic joints.[38] Furthermore, TNF-α concentrations are higher in osteoarthritis than in rheumatoid arthritis.[38] TNF-α shares the same cell-surface receptor as TNF-ß,[58] a cytokine produced by activated lymphocytes. Although TNF-ß also stimulates cartilage breakdown,[52] its synovial fluid concentration is extremely low.[57]

Experiments on the activation of cultured articular chondrocytes by synovial factors[29,43,54] suggest that the synovial regulation of cartilage catabolism may involve more than the actions of just IL-1 and TNF-α. Synovial factors induce the synthesis of plasminogen activator,[42] NMPs,[43] and PGE$_2$,[42,43] as part of a process called chondrocyte activation. As a result, we have termed the synovial agents which provoke this response "chondrocyte activating factors" (CAF).[59] To expedite studies of CAF, we have established an immortalized line of lapine synovial fibroblasts, designated HIG-82.[60] CAF prepared from HIG-82 cells proved more potent than IL-1 as an inducer of NMPs in lapine articular chondrocytes.[61] Further analysis provided two explanations of this phenomenon: synovial fibroblasts produced potentiators of chondrocyte activation by IL-1 and also an additional activator (or activators), which was neither IL-1 nor TNF-α.[54]

Possible potentiators include basic fibroblast growth factor (bFGF) and platelet-derived growth factor (PDGF). Basic FGF is a powerful potentiator of the *in vitro* activation of lapine articular chondrocytes by IL-1.[61,62] It also increases cartilage breakdown *in vivo,* when injected intraarticularly with IL-1.[63] Unlike IL-1, whose secretion from synovial fibroblasts requires induction, cultures of synovial fibroblasts synthesize bFGF constitutively.[54,64] In human synoviocytes, the bFGF remains cell associated,[64] although we have measured bFGF in the conditioned medium of lapine synoviocytes.[54] Human synovial fluid contains bFGF.[65] Acidic FGF, which also potentiates the actions of IL-1,[61,66] is produced by the rheumatoid synovium.[67]

Another cytokine which potentiates the activation of chondrocytes by IL-1 is platelet-derived growth factor (PDGF).[68] In this respect, lapine articular chondrocytes differ from human rheumatoid synovial fibroblasts whose induction of collagenase by IL-1 is inhibited by PDGF.[69] This growth factor is present in synovial fluid and is synthesized by rheumatoid synovium.[67] At least part of the effects of bFGF and PDGF on lapine articular chondrocytes may be related to their ability to increase the number of cell surface IL-1 receptors.[66,68] However, McCollum et al.[70] have reported that bFGF reduces by half the number of IL-1 receptors on human articular chondrocytes. This discrepancy may reflect differences of age and species. Further complexity is introduced by the observation that intraarticular infusions of bFGF promote cartilage repair[71,72] (see Section V.B.).

Although bFGF and PDGF potentiate the activation of chondrocytes by IL-1, they do not alone induce the synthesis of NMPs. However, if the IL-1 present in partially purified preparations of lapine CAF is inhibited, the ability of these preparations to activate chondrocytes is only partially blocked.[54] The remaining activity presumably represents an activator which is distinct from IL-1. Of the large number of cytokines which have been screened for their ability to induce the synthesis of NMPs in chondrocytes, only IL-1 and TNF-α are active (Table 1). However, synovial fibroblasts are a poor source of TNF-α and, in any case,

TABLE 1
Major Cytokines Produced by Synoviocytes and Their Effects on Prominent Aspects of Chondrocyte Metabolism

	Producer synoviocyte		Articular chondrocyte response				
			Synthesis of:			Matrix	Matrix
Cytokine	A	B	NMPs	PA	PGE$_2$	catabolism	anabolism
IL-1α,ß	+	+	+	+	+	+	–
TNF-α	+	±	{+	+	+	+}[a]	–
TGF-ß	+	+	–		+	–	+
IL-6	±	+	0		0	0,+,–	0
IL-8		+	0		0		
bFGF	+	+	Pot		Pot	Pot	+
aFGF	+	+	Pot		Pot		
GM-CSF[b]	+	+	0		0		
ß$_2$-Microglobulin		+	0		0		
PDGF	+	±	Pot		Pot		

Note: Cytokines shown to be produced either *in vivo* or *in vitro* by A or B type synoviocytes are indicated by a +; ± indicates a low level of production, but there is no other attempt to be quantitative. The responses of articular chondrocytes to these cytokines are indicated as follows. A + sign shows that the response is increased; a – sign that it is decreased, a 0, that the cytokine has no effect; Pot means that the cytokine has no effect alone, but potentiates the action of IL-1. If more than one sign appears at a given place, it indicates conflicting reports in the literature. If no sign appears, the relevant information is not in the literature. GM-CSF, Granulocyte macrophage-colony stimulating factor; NMPs, Neutral metalloproteinases; and PA, Plasminogen activator.

[a] TNF-a has the indicated effects on human chondrocytes, but not on lapine articular chondrocytes where TNF-a has '0' effect.

TNF-α does not induce these enzymes in lapine articular chondrocytes. Thus, it seems that synovial fibroblasts produce a factor, distinct from IL-1 and TNF-α, which induces the synthesis of NMPs by lapine articular chondrocytes. This is a potentially novel cytokine, about which little is presently known. The factor elutes from a size-exclusion column with an apparent molecular weight of 20,000 Da, and, unlike IL-1, is completely destroyed by heating to 60°C for 1 h.[54]

C. SYNOVIAL FACTORS WHICH INHIBIT CATABOLISM

Although crude, partially fractionated preparations of synovial cytokines have an overall catabolic effect, there are within this mixture substances which inhibit the activation of chondrocytes and the depletion of the matrix. One such agent is transforming growth factor-ß (TGF-ß), which inhibits the induction of neutral metalloproteinases in cultured chondrocytes by IL-1.[54,73] At least part of this inhibition occurs at the level of mRNA abundance.[54] It is also possible that the synthesis of proteinase inhibitors is increased.[74] TGF-ß protects fragments of cartilage from matrix depletion by IL-1,[75] as well as decreasing basal levels of matrix catabolism by unstimulated cartilage *in vitro.*[76] In addition, TGF-ß promotes synthesis of the cartilaginous matrix (Section V.B.), promotes chondrocyte differentiation (Section VI.C.), and, depending upon the presence of other growth factors, mitosis (Section VI.A.). TGF-ß is produced constitutively by normal synovial fibroblasts[54,77] and by rheumatoid synovium[78] *in vitro,* and can be detected by immunohistology of *ex vivo* synovium.[77] TGF-ß is present in both nonrheumatoid and rheumatoid synovial fluids, where concentrations as high as 46 ng/ml have been measured.[78] Concentrations are higher in rheumatoid than osteoarthritic fluids.[78,79] Systemic administration of TGF-ß has recently been shown to antagonize the development of experimental arthritis in rats.[80]

In addition to TGF-ß, there is evidence of other synovial inhibitors of the activation of articular chondrocytes by IL-1.[34] It remains to be seen whether synoviocytes synthesize the recently identified antagonist of the 80-kDa IL-1 receptor[81] or any other of the documented, naturally occurring inhibitors of IL-1.[82]

V. SYNOVIAL INFLUENCE OVER THE SYNTHESIS OF THE CARTILAGINOUS MATRIX

A. SYNOVIAL FACTORS WHICH INHIBIT MATRIX SYNTHESIS

IL-1 decreases the synthesis of proteoglycan molecules by fragments of cartilage in culture.[83] There is evidence that, in certain systems, this effect may be more damageing to cartilage than the direct catabolic effect discussed in Section IV.A. Adult human cartilage, for example, may mount a poor catabolic response to IL-1, while remaining responsive to its suppression of matrix synthesis.[25,26] Furthermore, greater concentrations of IL-1 are required to induce a catabolic response than to inhibit the synthesis of matrix. However, under physiological conditions where chondrocytes are probably exposed to complex mixtures of cytokines, this difference may disappear.

The recent finding[84] that anti-IL-6 antibodies block the inhibition of proteoglycan synthesis by IL-1 suggests that suppression of matrix synthesis by IL-1 is secondary to induction of IL-6. Although IL-1 indeed induces the synthesis of IL-6 by fragments of human articular cartilage,[84] exogenous IL-6 reduces proteoglycan synthesis only slightly,[84] if at all.[85] This may simply reflect poor penetration of cartilage by IL-6 or indicate an underlying mechanistic complexity to the phenomenon. However, IL-6 fails to influence proteoglycan synthesis by cell cultures of bovine articular chondrocytes,[86] where penetration is not an issue.

TNF-α also suppresses matrix synthesis by chondrocytes.[51,55] As with its catabolic effect, it does so more weakly than IL-1.

Another product of synoviocytes which inhibits matrix synthesis is PGE_2.[87,88] Although resting synoviocytes secrete only basal levels of PGE_2, activated synoviocytes secrete high amounts of this ecosanoid,[89] which is secreted by the synovium of arthritic joints[90] and accumulates to a considerable degree in synovial fluid.

Type A synoviocytes within the synovium of rats with adjuvant arthritis synthesizes hydrogen peroxide[91] which inhibits matrix synthesis by chondrocytes.[92] This appears to be linked to lowered intracellular concentrations of ATP and a general depression of protein synthesis and other cellular functions.[93] Synovium also synthesizes superoxide[94] and possibly other oxygen-derived free radicals with the potential to inhibit matrix synthesis. Production of such radicals may be particularly high following transient ischemia and reperfusion of the synovium as a result of the flexion of joints with effusions.[95] Whether the half-life of such radicals is long enough for them to diffuse from synovium to the cartilage and influence chondrocyte metabolism is unknown. This may only occur in areas of close apposition of synovium and cartilage. Such free radicals can directly damage cartilage macromolecules[96] in the absence of a chondrocytic response.

B. SYNOVIAL FACTORS WHICH INCREASE MATRIX SYNTHESIS

The major cytokines thus far identified as physiological enhancers of matrix synthesis by adult articular chondrocytes are insulin, insulin-like growth factor I (IGF-I), insulin-like growth factor II (IGF-II), and TGF-ß.[97] Of these, only TGF-ß has been identified as a product of synoviocytes (Section IV.C.), although the other factors are present in synovial fluid.[98] Monolayer cultures of lapine articular chondrocytes constitutively secrete IGF-I; TGF-ß inhibits, but bFGF enhances their production of this factor.[99]

TGF-ß$_1$ increases the synthesis of proteoglycans by articular cartilage in organ culture[97] and by monolayers of articular[100] and growth plate[101] chondrocytes. However, the synthesis of GAGs by lapine articular chondrocytes cultured in soft agar is reduced by TGF-ß.[102] A similar discrepancy is observed for collagen synthesis by articular chondrocytes, which TGF-ß inhibits in soft agar[102] but enhances in monolayer cultures.[100] The synthesis of noncollagenous proteins is also increased in the latter cultures. According to O'Keefe et al.,[101] TGF-ß inhibits the synthesis of collagen by growth plate chondrocytes, but enhances the synthesis of noncollagenous proteins. In this system, these effects were increased by a dialyzable serum factor. The anabolic effects of TGF-ß reinforced, additively or synergistically, those of IGF-I and FGF.

As we have seen (Section IV.B.), FGF potentiates the ability of IL-1 to activate chondrocytes *in vitro* and provoke cartilage breakdown *in vivo*. However, there is evidence that FGF can also increase matrix synthesis by fragments of articular cartilage *in vitro*[103] and promote the healing of articular cartilage *in vivo*.[71,72] A combination of FGF and insulin or epidermal growth factor (EGF) produces a markedly synergistic synthetic response *in vitro*.[103]

Synthesis of GAG by cell cultures of articular chondrocytes is also increased by FGF, a marked synergy occurring with TGF-ß.[104] However, FGF has been reported to inhibit GAG synthesis by fetal articular cartilage, while enhancing hyaluronate synthesis.[105]

Insulin-like growth factors, formally known as somatomedins, are of additional interest as they to some degree counteract the inhibitory effects of IL-1 and

TNF upon proteoglycan synthesis.[106] Of possible pathophysiological importance is evidence that the chondrocytes of arthritic joints are unresponsive to IGF-I.[107]

VI. SYNOVIAL INFLUENCE OVER OTHER ASPECTS OF CHONDROCYTE METABOLISM

A. MITOSIS

Chondrocytes in the growth plate divide frequently *in vivo,* and cultures of these cells respond mitotically to insulin, IGF-I, FGF, and EGF. Adult articular chondrocytes do not normally divide, but they do so in response to depletion of their extracellular matrix. This occurs in osteoarthritis and experimentally when the matrix is digested to release articular chondrocytes for tissue culture.

FGF, but not insulin, IGF-I, or EGF, increases the incorporation of [^{3}H]-thymidine by fragments of adult bovine articular cartilage in organ culture.[103] Although insulin alone is inactive in this regard, it strongly potentiates the effects of FGF.[103] Monolayers of articular chondrocytes also divide in response to FGF.[108]

IL-1 stimulates DNA synthesis by cultures of rat epiphyseal chondrocytes[109] and human articular chondrocytes.[27] In the latter case, γ-interferon synergized strongly with IL-1 as a mitogen. The mitogenic response of human chondrocytes to IL-1 and γ-interferon was preceded by a transient inhibition of DNA synthesis during the first 24 h of culture. This raises the possibility that the mitogenic effect of these cytokines is a secondary response to the induction of additional autocrine growth factors. If so, this would explain the discrepancy between this study and that of Chin and Lin,[110] who noted a very strong inhibition of DNA synthesis in lapine articular chondrocytes exposed to IL-1 for 24 h. TGF-ß inhibits the growth of lapine articular chondrocytes cultured in 2% serum, but enhances growth in 10% serum.[111] TGF-ß produces a delayed increase in DNA synthesis by epiphyseal chondrocyte cultures, the effect being greatly enhanced by dialyzable factors present in the serum.[112] The mitogenic responses of chick growth plate chondrocytes increases with increasing cellular maturity, the highest stimulation occurring in proliferating and early hypertrophic cells.[113]

PDGF is also a mitogen for articular chondrocytes,[114] but H_2O_2, on the contrary, causes an immediate inhibition of DNA synthesis.[93]

B. EICOSANOID PRODUCTION

Following activation, articular chondrocytes synthesize, in decreasing order, PGE_2,[42,43] $PGF_{2\alpha}$,[110] prostacyclin I_2,[110] and thromboxane A_2.[110] However, no leukotriene C_4 can be detected.[115] A number of synovial cytokines, including IL-1, TGF-ß,[54] and, to a lesser degree, bFGF[61] enhance PGE_2 synthesis by cultures of articular chondrocytes. TNF-α is also active on human,[53] but not lapine[54] articular chondrocytes.

The elevation in prostanoid production is, at least partly, due to increased activities of phospholipase A_2[116] and cyclooxygenase.[117]

Results from cell culture experiments strongly suggest that the induction of NMPs and PGE_2 by synovial CAF is independently regulated.[43,61] A similar conclusion has been reached for the effects of IL-1 on proteoglycan breakdown, proteoglycan synthesis, and PGE_2 synthesis by cartilage in organ culture.[118]

C. DIFFERENTIATION

Chondrocytes are highly differentiated cells which rapidly, but reversibly, become dedifferentiated in culture. As they provide convenient biochemical markers of differentiation, particularly the synthesis of type II collagen and cartilage-specific proteoglycans, chondrocytes have been widely used in studies of this phenomenon. Particular attention has been paid to TGF-ß, which was originally identified as "cartilage inducing factor",[119] a protein from bone with the ability to promote the chondrogenesis of prechondrogenic mesenchymal cells and to induce the synthesis of cartilage-specific macromolecules.

FGF also promotes the differentiated phenotype of chondrocytes. A combination of FGF and TGF-ß is particularly powerful in this regard such that growth plate chondrocytes cultured in their presence continue to express their differentiated proteoglycan phenotype despite displaying aberrant morphologies which normally would indicate dedifferentiation.[104]

These matters are of some interest in osteoarthritis where dedifferentiation of articular chondrocytes may occur. They are also of relevance to cartilage or chondrocyte transplantation where it would be necessary for the transplanted tissue to behave in an appropriately differentiated manner.

D. OTHER EFFECTS

Cartilage is avascular and has a correspondingly low oxygen tension.[120] Energy metabolism is predominantly anaerobic, although chondrocytes contain mitochondria and have some ability to respire aerobically.[121,122] In cell culture, IL-1 and synovial factors strongly increase the production of lactate by chondrocytes,[122] a finding of possible relevance to the acidosis that occurs in inflamed joints. This response may be related to the production of nitric oxide by lapine articular chondrocytes exposed to synovial factors and IL-1; TNF-α does not induce the synthesis of nitric oxide by these cells.[115] Lapine articular chondrocytes also produce hydrogen peroxide in response to certain cytokines.[123]

Human articular chondrocytes obtained from osteoarthritic joints produce superoxide in response to TNF-α. This effect is biphasic. At a dose of 23 U/ml, TNF-α suppresses superoxide production. However, at 1500 U/ml, TNF-α enhances superoxide production by chondrocytes.[124] Unlike normal cells, articular chondrocytes in arthritic joints possess I_a antigens.[125] This has important pathophysiological consequences as it implicates chondrocytes in antigen presentation. I_a antigens can be induced on chondrocytes by the lymphokine γ-interferon,[126] but it is not known whether cytokines produced by synoviocytes can do this. Stromelysin production in response to IL-1 is considerably greater in I_a positive chondrocytes, although collagenase induction is unaltered.[127] This result is a little surprising, as γ-interferon, which was used to induce I_a expression in these experiments, has been reported to suppress stromelysin synthesis by activated chondrocytes.[27,53]

VII. OTHER SYNOVIAL FACTORS

In addition to the synovial substances discussed in the foregoing sections, synoviocytes secrete a number of other agents whose possible effects upon chondrocyte metabolism, if any, remain to be evaluated. These include amyloid A,[128] *gro* protein,[129] IL-8,[130] and proteins of the complement system[131] produced by synovial fibroblasts *in vitro*. It would be teleologically satisfying to identify a role for IL-6, which is produced in large amounts by both synovial fibroblasts and chondrocytes. In addition, if the type A synoviocytes are like other macrophages, they may be expected to secrete the full spectrum of macrophage cytokines, such as platelet activating factor, and the macrophage inflammatory proteins α and β,[132] whose effects on chondrocytes are, as yet, unknown. Furthermore, synoviocytes produce a number of eicosanoids in addition to PGE_2, such as leukotriene B_4, and thromboxane A_2 which could influence chondrocyte metabolism.

Clearly, considerable work remains to be done in this important area of research.

VIII. SUMMARY AND CONCLUSIONS

Between them, types A and B synoviocytes secrete a variety of polypeptide cytokines (Table 1), eicosanoids, free radicals, and perhaps other types of molecules, which profoundly influence the metabolism of chondrocytes. Substances secreted by activated synoviocytes collectively suppress the synthesis of the cartilaginous matrix and accelerate its degradation. These responses are mediated by the articular chondrocytes which, at the cellular level, undergo a number of complex metabolic changes (Figure 3). Although these changes compromise the integrity of the cartilaginous matrix, synovial CAF contains individual cytokines which inhibit cartilage catabolism and promote matrix synthesis. Conditions may exist where these factors are preferentially secreted by synoviocytes, in which case synovium would play a role in maintaining the normal metabolic equilibrium of cartilage and promoting its repair.

It is clear that these responses do not occur in isolation. Not only do cytokines influence each other's activity at the level of the chondrocyte, but they influence each other's synthesis. It should also be emphasized that chondrocytes are themselves capable of synthesizing many of the synovial factors discussed in this chapter. These include IL-6, bFGF, TGF-ß, IL-1, PGE_2, and free radicals. This raises the likelihood of autocrine regulation by the chondrocytes and gives chondrocytes the potential to talk back to synoviocytes via chondrocytic factors diffusing from the cartilage to the synovium (Figure 1). As synovial factors modulate the synthesis of these chondrocytic factors, the conversations between synovium and cartilage may become very complicated.

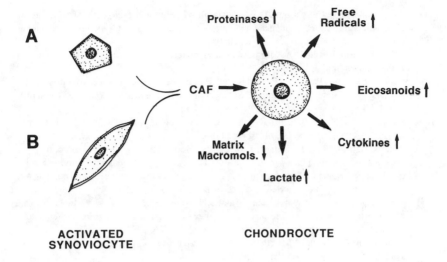

FIGURE 3. Responses of articular chondrocytes to synovial factors. Activated synoviocytes secrete a cocktail of cytokines and other mediators, known collectively as "chondrocyte activating factors" (CAF). As indicated, articular chondrocytes mount a pleiotropic response to CAF.

Under inflammatory conditions, additional potential cellular sources of cytokines exist (Figure 1). As well as directly affecting chondrocyte metabolism, the products of inflammatory cells could also alter the responses of chondrocytes to synovial factors and help regulate the synthesis of synovial factors by synoviocytes themselves. Thus, the interaction between the synoviocyte and the chondrocyte is but one component of a complex network of intercellular communication that occurs in the joint. Understanding this system is vital to the complete understanding of how the joint functions in health and disease.

REFERENCES

1. **Fell, H. B. and Jubb, R. W.,** The effect of synovial tissue on the breakdown of articular cartilage in organ culture, *Arthritis Rheum.,* 20, 1359, 1977.
2. **Jubb, R. W. and Fell, H. B.,** The effect of synovial tissue on the synthesis of proteoglycan by the articular cartilage of young pigs, *Arthritis Rheum.,* 23, 545, 1980.
3. **Simkin, P. A.,** Joints: structure and function, in *Primer on the Rheumatic Diseases,* 9th ed., Schumacher, H. R., Klippel, J. H., and Robinson, D. R., Eds., Arthritis Foundation, Atlanta, 1988, 18.
4. **Barland, P., Novikoff, A. B., and Hammerman, D.,** Electron microscopy of the human synovial membrane, *J. Cell Biol.,* 14, 207, 1962.
5. **Ghadially, F. N. and Roy, S.,** Ultrastructure of rabbit synovial membrane, *Ann. Rheum.Dis.,* 25, 315, 1966.

6. **Theofilopoulos, A. N., Carson, D. A., Tavassoli, M., Slovin, S. F., Speers, W. C., Jenson, F. B., and Vaugh, J. H.,** Evidence for the presence of receptors for C_3 and IgG F_c on human synovial cells, *Arthritis Rheum.,* 23, 1, 1980.

7. **Mapp, P. I. and Revell, P. A.,** Ultrastructural characterisation of macrophages (type A cells) in the synovial lining, *Rheumatol. Int.,* 8, 171, 1988.

8. **Pollock, L., Lalor, P., Mapp, P., and Revell, P.,** The presence and basal distribution of type IV collagen and laminin in the intimal cell layer of synovium, *Arthritis Rheum.,* 32 (Suppl.), 153, 1989.

9. **Stevens, C. R., Mapp, P. I., and Revell, P. A.,** A monoclonal antibody (Mab 67) marks type B synoviocytes, *Rheumatol. Int.,* 10, 103, 1990.

10. **Clarris, B. J., Leizer, T., Fraser, J. R. E., and Hamilton, J. A.,** Diverse morphological responses of normal human synovial fibroblasts to mononuclear leukocyte products: relationship to prostaglandin production and plasminogen activator activities, and comparison to the effects of purified interleukin-1, *Rheumatol. Int.,* 7, 35, 1987.

11. **Goto, M., Sasano, M., Yamanaka, H., Miyasaka, N., Kamatani, N., Inoue, K., Nishioka, K., and Miyamoto, T.,** Spontaneous production of an interleukin-1-like factor by cloned rheumatoid synovial cells in long term culture, *J. Clin. Invest.,* 80, 786, 1987.

12. **Brinckerhoff, C. E. and Mitchell, T. I.,** Autocrine control of collagenase synthesis by synovial fibroblasts, *J. Cell. Physiol.,* 136, 72, 1988.

13. **Baratz, M. E., Georgescu, H. I., and Evans, C. H.,** Studies on the autocrine activation of a synovial cell line, *J. Orthop. Res.,* 9, 651, 1991.

14. **Evans, C. H., Mears, D. C., and Cosgrove, J. L.,** Release of neutral proteinases from mononuclear phagocytes and synovial cells in response to cartilaginous wear particles in vitro, *Biochim. Biophys. Acta,* 677, 287, 1981.

15. **Evans, C. H., Mazzocchi, R. A., Nelson, D. D., and Rubash, H. E.,** Experimental arthritis induced by intraarticular injection of allogenic cartilaginous particles into rabbit knees, *Arthritis Rheum.,* 27, 200, 1984.

16. **Hasselbacher, P., McMillan, R. M., Vater, C. A., Hahn, J., and Harris, E. D.,** Stimulation of secretion of collagenase and prostaglandin E_2 by synovial fibroblasts in response to crystals of monosodium urate monohydrate: a model for joint destruction in gout, *Trans. Assoc. Am. Phys.,* 94, 243, 1981.

17. **Vance, B. A., Kowalski, C. G., and Brinckerhoff, C. E.,** Heat shock of rabbit synovial fibroblasts increases expression of mRNAs for two metalloproteinases, collagenase and stromelysin, *J. Cell Biol.,* 108, 2037, 1989.

18. **Aydelotte, M. B. and Kuettner, K. E.,** Differences between sub-populations of cultured bovine articular chondrocytes. I. Morphology and cartilage matrix production, *Connect. Tissue Res.,* 18, 205, 1988.

19. **Adams, M. E. and Huang, D. Q.,** Measurement of proteoglycan mRNA in articular cartilage — preliminary investigation, *J. Rheumatol.,* 18 (Suppl. 27), 55, 1991.

20. **Green, W. T.,** Behavior of articular chondrocytes in cell culture, *Clin. Orthop.,* 75, 248, 1971.

21. **Dingle, J. T., Saklatvala, J., Hembry, R., Tyler, J., Fell, H. B., and Jubb, R.,** A cartilage catabolic factor from synovium, *Biochem. J.,* 184, 177, 1979.

22. **Saklatvala, J., Curry, V. A., and Sarsfield, S. J.,** Purification to homogeneity of pig leucocyte catabolin, a protein that causes cartilage resorption in vitro, *Biochem. J.,* 215, 385, 1983.

23. **Saklatvala, J., Pilsworth, L. M. C., Sarsfield, S. J., Gavrilovic, J., and Heath, J. K.,** Pig catabolin is a form of interleukin-1, *Biochem. J.,* 224, 461, 1984.

24. **March, C. J., Mosley, B., Larsen, A., and Cerretti, D. P., et al.,** Cloning, sequence and expression of two distinct human interleukin-1 complementary DNAs, *Nature (London),* 315, 641, 1985.

25. **Nietfeld, J. J., Wilbrink, B., Den Otter, W., Huber, J., and Huker-Bruning, O.,** The effect of human interleukin-1 on proteoglycan metabolism in human and porcine cartilage explants, *J. Rheumatol.,* 17, 818, 1990.

26. **Bayliss, M. T., Hickery, M. S., and Hardingham, T. E.,** Effects of IL-1 and TNF-α on human articular cartilage: differences in biosynthetic and degradative responses, *Trans. Orthop. Res. Soc.,* 16, 147, 1991.

27. **Andrews, H. J., Bunning, R. A. d., Dinarello, C. A., and Russell, R. G. G.,** Modulation of human chondrocyte metabolism by recombinant human interferon gamma: in-vitro effects on basal and IL-1-stimulated proteinase production, cartilage degradation and DNA synthesis, *Biochim. Biophys. Acta,* 102, 128, 1989.

28. **Hubbard, J. R., Steinberg, J. J., Bednar, M. S., and Sledge, C. B.,** Effect of purified human interleukin-1 on cartilage degradation, *J. Orthop. Res.,* 6, 180, 1988.

29. **Campbell, I. K., Piccoli, D. S., Butler, C. M., Singleton, D. K., and Hamilton, J. A.,** Recombinant human interleukin-1 stimulates human articular cartilage to undergo resorption and human chondrocytes to produce both tissue and urokinase-type plasminogen activator, *Biochim. Biophys. Acta,* 967, 183, 1988.

30. **Pettipher, E. R., Higgs, G. A., and Henderson, B.,** Interleukin-1 induces leukocyte infiltration and cartilage proteoglycan degradation in the synovial joint, *Proc. Natl. Acad. Sci. U.S.A.,* 83, 8749, 1986.

31. **Dalton, B. J., Connor, J. R., and Johnson, W. J.,** Interleukin-1 induces interleukin-1 α and interleukin-1 ß gene expression in synovial fibroblasts and peripheral blood monocytes, *Arthritis Rheum.,* 32, 279, 1989.

32. **Johnson, W. J., Breton, J., Newman-Tarr, T., Connor, J. R., Meunier, P. C., and Dalton, B. J.,** Interleukin-1 release by rate synovial cells is dependent on sequential treatment with γ-interferon and lipopolysaccharide, *Arthritis Rheum.,* 33, 261, 1990.

33. **Hazuda, D. J., Lee, L. C., and Young, P. R.,** The kinetics of interleukin-1 secretion from activated monocytes. Differences between interleukin-1 α and interleukin-1 ß, *J. Biol. Chem.,* 263, 8473, 1988.

34. **Bandara, G., Georgescu, H. I., McIntyre, L., and Evans, C. H.,** Preferential secretion of IL-1α and retention of IL-1ß by human synovial fibroblasts, *Calc. Tiss. Int.,* 50 (Suppl. 1), A33, 1992.

35. **Firestein, G. S., Alvaro-Garcia, J. M., and Maki, R.,** Quantitative analysis of cytokine gene expression in rheumatoid arthritis, *J. Immunol.,* 44, 3347, 1990.

36. **Buchan, G., Barrett, K., Turner, M., Chantry, D., Baini, R. N., and Feldmann, M.,** Interleukin-1 and tumour necrosis factor mRNA expression in rheumatoid arthritis: prolonged production of IL-1α, *Clin. Exp. Immunol.,* 73, 449, 1988.

37. **Miyasaka, N., Sato, K., Goto, M., Sasano, M., Natsuyama, M., Inoue, K., and Nishioka, K.,** Augmented interleukin-1 production and HLA-DR expression in the synovium of rheumatoid arthritis patients: possible involvement in joint destruction, *Arthritis Rheum.,* 31, 480, 1988.

38. **Westacott, C. I., Whincher, J. T., Barnes, I. C., Thompson, D., Swan, A. J., and Dieppe, P. A.,** Synovial fluid concentration of five different cytokines in rheumatic diseases, *Ann. Rheum. Dis.,* 49, 676, 1990.

39. **Ollivierre, F., Gubler, U., Towle, C. A., Laurencin, C., and Treadwell, B. V.,** Expression of IL-1 genes in human and bovine chondrocytes: a mechanism for autocrine control of cartilage matrix degradation, *Biochem. Biophys. Res. Commun.,* 141, 904, 1986.

40. **Tiku, K., Tiku, M. L., and Skosey, J. L.,** Normal human neutrophils are a source of a specific interleukin-1 inhibitor, *J. Immunol.,* 136, 3677, 1986.

41. **Watanabe, S., Georgescu, H. I., Kuhns, D. B., and Evans, C. H.,** Chondrocyte activation by a putative interleukin-1 derived from lapine polymorphonuclear leukocytes, *Arch. Biochem. Biophys.,* 270, 69, 1989.

42. **Meats, J. E., McGuire, M. B., and Russell, R. G. G.,** Human synovium releases a factor which stimulates chondrocyte production of PGE and plasminogen activator, *Nature (London),* 286, 891, 1980.

43. **Watanabe, S., Georgescu, H. I., Mendelow, D., and Evans, C. H.,** Chondrocyte activation in response to factor(s) produced by a continuous line of lapine synovial fibroblasts, *Exp. Cell Res.,* 167, 218, 1986.

44. **Evans, C. H.,** The role of proteinases in cartilage destruction, in *Drugs in Inflammation,* Parnham, M. J., Bray, M. A., and Van den Berg, W. B., Eds., Birkhauser Verlag, Basel, 1991, 135.

45. **Lin, C. W., Georgescu, H. I., Phillips, S. I., and Evans, C. H.,** Cycloheximide inhibits the induction of collagenase mRNA in chondrocytes exposed to synovial factors or recombinant interleukin-1, *Agents Actions,* 27, 445, 1989.

46. **Stephenson, M. L., Goldring, M. B., Birkhead, J. R., Krane, S. M., Rahmsdorf, H. J., and Angel, P.,** Stimulation of procollagenase synthesis parallels increases in cellular procollagenase mRNA in human articular chondrocytes exposed to recombinant interleukin-1ß or phorbol esters, *Biochem. Biophys. Res. Commun.,* 144, 583, 1987.

47. **Lin, C. W., Phillips, S. L., Brinckerhoff, C. E., Georgescu, H. I., Bandara, G., and Evans, C. H.,** Induction of collagenase mRNA in lapine articular chondrocytes by synovial factors and interleukin-1, *Arch. Biochem. Biophys.,* 261, 351, 1988.

48. **Martel-Pelletier, J., Zafarullah, M., Kodama, S., and Pelletier, J. P.,** In vitro effects of interleukin 1 on the synthesis of metalloproteinases, TIMP, plasminogen activators and inhibitors in human articular cartilage, *J. Rheumatol.,* 18 (Suppl. 27), 80, 1991.

49. **Murphy, G., Reynolds, J. J., and Werb, Z.,** Biosynthesis of tissue inhibitor of metalloproteinases by human fibroblasts in culture: stimulation by 12-0-tetradecanoylphorbol-13-acetate and interleukin-1 in parallel with collagenase, *J. Biol. Chem.,* 260, 3079, 1985.

50. **Tyler, J. A.,** Chondrocyte-mediated depletion of articular cartilage proteoglycans in vitro, *Biochem. J.,* 225, 493, 1985.

51. **Saklatvala, J.,** Tumour necrosis factor α stimulates resorption and inhibits synthesis of proteoglycan in cartilage, *Nature (London),* 322, 547, 1986.

52. **Campbell, I. K., Piccoli, D. S., Roberts, M. J., Muirden, K. D., and Hamilton, J. A.,** Effects of tumor necrosis factor α and ß on resorption of articular cartilage and production of plasminogen activator by human articular chondrocytes, *Arthritis Rheum.,* 33, 542, 1990.

53. **Bunning, R. A. D. and Russell, R. G. G.,** The effect of tumor necrosis factor α and γ-interferon on the resorption of human articular cartilage and on the production of prostaglandin E and of caseinase activity by human articular chondrocytes, *Arthritis Rheum.,* 32, 780, 1989.

54. **Bandara, G., Lin, C. W., Georgescu, H. I., and Evans, C. H.,** The synovial activation of chondrocytes: evidence for complex cytokine interactions involving a possible novel factor, *Biochim. Biophys. Acta,* in press, 1992.

55. **Lefebvre, V., Peeters-Joris, C., and Vaes, G.,** Modulation by interleukin-1 and tumor necrosis factor α of production of collagenase, tissue inhibitor of metalloproteinases and collagen types in differentiated and dedifferentiated articular chondrocytes, *Biochim. Biophys. Acta,* 1052, 366, 1990.

56. **Henderson, B. and Pettipher, E. R.,** Arthritogenic actions of recombinant IL-1 and tumour necrosis factor α in the rabbit: evidence for synergistic interactions between cytokines in vivo, *Clin. Exp. Immunol.,* 75, 306, 1989.

57. **Sazne, T., Palladino, M. A., Heinegard, D., Talal, N., and Wollheim, F. A.,** Detection of tumor necrosis factor α but not tumor necrosis factor ß in rheumatoid arthritis synovial fluid and serum, *Arthritis Rheum.,* 30, 864, 1988.

58. **Aggarwal, B. B., Eessalu, T. E., and Hass, P. E.,** Characterization of receptors for human tumor necrosis factor and their regulation by γ-interferon, *Nature (London),* 318, 665, 1985.

59. **Sung, K., Mendelow, D., Georgescu, H. I., and Evans, C. H.,** Characterisation of chondrocyte activation in response to cytokines synthesised by a synovial cell line, *Biochim. Biophys. Acta,* 971, 148, 1988.

60. **Georgescu, H. I., Mendelow, D., and Evans, C. H.,** HIG-82: an established line from rabbit periarticular soft tissue, which retains the "activatable" phenotype, *In Vitro,* 24, 1015, 1988.

61. **Bandara, G., Lin, C. W., Georgescu, H. I., Mendelow, D., and Evans, C. H.,** Chondrocyte activation by interleukin-1: analysis of the synergistic properties of fibroblast growth factor and phorbol myristate acetate, *Arch. Biochem. Biophys.,* 274, 539, 1989.

62. **Phadke, K.,** Fibroblast growth factor enhances the interleukin-1 mediated chondrocyte protease release, *Biochem. Biophys. Res. Commun.,* 142, 448, 1987.

63. **Stevens, P. and Shatzen, E. M.,** Synergism of basic fibroblast growth factor and interleukin-1ß to induce articular cartilage - degradation in the rabbit, *Agents Actions,* 34, 217, 1991.

64. **Melnyk, V. O., Shipley, G. D., Sternfeld, M. D., Sherman, L., and Rosenbaum, J. T.,** Synoviocytes synthesize, bind, and respond to basic fibroblast growth factor, *Arthritis Rheum.,* 33, 493, 1990.

65. **Hamerman, D., Taylor, S., Kirschenbaum, I., Klagsbrun, M., Raines, E. W., Ross, R., and Thomas, K. A.,** Growth factors with heparin binding affinity in human synovial fluid, *Proc. Soc. Exp. Biol. Med.,* 186, 384, 1987.

66. **Chandrasekhar, S. and Harvey, A. K.,** Induction of interleukin-1 receptors on chondrocytes by fibroblast growth factor: a possible mechanism for modulation of interleukin-1 activity, *J. Cell Physiol.,* 138, 236, 1989.

67. **Remmers, E. F., Sano, H., Lafyatis, R., Case, J. P., Kumkumian, G. K., Hla, T., Maciag, T., and Wilder, R. L.,** Production of platelet derived growth factor ß chain (PDGF-ß/c-sis) mRNA and immunoreactive PDGF ß-like polypeptide by rheumatoid synovium: co-expression with heparin binding acidic fibroblast growth factor-1, *J. Rheumatol.,* 18, 7, 1991.

68. **Smith, R. J., Justen, J. J., Sam, L. M., Rohloff, N. A., Ruppel, P. L., Brunden, M. N., and Chin, J. E.,** Platelet-derived growth factor potentiates cellular responses in articular chondrocytes to interleukin-1, *Arthritis Rheum.,* 34, 697, 1991.

69. **Kumkumian, G. K., Lafyatis, R., Remmers, E. F., Case, J. P., Kim, S.-J., and Wilder, R. L.,** Platelet-derived growth factor and IL-1 interactions in rheumatoid arthritis: regulation of synoviocyte proliferation, prostaglandin production and collagenase transcription, *J. Immunol.,* 143, 833, 1989.

70. **McCollum, R., Martel-Pelletier, J., DiBattista, J., and Pelletier, J. P.,** Regulation of interleukin-1 receptors in human articular chondrocytes, *J. Rheumatol.,* 18 (Suppl. 27), 85, 1991.

71. **Wellmitz, G., Petzold, E., Jentzsch, K. D., Heder, G., and Buntrock, P.,** The effect of brain factors with fibroblast growth factor activity on regeneration and differentiation of articular cartilage, *Exp. Pathol.,* 18, 282, 1980.

72. **Cuevas, P., Burgos, J., and Baird, A.,** Basic fibroblast growth factor (FGF) promotes cartilage repair in vivo, *Biochem. Biophys. Res. Commun.,* 156, 611, 1988.

73. **Chandrasekhar, S. and Harvey, A. K.,** Transforming growth factor-beta is a potent inhibitor of IL-1 induced protease activity and cartilage proteoglycan degradation, *Biochem. Biophys. Res. Commun.,* 157, 1352, 1988.

74. **Overall, C. M., Wrana, J. L., and Sodek, J.,** Independent regulation of collagenase, 72-kDa progelatinase, and metalloendoproteinase inhibitor expression in human fibroblasts by transforming growth factor-beta, *J. Biol. Chem.,* 264, 1860, 1989.

75. **Andrews, H. J., Edwards, T. A., Cawston, T. E., and Hazelman, B. L.,** Transforming growth factor-beta causes partial inhibition of interleukin-1-stimulated cartilage degradation in vitro, *Biochem. Biophys. Res. Commun.,* 162, 144, 1989.

76. **Morales, T. I. and Roberts, A. B.,** Transforming growth factor ß regulates the metabolism of proteoglycans in bovine cartilage organ cultures, *J. Biol. Chem.,* 263, 12828, 1988.

77. **Lafyatis, R., Thompson, N. L., Remmers, E. F., Flanders, K. C., Roche, N. S., Kim, S. J., Case, J. P., Spron, M. B., Roberts, A. B., and Wilder, R. L.,** Transforming growth factor-ß production by synovial tissues from rheumatoid patients and streptococcal cell wall arthritic rats: studies on secretion by synovial fibroblast-like cells and immunohistologic localization, *J. Immunol.,* 143, 1142, 1989.

78. **Brennan, F. M., Chantry, D., Turner, M., Foxwell, B., Maini, R., and Feldmann, M.,** Detection of transforming growth factor-beta in rheumatoid arthritis synovial tissue: lack of effect on spontaneous cytokine production in joint cell cultures, *Clin. Exp. Immunol.,* 81, 278, 1990.

79. **Fava, R., Olsen, N., Keski-Oja, J., Moses, H., and Pincus, T.,** Active and latent forms of TGF-ß activity in synovial effusions, *J. Exp. Med.,* 169, 291, 1989.

80. **Brandes, M. E., Aller, J. B., Ogawa, Y., and Wahl, S. M.,** Transforming growth factor ß$_1$ suppresses acute and chronic arthritis in experimental animals, *J. Clin. Invest.,* 87, 1108, 1991.

81. **Eisenberg, S. P., Evans, R. J., and Arend, W. P., et al.,** Primary structure and functional expression from complementary DNA of a human interleukin-1 receptor antagonist, *Nature (London),* 343, 341, 1990.
82. **Larrick, J.,** Native interleukin-1 inhibitors, *Immunol. Today,* 10, 61, 1990.
83. **Tyler, J. A.,** Articular cartilage cultured with catabolin (pig interleukin-1) synthesizes a decreased number of normal proteoglycan molecules, *Biochem. J.,* 227, 869, 1985.
84. **Nietfeld, J. J., Wilbrink, B., Helle, M., Van Roy, J. L. A. M., Denootter, W., Swaak, A. J. G., and Huber-Bruning, O.,** Interleukin-1 induced interleukin-6 is required for the inhibition of proteoglycan synthesis by interleukin-1 in human cartilage, *Arthritis Rheum.,* 33, 1695, 1990.
85. **Seckinger, P., Yaron, I., Meyer, F. A., Yaron, M., and Dayer, J. M.,** Modulation of the effects of interleukin-1 on glycosaminoglycan synthesis by the urine-derived interleukin-1 inhibitor but not by interleukin-6, *Arthritis Rheum.,* 33, 1807, 1990.
86. **Kandel, R. A., Petelycky, M., Dinarello, C. A., Minden, M., Pritzker, K. P. H., and Cruz, T. E.,** Comparison of the effect of interleukin-6 and interleukin-1 on collagenase and proteoglycan production by chondrocytes, *J. Rheumatol.,* 17, 953, 1990.
87. **Lippiello, L., Yamamoto, K., Robinson, D., and Mankin, H. J.,** Involvement of prostaglandins from rheumatoid synovium in inhibition of articular cartilage metabolism, *Arthritis Rheum.,* 21, 909, 1978.
88. **Fulkerson, J. P. and Damiano, P.,** Effect of prostaglandin E_2 on adult pig articular cartilage slices in culture, *Clin. Orthop. Rel. Res.,* 179, 266, 1983.
89. **Brinckerhoff, C. E., McMillan, R. M., Fahey, J. V., and Harris, E. D.,** Collagenase production by synovial fibroblasts treated with phorbol myristate acetate, *Arthritis Rheum.,* 22, 1109, 1979.
90. **Pettipher, E. R., Henderson, B., Moncada, S., and Higgs, G. A.,** Leucocyte infiltration and cartilage proteoglycan loss in immune arthritis in the rabbit, *Br. J. Pharmacol.,* 95, 169, 1988.
91. **Hoffstein, S. T., Gennaro, D. E., and Meunier, P. A.,** Cytochemical demonstration of constitutive H_2O_2 production by macrophages in synovial tissue from rates with adjuvant arthritis, *Am. J. Pathol.,* 130, 120, 1988.
92. **Bates, E. J., Johnson, C. C., and Lowther, D. A.,** Inhibition of proteoglycan biosynthesis by hydrogen peroxide in cultured bovine articular cartilage, *Biochim. Biophys. Acta,* 838, 221, 1985.
93. **Baker, M. S., Feigan, J., and Lowther, D. A.,** The mechanism of chondrocyte hydrogen peroxide damage. Depletion of intracellular ATP due to suppression of glycolysis caused by oxidation of glyceraldehyde-3-phosphate dehydrogenase, *J. Rheumatol.,* 16, 7, 1989.
94. **Allen, R. E., Blake, D. R., Nazhat, N. B., and Jones, P.,** Superoxide radical generation by inflamed human synovium after hypoxia, *Lancet,* 1, 282, 1989.
95. **Levick, J. R.,** Hypoxia and acidosis in chronic inflammatory arthritis: relation to vascular supply and dynamic effusion pressure, *J. Rheumatol.,* 17, 579, 1990.
96. **Burkhardt, H., Schwingel, M., Menninger, H., McCartney, H. W., and Tschesche, H.,** Oxygen radicals as effectors of cartilage destruction, *Arthritis Rheum.,* 29, 379, 1986.
97. **Morales, T. I. and Hascale, V. C.,** Factors involved in the regulation of proteoglycan metabolism in articular cartilage, *Arthritis Rheum.,* 32, 1197, 1989.
98. **Schalkwijk, J., Joosten, L. A. B., Van Der Berg, W. B., Van Wyk, J. J., and Van de Putte, L. B. A.,** Insulin-like growth factor stimulation of chondrocyte proteoglycan synthesis by human synovial fluid, *Arthritis Rheum.,* 32, 66, 1989.
99. **Elford, P. R. and Lamberts, S. W. J.,** Contrasting modulation by transforming growth factor-ß-1 of insulin-like growth factor-1 production in osteoblasts and chondrocytes, *Endocrinology,* 127, 1635, 1990.
100. **Redini, F., Galera, P., Mauriel, A., Loyau, G., and Pijol, J. P.,** Transforming growth factor ß stimulates collagen and glycosaminoglycan biosynthesis in cultured rabbit articular chondrocytes, *FEBS Lett.,* 234, 172, 1988.

101. **O'Keefe, R. J., Puzas, J. E., Brand, J. S., and Rosier, R. N.,** Effects of transforming growth factor-ß on matrix synthesis by chick growth plate chondrocytes, *Endocrinology,* 122, 2953, 1988.

102. **Skantze, K. A., Brinckerhoff, C. E., and Collier, J. P.,** Use of agarose culture to measure the effect of transforming growth factor ß and epidermal growth factor on rabbit articular chondrocytes, *Cancer Res.,* 45, 4416, 1985.

103. **Osburn, K. D., Trippel, S. B., and Mankin, H. J.,** Growth factor stimulation of adult articular cartilage, *J. Orthop. Res.,* 7, 35, 1989.

104. **Inoue, H., Kato, Y., Iwanoto, M., Hiraki, Y., Sakuda, M., and Suzuki, F.,** Stimulation of cartilage-matrix proteoglycan synthesis by morphologically transformed chondrocytes grown in the presence of fibroblast growth factor and transforming growth factor-beta, *J. Cell. Physiol.,* 138, 329, 1989.

105. **Hamerman, D., Sasse, J., and Klagsbrun, M. A.,** Cartilage-derived growth factor enhances hyaluronate synthesis and diminishes sulfated glycosaminoglycan synthesis in chondrocytes, *J. Cell. Physiol.,* 127, 317, 1986.

106. **Tyler, J. A.,** Insulin-like growth factor 1 can decrease degradation and promote synthesis of proteoglycan exposed to cytokines, *Biochem. J.,* 260, 543, 1989.

107. **Van den Berg., W. B., Joosten, L. A. B., Schalkwijk, J., Van de Loo, F. A. J., and Van Beuningen, H. M.,** Mechanisms of cartilage destruction in experimental arthritis: lack of IGF-1 responsiveness, in *Therapeutic Approaches to Inflammatory Diseases,* Lewis, A. J., Doherty, N. S., and Ackerman, N. R., Eds., Elsevier, Amsterdam, 1989, 47.

108. **Jones, K. L. and Addison, J.,** Pituitary fibroblast growth factor as a stimulator of growth in cultured rabbit articular chondrocytes, *Endocrinology,* 9, 359, 1975.

109. **Soder, O. and Madsen, K.,** Stimulation of chondrocyte DNA synthesis by interleukin-1, *Br. J. Rheumatol.,* 27, 21, 1988.

110. **Chin, J. E. and Lin, Y.,** Effects of recombinant human interleukin-1ß on rabbit articular chondrocytes, *Arthritis Rheum.,* 31, 1290, 1988.

111. **Vivien, D., Galera, P., Lebrun, E., Loyau, G., and Pujol, J.-P.,** Differential effects of transforming growth factor-ß and epidermal growth factor on the cell cycle of cultured rabbit articular chondrocytes, *J. Cell. Physiol.,* 143, 534, 1990.

112. **O'Keefe, R. J., Puzas, J. E., Brand, J. S., and Rosier, R. N.,** Effect of transforming growth factor-ß on DNA synthesis by growth plate chondrocytes: modulation by factors present in serum, *Calcif. Tissue Int.,* 43, 352, 1988.

113. **Rosier, R. N., O'Keefe, R. J., Crabb, I. D., and Puzas, J. E.,** Transforming growth factor beta: an autocrine regulator of chondrocytes, *Connect. Tissue Res.,* 20, 295, 1989.

114. **Prins, A. P. A., Lipman, J. M., and Sokoloff, L.,** Effects of purified growth factors on rabbit articular chondrocytes in monolayer culture, *Arthritis Rheum.,* 25, 1217, 1982.

115. **Stadler, J., Stefanovic-Racic, M., Billiar, T. R., Curran, R. D., McIntyre, L. A., Georgescu, H. I., Simmons, R. L., and Evans, C. H.,** Articular chondrocytes synthesize nitric oxide in response to cytokines and lipopolysaccharide, *J. Immunol.,* 147, 3915, 1991.

116. **Chang, J., Gilman, S. C., and Lewis, A. J.,** Interleukin 1 activates phospholipase A_2 in rabbit chondrocytes: a possible signal for IL-1 action, *J. Immunol.,* 136, 1283, 1986.

117. **Raz, A., Wyche, A., Siegel, N., and Needleman, P.,** Regulation of fibroblast cyclooxygenase synthesis by interleukin-1, *J. Biol. Chem.,* 263, 3022, 1988.

118. **Arner, E. and Pratta, M. A.,** Independent effects of interleukin-1 on proteoglycan breakdown, proteoglycan synthesis, and prostaglandin E_2 release from cartilage in organ culture, *Arthritis Rheum.,* 32, 288, 1989.

119. **Ellingsworth, L. R., Brennan, J. E., Fok, K., Rosen, D. M., Bentz, H., Piez, K. A., and Seyedin, S. M.,** Antibodies to the N-terminal portion of cartilage-inducing factor A and transforming growth factor ß, *J. Biol. Chem.,* 261, 12362, 1986.

120. **Stockwell, R. A.,** *Biology of Cartilage Cells,* Cambridge University Press, Cambridge, 1979, 81.

121. **Yamamoto, T. and Gay, C. V.,** Ultrastructural analysis of cytochrome oxidase in chick epiphyseal growth plate cartilage, *J. Histochem. Cytochem.,* 36, 1161, 1988.
122. **Stefanovic-Racic, M., Stadler, J., and Evans, C. H.,** unpublished observations.
123. **Tiku, M. L.,** Production of hydrogen peroxide by rabbit articular chondrocytes, *J. Immunol.,* 145, 690, 1990.
124. **Ahmadzadeh, N., Shingu, M., and Nobunaga, M.,** The effect of recombinant tumor necrosis factor-α on superoxide and metalloproteinase production by synovial cells and chondrocytes, *Clin. Exp. Rheumatol.,* 8, 387, 1990.
125. **Burmester, G. R., Menche, D., Merryman, P., Klein, M., and Winchester, R.,** Application of monoclonal antibodies to the characterization of cells eluted from human articular cartilage: expression of I_a antigens in certain diseases and identification of an 85-kd cell surface molecule accumulated in the pericellular matrix, *Arthritis Rheum.,* 26, 1187, 1983.
126. **Jahn, B., Burmester, G. R., Schmid, H., Weseloh, G., Rohwer, P., and Kalder, J. R.,** Changes in cell surface antigen expression on human articular chondrocytes induced by gamma-interferon: induction of I_a antigens, *Arthritis Rheum.,* 30, 64, 1987.
127. **Quintavalla, J. C., Robertson, F. M., and Beaauris, A. J., et al.,** Interferon-γ differentially regulates IL-1 induction of stromelysin in HLA-DR(+) vs. HLA-DR(–) chondrocytes, Abstr. 63, *5th Int. Conf. Inflamm. Res. Assoc.,* New Haven, PA, 1990.
128. **Brinckerhoff, C. E., Mitchell, T. I., Karmilowicz, M. J., Klure-Beckerman, B., and Benson, M. D.,** Autocrine induction of collagenase by serum amyloid A-like and ß_2-microglobulin-like proteins, *Science,* 243, 655, 1989.
129. **Golds, E. E., Mason, P., and Nyirkos, P.,** Inflammatory cytokines induce synthesis and secretion of gro protein and a neutrophil chemotactic factor but not ß_2-microglobulin in human synovial cells and fibroblasts, *Biochem. J.,* 259, 585, 1989.
130. **Watson, M. L., Lewis, G. P., and Westwick, J.,** Neutrophil stimulation by recombinant cytokines and a factor produced by IL-1 treated synovial cell cultures, *Immunology,* 65, 567, 1988.
131. **Katz, Y., and Strunk, R. C.,** Synovial fibroblast-like cells synthesize seven proteins of the complement system, *Arthritis Rheum.,* 31, 13651, 1988.
132. **Davatelis, G., Wolpe, S. D., Sherry, B., Dayer, J. M., Chicheportiche, R., and Cerami, A.,** Macrophage inflammatory protein-1: a prostaglandin-independent endogenous pyrogen, *Science,* 243, 1066, 1989.

Chapter 9

ARTICULAR CARTILAGE REPAIR: POTENTIAL ROLE OF GROWTH AND DIFFERENTIATION FACTORS

Frank P. Luyten and A.H. Reddi

TABLE OF CONTENTS

I. INTRODUCTION

Osteoarthritis is the most common of the various arthritic disorders.[1,2] The tissues involved in this slow and destructive process are mainly articular cartilage and subchondral bone. While it is well known that bone has an impressive potential for regeneration, cartilage is considered to have a limited capacity for repair. Thus, the study of the biological basis of cartilage repair and regeneration is a scientific challenge with direct implications for potential therapeutic approaches for resurfacing of joints. The aim of this chapter is to present a concise current status and explore potential future approaches.

II. CARTILAGE INJURY AND IMPERFECT REPAIR

Injury to cartilage initiates a specific reparative response. Partial thickness defects, restricted to the articular cartilage layer, result in the loss of noncollagenous matrix. These lesions can lead to complete repair of the damaged matrix. In the more severe cases, where there is damage of the fibrillar network and cell death, the articular cartilage does not heal. Local limited cell proliferation and increased matrix synthesis are attempts of the chondrocytes to repair their matrix integrity. However, with disruption of the fibrillar network, they never result in the restoration *ad integrum* of the tissue.[3-8]

Full-thickness defects involve both cartilage and subchondral bone and marrow.[4,5,9-11] As demonstrated by drilling holes in the articular cartilage of rabbit knee joints, these defects undergo repair which is satisfactory 2 to 3 months following injury. A new layer of bone and cartilage is formed and, although the replacement tissue appears normal histologically, the macromolecular organization and the biochemical characteristics of the matrix are imperfect. The persistence of high levels of type I collagen[12,13] and the substitution of the cartilage specific proteoglycans by other types such as dermatan sulfate containing proteoglycans[8] illustrates this imperfect healing. This culminates in a repair tissue with fibrillations and extensive degenerative changes after 3 to 6 months, and complete loss of the integrity of the tissue 10 to 12 months after the initiation of the repair process.[14]

Nevertheless, the repair phenomena in both partial- and full-thickness lesions are an indication that a detailed study of these processes will provide new insights and potential methods of inducing the restoration of articular surfaces. One approach is the study of the role of growth and differentiation factors in articular cartilage repair. Growth and differentiation factors stimulate cell proliferation and matrix synthesis, and decrease matrix degradation. They also have the potential to influence cellular migration and differentiation, enabling progenitor cells (from marrow, synovial membrane, periost, underlying bone, or cartilage) to develop into mature articular chondrocytes capable of producing a functional repair tissue. It is likely that factors emanating from the subchondral bone may trigger the responding cells into cartilage differentiation. We have therefore focused on differentiation factors of the underlying bone, especially osteogenin and related bone morphogenetic proteins.

III. CHARACTERIZATION OF OSTEOGENIN/BONE MORPHOGENETIC PROTEINS IN SUBCHONDRAL BONE

It is well known that demineralized bone matrix is a repository of growth and differentiation factors and has the potential to induce locally, in heterotopic sites, new cartilage and bone formation.[15,16] This cartilage and bone inductive activity, named osteogenin, can be dissociatively extracted from demineralized bone matrix, and reconstituted with the collagenous bone matrix (from which all the activity has been removed) to induce *in vivo de novo* cartilage and bone.[17] This reconstitution assay opened the way to the further purification and characterization of cartilage and bone-inducing proteins from bone matrix. Osteogenin was isolated by heparin affinity, hydroxylapatite, and molecular sieve chromatography.[18] The activity was purified to homogeneity by SDS-polyacrylamide gel electrophoresis followed by gel elution, and the sequences of several tryptic peptides have been determined.[19] The sequences were similar to the amino acid sequence deduced from the cDNA clone of bone morphogenetic protein-3 (BMP-3).[20] The carboxy-terminal quarter of osteogenin (BMP-3) has sequence identity to the corresponding regions of two related proteins BMP-2A and BMP-2B (renamed BMP-4), showing 49 and 48% sequence identity, respectively.[20] The bone morphogenetic proteins (now containing 6 members)[21] are members of the TGF-ß superfamily, by virtue of the location of seven highly conserved cysteines in their carboxy-terminal quarter. The recently described osteogenic proteins OP-1 and OP-2 appear to be identical to BMP-7 and BMP-2, respectively.[22]

IV. *IN VITRO* ACTIVITY OF OSTEOGENIN AND RELATED BONE MORPHOGENETIC PROTEINS

A. MONOLAYER CULTURES
Osteogenin and related BMPs initiate the formation of cartilage and bone formation *in vivo*. To better understand the mechanisms of action of this family of proteins, and particularly to define their possible role in joint repair, we have investigated the effects of osteogenin (BMP-3) on primary cultures of skeletal cells *in vitro*. Osteogenin stimulated alkaline phosphatase activity and collagen synthesis in rat periosteal cells, rat calvarial osteoblasts, and mouse MC3T3-E1 cells.[23,24] A slight increase in cell proliferation and a profound stimulation of proteoglycan synthesis in fetal rat chondroblasts and in rabbit articular chondrocytes was observed.[23] The increased cell proliferation may be in a subpopulation of chondrocytes. In addition, osteogenin and BMP-2B stimulate cartilage matrix synthesis in micromass cultures of chick limb bud cells.[25,26]

The monolayer culture system has a number of drawbacks, the major one being the rapid dedifferentiation of the chondrocytes.[27] The cartilage phenotype is unstable, with decline and eventual loss of type II collagen synthesis[28] and of cartilage proteoglycan.[29] For this reason, one has to study the response of chondrocytes to exogenous factors in other systems. The agarose cultures and the explant culture system are useful to minimize the usual phenotypic drift of cells in monolayers.

B. AGAROSE CULTURE SYSTEM

Chondrocytes have been shown to be capable of growth in soft agar and agarose.[30,31] Chondrocytes which lose their phenotype as a result of passaging in culture regain their cartilage characteristics after transfer to agarose-containing medium supplemented with 10% fetal bovine serum. We have investigated the possibility that specific growth factors can replace this serum effect. Reexpression of the cartilage phenotype and clonal growth in agarose cultures are relevant to cartilage repair, as fibrocartilaginous degeneration and cluster formation have been described in osteoarthritic cartilage.[27]

The original culture technique in agarose has been described previously.[31-33] Briefly, the bottom of 24-well tissue culture plates are coated with 1% aqueous agarose (high Tm agarose) which, after gelation, is overlaid with 0.5% upper gel solution (low Tm agarose) in basal medium containing chondrocytes or dedifferentiated chondrocytes. Medium containing the appropriate agents is then added on top of this culture. The formation of colonies and the production of cartilage-specific collagen (type II) and proteoglycans are determined as markers for expression of the chondrocytic phenotype. We have been using this model to study the reexpression of the cartilage phenotype by dedifferentiated rabbit articular chondrocytes obtained after serial passage for four times. A serum-free, chemically defined medium was developed,[33] allowing the study of the effect of individual growth factors in the system. The data showed that highly purified osteogenin, in the presence of platelet-derived growth factor (PDGF), epidermal growth factor (EGF), and fibroblast growth factor (FGF), induced anchorage-independent growth and colony formation with the production of an extracellular matrix containing both metachromatic matrix and collagen type II (Table 1). When osteogenin was combined with transforming growth factor-ß (TGF-ß), the number of colonies formed was considerably greater than that formed in the presence of osteogenin alone. It is noteworthy that the effects were insulin dependent, and that the insulin in the basal medium could not be replaced by insulin-like growth factor I (IGF-I).[33] The number of colonies obtained in the cultures treated with the growth factor combinations reached higher values than control cultures grown in the presence of 10% fetal bovine serum.

C. THE ARTICULAR CARTILAGE EXPLANT CULTURE SYSTEM

It may be necessary to ascertain the effect of growth and differentiation factors on chondrocytes surrounded by extracellular matrix *in situ.* Explant or organ cultures of articular cartilage allows one to study the homeostatic mecha-

TABLE 1
Osteogenin Promotes the Reexpression of the
Cartilage Phenotype in Dedifferentiated Chondrocytes

	Number of colonies	
Treatment	Fresh cells	Stored cells
Basal medium	0	0
10% FBS	22 ± 2	13 ± 0
Ogn (1 µg/ml)	1.5 ± 0.3	1.8 ± 0.2
Ogn + EGF(10 ng/ml)	30 ± 1.3	0
Ogn + PDGF(4 ng/ml)	28 ± 2	6.3 ± 0.7
Ogn + FGF(10 ng/ml)	23 ± 0.9	3.3 ± 0.2
Ogn + EGF + FGF	83 ± 2	14.5 ± 2.5
Ogn + PDGF + FGF	122 ± 2	11.5 ± 0.3
Ogn + EGF + PDGF	77 ± 2	16.5 ± 0.4
Ogn + EGF + PDGF + FGF	178 ± 0.9	25.5 ± 1.7

Note: Values are means ± standard error. Ogn, osteogenin; FBS, fetal bovine serum; EGF, epidermal growth factor; PDGF, platelet derived growth factor; FGF, basic fibroblast growth factor.

From Harrison, E. T., Luyten, F. P., and Reddi, A. H., *Exp. Cell Res.*, 192, 340, 1991. With permission.

nisms of the articular chondrocytes in more physiological conditions.[27] Long-term organ culture of human normal and osteoarthritic articular cartilage has been reported,[34-37] but it is obvious that reproducible sampling of human articular tissues is a major limitation.[38] To improve the reproducibility of the studies, we have been using articular cartilage dissected from the bovine metacarpophalangeal joints as described.[39] The tissues are minced, washed in serum-free Dulbecco's Modified Eagle's Medium (DMEM), and about 100 mg of wet tissue (6 to 8 pieces) is distributed at random in 12-well tissue culture plates. The explant cultures are maintained for 7 d in DMEM supplemented with 0.2% albumin to bring them to a steady-state basal level of proteoglycan synthesis and to deplete the growth factors bound or stored in the matrix.[39,40] Thereafter, the tissues are cultured in the presence of serum, growth factors, or other experimental conditions, and proteoglycan (the major constituent of the cartilage matrix) biosynthesis and catabolism are analyzed.

In this model, 20 ng/ml IGF-I increased proteoglycans synthesis to levels comparable with media containing 20% fetal bovine serum;[40,41] higher synthetic and lower catabolic rates are achieved under these conditions such that a constant amount of total proteoglycans is maintained in the matrix for up to 6 weeks in culture.[41] The amounts of glycosaminoglycans, DNA, and hydroxyproline content on an initial wet weight basis are constant in these IGF-I-treated cultures. Analysis of the size of the newly synthesized proteoglycan, the glycosaminoglycan chain size, and the glycosaminoglycan type did not reveal any difference when

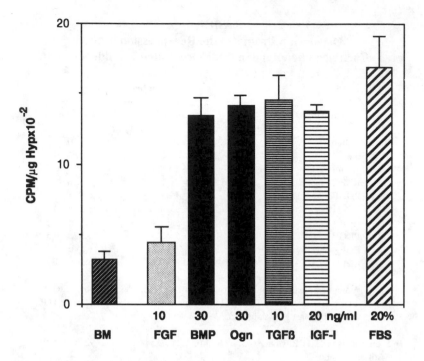

FIGURE 1. Effect of growth and differentiation factors on proteoglycan synthesis in cartilage explants. Triplicate cultures of bovine articular cartilage explants were grown for 7 d in Basal Medium (BM = Dulbecco's Modified Eagle's Medium with 0.2% albumin), in BM in the presence of basic fibroblast growth factor (FGF), of bone morphogenetic protein-2B (BMP), of Osteogenin (Ogn), of transforming growth factor-ß (TGF-ß), insulin-like growth factor-I (IGF-I) at the indicated doses, or 20% fetal bovine serum (FBS). The tissues were labeled with [^{35}S]sulfate and the amount of radiolabeled macromolecules was determined per hydroxyproline content (Hyp). The bars represent means and standard error of the mean.

compared with tissues cultured in the presence of 20% fetal bovine serum.[42] This response is achieved with tissue from immature (4 to 6 weeks) and adolescent (4 to 6 months) animals. Tissue obtained from adult steers (18 to 24 months) does respond in the same manner to IGF-I, but the response is less dramatic.[42] Additional experiments indicated that if the chondrocytes synthesize IGF-I, they do so at levels too low for the maintenance of a steady-state condition; indeed, articular cartilage cultured for weeks in serum free DMEM in the presence of 0.2% albumin progressively loses proteoglycans. A similar effect has been described in the same explant culture system with TGF-ß1.[43] TGF-ß1, at doses of 10 ng/ml, maintains steady-state conditions of proteoglycans by increasing the synthesis and decreasing the catabolic rate of proteoglycans for two weeks, the maximum period during which these two parameters have been studied. Recent experiments have shown that bFGF does not affect proteoglycan synthesis or degradation in this model (Figure 1), while osteogenin (bone morphogenetic protein-3) and bone morphogenetic protein 2B stimulate proteoglycan synthesis and inhibit the degradation in a dose-dependent manner.[42] Analysis of the size of the newly synthesized proteoglycan, the glycosaminoglycan chain size, and the glycosaminoglycan type did not show any difference when compared with cartilage treated with either IGF-I or fetal bovine serum for two weeks of culture.[42]

While several growth factors, such as IGF-I and TGF-ß1, are able to increase proteoglycan synthesis to levels similar to those attained with medium containing 20% fetal bovine serum, the decrease in catabolic rate is not quite comparable with that observed in a medium with 20% fetal bovine serum. This difference could be the result of the presence of proteinase inhibitors and/or a combination of several growth factors in the serum.

The bovine articular cartilage explant culture system is a useful model to study the mechanisms underlying homeostasis.[44] It is close to *in vivo* conditions and it allows the design of a variety of experimental conditions in serum-free, chemically defined media. On the other hand, growth factors and cytokines that interact or bind to anionic cartilage proteoglycans or other glycoproteins in the matrix may potentially complicate interpretation of the results. For example, matrix-bound growth factors may be released slowly over time from a depot. Dissection of the tissue results in the disruption of the collagen network and a continuous slow release of PG into the culture medium. The model can become a more "stringent" repair system if one "pretreats" the tissues with enzymes such as trypsin[35] or interleukin-1.[45] This model can potentially be used as a screening method for possible therapeutic agents prior to *in vivo* studies in animal cartilage repair models which are cumbersome and costly.

V. CHALLENGES FOR THE FUTURE

Biological repair and regeneration of tissues is both a scientific and clinical challenge. The growing information on the processes of cellular proliferation and acquisition of the specialized phenotype has triggered the interest of both clinicians and basic scientists. It has become evident that tissue regeneration recapitulates the developmental sequences of tissue formation. These developmental processes can be influenced by several variables such as mechanical forces, extracellular matrix components, cell adhesion molecules, and, certainly, soluble signals such as growth and differentiation factors.

The study and characterization of growth and differentiation factors involved in the formation of cartilage and in the maintenance of the integrity of the cartilaginous matrix is exiting. With respect to the formation of new articular cartilage, and eventually new underlying bone, the recently discovered family of osteogenin and related bone morphogenetic proteins are certainly major candidates for a crucial role in this process. As the cartilaginous joint surface is eroded and degraded and subchondral bone sclerosis occurs, as is typically seen in osteoarthritic disease, osteophytic formation is believed to be a key phenomenon in the attempt for regeneration of the joint. It is very likely that factors such as osteogenin and related BMPs are central players in these events. Bone is a repository of growth and differentiation factors including TGF-ß1 and TGF-ß2, IGF-I and IGF- II, PDGF, EGF, aFGF and bFGF,[46-49] osteogenin, and related BMPs.[19-22,50] Most of these factors have a direct effect on cartilage metabolism, as described above. The potential of the cartilage and bone-inducing proteins is promising. Not only can they initiate chondrogenesis *in vivo,* but these differentiation factors are stimulatory in *in vitro* systems for chondrogenesis. They are crucial in the reexpression of the chondrocytic phenotype of dedifferentiated chondrocytes, and they are able to stabilize the cartilage phenotype and demonstrate a maintenance function in articular cartilage explants. Further study of these differentiation factors is necessary to discriminate between their chondrogenic, as opposed to their osteogenic potential. The systematic analysis of chondrogenic factors present in cartilage or produced by chondrocytes will certainly add important new information.

REFERENCES

1. **Howell, D. S., Mankin, H. J., Moskowitz, R. W., and Saunders, W. B.,** *Osteoarthritis: Diagnosis and Management,* 2nd ed., W. B. Saunders, Philadelphia, 1990, 1.
2. **Kelley, W. N., Harris, E. D., Ruddy, S., and Sledge, C. B.,** *Textbook of Rheumatology,* 3rd ed., W. B. Saunders, Philadelphia, 1989, 1480.
3. **Meachim, G.,** The effect of scarification on articular cartilage in the rabbit, *J. Bone Joint Surg.,* 45B, 150, 1963.
4. **Mankin, H. J.,** The reaction of articular cartilage to injury and osteoarthritis. Part I. *N. Engl. J. Med.,* 291, 1285, 1974a.
5. **Mankin, H. J.,** The reaction of articular cartilage to injury and osteoarthritis. Part II. *N. Engl. J. Med.,* 291, 1335, 1974b.
6. **Fuller, J. A. and Gadhially, F. N.,** Ultrastructural observations on surgically produced partial-thickness defects in articular cartilage, *Clin. Orthop.,* 86, 193, 1972.
7. **Gadhially, F. N., Thomas, I., Oryschak, A. F., and Lalonde, J. M.,** Long-term results of superficial defects in articular cartilage: a scanning electron-microscopic study, *J. Pathol.,* 121, 213, 1977.
8. **Rosenberg, L.,** Biological basis for the imperfect repair of articular cartilage following injury, in *Soft and Hard Tissue Repair. Biological and Clinical Aspects,* Hunt, T. K., Heppenstall, R. B., Pines, E., and Rovee, D., Eds., Praeger, New York, 1984, 143.
9. **Meachim, G. and Roberts, C.,** Repair of the joint surface from subarticular tissue in the rabbit knee, *J. Anat.,* 109, 317, 1971.
10. **Mitchell, N. and Shepard, N.,** The resurfacing of adult rabbit articular cartilage by multiple perforations through the subchondral bone, *J. Bone Joint Surg.,* 58A, 230, 1976.
11. **Mitchell, G. and Shepard, N.,** Healing of articular cartilage in intra-articular fractures in rabbits, *J. Bone Joint Surg.,* 62A, 628, 1980.
12. **Chueng, H. S., Cottrell, W. H., Stephenson, K., and Nimni, M. E.,** In vitro collagen biosynthesis in healing and normal rabbit articular cartilage, *J. Bone Joint Surg.,* 60A, 1076, 1978.
13. **Furukawa, T., Eyre, D., Koide, S., and Glimcher, M. J.,** Biochemical studies on repair cartilage resurfacing experimental defects in the rabbit knee, *J. Bone Joint Surg.,* 62A, 79, 1980.

14. **Buckwalter, J. A., Rosenberg, L., and Hunziker, E. B.**, Articular cartilage: composition structure, response to injury and methods of facilitating repair, in *Articular Cartilage and Knee Joint Function: Basic Science and Arthroscopy,* Ewing, J. E., Ed., Raven Press, New York, 1990, 19.

15. **Urist, M. R.**, Bone: formation by autoinduction, *Science,* 150, 893, 1965.

16. **Reddi, A. H. and Huggins, C. B.**, Biochemical sequences in the transformation of normal fibroblasts in adolescent rats, *Proc. Natl. Acad. Sci. U.S.A.,* 69, 1601, 1972.

17. **Sampath, T. K. and Reddi, A. H.**, Dissociative extraction and reconstitution of extracellular matrix components involved in local bone differentiation, *Proc. Natl. Acad. Sci. U.S.A.,* 78,7599, 1981.

18. **Sampath, T. K., Muthukumaran, N., and Reddi, A. H.**, Isolation of osteogenin, an extracellular matrix-associated, bone inductive protein, by heparin affinity chromatography, *Proc. Natl. Acad. Sci. U.S.A.,* 84, 7109, 1987.

19. **Luyten, F. P., Cunningham, N., Ma, S., Muthukumaran, N., Hammonds, R. G., Nevins, W. B., Wood, W. I., and Reddi, A. H.**, Purification and partial amino acid sequence of osteogenin, a protein initiating bone differentiation, *J. Biol. Chem.,* 264, 13777, 1989.

20. **Wozney, J. M., Rosen, V., Celeste, A. J., Mitsock, L. M., Whitters, M. J., Kritz,R. W., Hewick, R. M., and Wang, E. A.**, Novel regulators of bone formation: molecular clones and activities, *Science,* 242, 1528, 1988.

21. **Celeste, A. J., Iannazzi, J. A., Taylor, R. C., Hewick, R. M., Rosen, V., Wang, E. A., and Wozney, J. M.**, Identification of transforming growth factor B family members in bone-inductive protein purified from bovine bone, *Proc. Natl. Acad. Sci. U.S.A.,* 87, 9843, 1990.

22. **Sampath, T. K., Coughlin, J. E., Whetstone, R. M., Banach, D., Corbett, C., Ridge, R., Ozkaynak, E., Opperman, H., and Rueger, D. C.**, Bovine osteogenic protein is composed of dimers of OP-1 and BMP-2A, two members of the transforming growth factor-B superfamily, *J. Biol. Chem.,* 265, 13198, 1990.

23. **Vukicevic, S., Luyten, F. P., and Reddi, A. H.**, Stimulation of the expression of osteogenic and chondrogenic phenotypes in vitro by osteogenin, *Proc. Natl. Acad. Sci. U.S.A.,* 86, 8793, 1989.

24. **Vukicevic, S., Luyten, F. P., and Reddi, A. H.**, Osteogenin inhibits proliferation and stimulates differentiation in mouse osteoblast-like cells (MC3T3-E1), *Biochem. Biophys. Res. Commun.,* 166, 750, 1990.

25. **Carrington, J. L., Chen, P., Yanagishita, M., and Reddi, A. H.**, Osteogenin (bone morphogenetic protein-3) stimulates cartilage formation by chick limb bud cells in vitro, *Dev. Biol.,* 146, 406, 1991.

26. **Chen, P., Carrington, J. L., Hammonds, R. G., and Reddi, A. H.**, Stimulation of chondrogenesis in limb bud mesoderm cells by recombinant BMP-2B and modulation by TGF-ß1 and TGF-ß2, *Exp. Cell Res.,* 195, 509, 1991.

27. **Sokoloff, L.**, In vitro culture of joints and articular tissues, in *The Joints and Synovial Fluid,* Vol. 2, Sokoloff, L., Ed., Academic Press, New York, 1980, chap. 1.

28. **Layman, D. L., Sokoloff, L., and Miller, E. J.**, Collagen synthesis by articular chondrocytes in monolayer culture, *Exp. Cell Res.,* 73, 107, 1972.

29. **Srivastava, V. M. L., Malemud, D. C. J., Hough, A. J., Bland, J., and Sokoloff, L.**, Preliminary experience with cell culture of human articular chondrocytes, *Arthritis Rheum.,* 17, 165, 1974.

30. **Sokoloff, L., Malemud, C. J., Srivastava, V. M. L., and Morgan, W. D.**, In vitro culture of articular chondrocytes, *Fed. Proc. Fed. Am. Soc. Exp. Biol.,* 32, 118, 1973.

31. **Benya, P. D. and Schaffer, J. E.**, Dedifferentiated chondrocytes reexpress the differentiated collagen phenotype when cultured in agarose gels, *Cell,* 30, 215, 1982.

32. **Thompson, A. Y., Piez, K. A., and Seyedin, S. M.**, Chondrogenesis in agarose gel culture. A model for chondrogenic induction, proliferation and differentiation, *Exp. Cell Res.,* 157, 483, 1985.

33. **Harrison, E. T., Luyten, F. P., and Reddi, A. H.,** Osteogenin promotes reexpression of cartilage phenotype by dedifferentiated articular chondrocytes in serum-free medium, *Exp. Cell Res.,* 192, 340, 1991.
34. **Fukae, M., Mechanic, G. L., Adamy, L., and Schwartz, E. R.,** Chromatographically different type II collagens from human normal and osteoarthritic cartilage, *Biochem. Biophys. Res. Commun.,* 67, 1575, 1975.
35. **Verbruggen, G., Luyten, F. P., and Veys, E. M.,** Repair function in organ-cultured human cartilage. Replacement of enzymatically removed proteoglycans during long term organ culture, *J. Rheumatol.,* 4, 665, 1985.
36. **Luyten, F. P., Verbruggen, G., Veys, E. M., Goffin, E., and De Pypere, H.,** In vitro repair potential of articular cartilage. Proteoglycan metabolism in the different areas of the femoral condyles in human cartilage explants, *J. Rheumatol.,* 2, 329, 1987.
37. **Luyten, F. P., Verbruggen, G., and Veys, E. M.,** Reparative response of human articular cartilage in tissue culture. Comparison between a normal and an osteoarthritic knee of the same donor, *Clin. Exp. Rheumatol.,* 5, 103, 1987.
38. **Bayliss, M. T.,** Sampling, characterization and handling ex vivo of tissue specimens, in *Methods in Cartilage Research,* Maroudas, A. and Keuttner, K., Eds., Academic Press, San Diego, 1990, Section 1.
39. **Hascall, V. C., Handley, C. J., McQuillan, D. J., Hascall, G. K., Robinson, H. C., and Lowther, D. A.,** Effect of serum on biosynthesis of proteoglycans by bovine articular cartilage in culture, *Arch. Biochem. Biophys.,* 224, 206, 1983.
40. **McQuillan, D. J., Handley, C. J., Campbell, M. A., Bolis, S., Milway, V. E., and Herington, A. C.,** Stimulation of proteoglycan biosynthesis by serum and insulin-like growth factor-1 in cultured bovine articular cartilage, *Biochem. J.,* 240, 423, 1986.
41. **Luyten, F. P., Hascall, V. C., Nissley, S. P., Morales, T. I., and Reddi A. H.,** Insulin-like growth factors maintain steady-state metabolism of proteoglycans in bovine articular cartilage explants, *Arch. Biochem. Biophys.,* 267, 416, 1988.
42. **Luyten, F. P., Yu M. Yu, Yanagishita, M., Vukicevic, S., Hammonds, R. G., and Reddi, A. H.,** Bovine osteogenin and recombinant human bone morphogenetic protein-2B are equipotent in the maintenance of proteoglycans in bovine articular cartilage explant cultures, *J. Biol. Chem.,* 267, 3691, 1992.
43. **Morales, T. I. and Roberts, A.,** Transforming growth factor B regulates metabolism of proteoglycans in bovine cartilage organ cultures, *J. Biol. Chem.,* 263, 12828, 1988.
44. **Hascall, V. C., Luyten, F. P., Plaas, A. H., and Sandy, J. D.,** Steady state metabolism of proteoglycans in bovine articular cartilage explants, in *Methods in Cartilage Research,* Maroudas, A. and Kuettner, K., Eds., Academic Press, Orlando, FL, 1991, 112.
45. **Tyler, J. A.,** Articular cartilage cultured with catabolin pig interleukin 10 synthesizes a decreased number of normal proteoglycan molecules, *Biochem. J.,* 227, 869, 1985b.
46. **Hauschka, P. V., Mavrakos, A. E., Iafrata, M. D., Doleman, S. E., and Klagsbrun, M.,** Growth factors in bone matrix: isolation of multiple types by affinity chromatography on heparin sepharose, *J. Biol. Chem.,* 261, 12665, 1986.
47. **Canalis, E., McCarthy, T., and Centrella, M.,** Isolation of growth factors from adult bovine bone, *Calcif. Tissue Int.,* 43, 346, 1988.
48. **Seyedin, S. M., Thomas, T. C., Thompson, A. Y., Rosen, D. M., and Piez, K. A.,** Purification and characterization of two cartilage-inducing factors from bovine demineralized bone, *Proc. Natl. Acad. Sci. U.S.A.,* 82, 2267, 1985.
49. **Mohan, S., Jennings, J. C., Linkhart, T. A., and Baylink, D. J.,** Primary structure of human skeletal growth factor: homology with human IGF-II, *Biochim. Biophys. Acta,* 966, 44, 1988.
50. **Reddi, A. H., Muthukumaran, N., Ma, S., Carrington, J. L., Luyten, F. P., Paralkar, V. M., and Cunningham, N. S.,** Initiation of bone development by osteogenin and promotion by growth factors, *Connect. Tissue Res.,* 20, 303, 1989.

Chapter 10

REGULATION OF CHONDROCYTES IN AGING

Martin Lotz

TABLE OF CONTENTS

I. INTRODUCTION

Studies on aging of chondrocytes are directed at defining mechanisms responsible for age-related changes in structure and function of adult cartilage. Such changes are of interest to the understanding of chondrocyte biology and may contribute to the pathogenesis osteoarthritis, the most frequent joint disease with a strongly age-related incidence.

The first part of this chapter will summarize information on age-related phenomena in cartilage. This will concern macroscopic changes in cartilage, changes in cartilage cellularity, and in the composition of extracellular matrix. Chondrocyte secretory and proliferative function in aging are then reviewed to examine the hypothesis that altered cartilage structure is the consequence of aging of chondrocyte functions and to create a background for the discussion of regulatory mechanisms in chondrocytes that are potentially responsible for these changes. Cytokines, growth factors, and neuropeptides are discussed as constituents of the cartilage regulatory network. Receptors for these factors and signal transduction pathways will be summarized, with particular attention to events that can contribute to the age-related decline in chondrocyte proliferative potential. Several important aspects of chondrocyte function have yet to be studied with respect to changes that may occur in aging. This is, in part, related to the relatively recent identification of many of the basic regulatory mechanisms that are operative in development and homeostasis of cartilage. These aspects will be discussed in order to define the questions on aging of chondrocytes that can now be tested experimentally.

II. AGE-RELATED CHANGES IN CARTILAGE

A. MACROSCOPIC CHANGES

Articular cartilage from older individuals has a xanthous appearance, as opposed to the white color of young cartilage.[1] This may be due to increased intracellular lipofuscin. In addition, cartilage extracellular matrix constituents may also undergo nonenzymatic advanced glycosylation.[2] In this process, exposure of proteins (including collagens as well as lipids) to ambient glucose leads to the formation of highly reactive, late addition products. This occurs on proteins with long half-lives during normal aging and at an accelerated rate in diabetes. It is unknown whether this impairs the function of collagen and other proteins within cartilage extracellular matrix. The implications of advanced glycosylation of proteins for cell function has been studied in detail in mononuclear phagocytes. These cells appear to express specific receptors for advanced glycosylation products, and occupancy of these receptors with glycosylated albumin or myelin has been shown to induce the release of platelet-derived growth factor.[2]

Cartilage from older donors has reduced thickness in weight-bearing and nonweight-bearing joints.[3] Areas that are exposed to mechanical stress (such as in the knee, the contact zones of the femoral condyles with the tibial plateau, or specific sites in the patellofemoral joint) are frequently denuded. A high number of samples show fibrillations and softening of the articular surface. Fibrillated areas are characterized by a loss of proteoglycans and a discontinuity in the collagen network.[4] Many of these cases with macroscopically abnormal cartilage, as examined at autopsy, can be classified as osteoarthritis (OA), although they may not have been clinically symptomatic.

B. CARTILAGE CELLULARITY

The most striking difference in cartilage cellularity is seen between fetal and adult cartilage. Fetal epiphyseal cartilage was found to contain 24×10^6 cells per gram of tissue, while the corresponding figure for adults was only 0.4×10^6.[5] Quintero et al. analyzed the femoral condyles from 77 autopsy specimens and observed an approximately 50% decrease in cellular density in specimens from donors older than 40 years as compared to donors younger than 40 years.[6] This difference was seen in both superficial, as well as in calcified zones. The number of empty lacunae also increased as a function of aging. The same investigators, in a subsequent study, correlated osteoarthritic changes, cellularity, and aging.[7] It was found that cellularity in fissured cartilage was lower than in normal cartilage. Apparently, normal cartilage in osteoarthritic joints was also hypocellular, as compared to nonosteoarthritic samples. Vignon et al. studied the femoral head and found an age-related decrease in cell numbers that was more pronounced in the superficial as compared to the deeper areas of cartilage.[8] They also observed a correlation between reduced cellularity and increased frequency of fibrillations. Reduced cellularity is not limited to weight-bearing articular cartilage, but has also been observed in the hyaline arytaenoid cartilage.[9] Collectively, these observations point to a predisposition to matrix degeneration as a consequence of loss of chondrocytes.

C. ALTERATIONS IN CARTILAGE EXTRACELLULAR MATRIX STRUCTURE

Profound age-related changes occur in the composition of cartilage extracellular matrix. Conceptually, at least two distinct mechanisms can be proposed to account for this. Impaired chondrocyte gene expression or posttranslational processing may lead to reduced amounts and qualitvely different matrix components. Alternatively, matrix may be degraded by proteases derived from chondrocytes or synovium. The age-related changes in cartilage collagens, proteoglycans, and link proteins have been extensively reviewed.[10-12] However, since altered cartilage matrix provides a different regulatory environment for chondrocytes and defects in chondrocyte function may be responsible for this, the following paragraphs will provide a brief summary.

Proteoglycans in aging cartilage are smaller in size.[10] The total glucosaminoglycan content of human articular cartilage may not significantly change with aging. However, there is a decrease in the relative amount of chondroitin sulfate and an increase in keratan sulfate content.[13] These changes can be the result of cleavage of the C-terminal chondroitin sulfate-rich region.

The changes in chondroitin sulfate are thought to account for the decreased water content in aged cartilage. Ageing cartilage also contains increased quantities of hyaluronic acid binding protein, which is generated by proteolysis of the proteoglycan core protein.[10] Keratan sulfate-rich and chondroitin sulfate-poor proteoglycans with the composition of older cartilage can also be synthesized by chondrocytes on the same core protein as proteoglycans from younger cartilage. This suggests that the age-related changes represent a distinct pattern of core protein glycosylation. The patterns of proteoglycan sulfation can be modulated by growth factors. In mammary epithelial cells, TGF-ß has been shown to alter the relative composition of glucosaminoglycan side chains on proteoglycans.[14] In cultured rabbit chondrocytes, TGF-ß and FGF promoted a change to fibroblastic morphology, but stimulated the production of cartilage-specific, large, chondroitin sulfate-rich proteoglycans and had no effect on the production of small proteoglycans or hyaluronic acid.[15]

Cartilage hyaloran content increased from 2% in young tissue to 10% in 60 to 80-year-old patients.[16] The size of hyaloronic acid in cartilage is reduced as a function of aging during adult life. The material isolated from cartilage was, in all age groups, smaller than the size of material newly synthesized by tissue from the different age groups, suggesting postsynthetic modifications.[10]

Link proteins which stabilize the interaction of proteoglycan subunits with hyaluronic acid show increased heterogeneity and fragmentation in human aging.[17] These changes may represent defects in biosynthesis and assembly of link proteins with hyaluronic acid and proteoglycan and/or protease effects. Ageing chondrocytes appear to assemble link protein-stabilized aggregates more slowly.[18] Flannery et al.[19] presented a detailed study on the age-related changes in link protein structure and content in rabbit cartilage. They found only a very small proportion of fragmented link protein (5%) and no age-associated decrease in the overall levels.[19] This difference in aging of the same set of proteins in rabbit and human cartilage suggest that defects in link proteins occur only in species with a longer lifespan.

D. SYNTHESIS OF EXTRACELLULAR MATRIX BY AGING CHONDROCYTES

Several *in vitro* studies presented evidence that aging chondrocytes produce qualitatively and quantitatively different matrix components. In a study on auricular chondrocytes from rabbits at different ages, Madsen[20] reported that the levels of elastin, collagen, and proteoglycan decreased to 50% or even lower levels in 60-month-old, as compared to 1-week-old animals. Chondrocytes also synthesized qualitatively different proteoglycans. Proteoglycans produced by cells from older animals showed decreased size and aggregation and an increase in the amount of 6-sulfate. Ageing of bovine chondrocytes *in vivo* was associated with the synthesis of smaller proteoglycans with more keratan sulfate.[21] Chondrocytes from aged donors or cells from young rabbits that had been extensively subcultured *in vitro* produced significantly lower levels of link protein-stabilized proteoglycan aggregates[22] and this was, in part, related to lower levels of link protein synthesis.[23] A more marked difference was observed with respect to the relative levels of secretion of link protein and proteoglycan. Mature cells secreted relatively more proteoglycan as compared to link protein, and this resulted in a relative enrichment of proteoglycans in the culture medium.[18] The *in vitro* formation of collagen lattices is a complex response that involves cell attachment to fibrils via specific receptors, protein synthesis, and the expression of tractional forces. Wang et al. studied femoral rat chondrocytes and found no age-related difference in this response.

Fibronectin is an extracellular matrix component which is involved with cell migration, spreading and attachment, and regulates cell differentiation. The fibronectin gene can be alternatively spliced at three different exons and give rise to proteins of different size and function. Splice patterns change during development and in malignant transformation of cells.

Fibronectin synthesis has been studied during *in vitro* senecence of fibroblasts. At the stage where fibroblasts reach *in vitro* senescence, their cell size increases and, simultaneously with this, there is a higher amount of fibronectin mRNA per cell, as demonstrated by *in situ* hybridization. Senescent cells also produced two- to threefold more fibronectin than younger cells.[25] The splicing patterns of fibronectin mRNAs showed significant age-related changes *in vivo* and *in vitro*.[26] In general, older cells and tissues showed an increase in fibronectin mRNAs, where all three alternatively spliced exons were spliced out. The exon that encodes the connecting sequence III (IIICS) can give rise to two distinct protein sequences (CS1 and CS5) which have differential affinities for integrin receptors. The expression of the different splice forms is regulated by growth factors. Fibronectin that contains the extra domain A (ED-A) is characteristic of fibroblasts. This splice form is selectively increased by TGF-ß in cultured normal human fibroblasts.[27] Serum deprivation of young fibroblasts resulted in a decrease in fibronectin mRNAs containing the three extra domains, a pattern similar to that seen in senescent cells. One possible interpretation of this similarity is that senescent cells are unresponsive to growth factor stimulation.[26]

ED-A fibronectin represents only a very small (<2%) proportion of total cartilage fibronectin. Canine chondrocytes secreted fibronectin in culture and the relative amount of ED-A increased with time in culture and reached up to 36% of total fibronectin. This shift was inhibited by the addition of dibutyryl cyclic AMP to the chondrocyte cultures.[28] Similar to the shift from synthesis of type II to type I collagen in culture, the increase in ED-A fibronectin in cultured chondrocytes may be an indicator of chondrocyte dedifferentiation. Changes in fibronectin splice forms during aging of human chondrocytes *in vivo* and the ED-A content of aging and osteoarthritis cartilage have not been studied.

E. FORMATION OF INORGANIC PHOSPHATE AND CARTILAGE CALCIFICATION

Deposition of calcium pyrophosphate dihydrate (CPPD) crystals in articular cartilage is a common finding in the elderly.[29] It is detectable radiographically in greater than 50% of individuals at ages greater than 90 years, but rare in individuals below the age of 50. Patients with CPPD disease (pseudo-gout) have increased synovial fluid levels of inorganic phosphate (PPi).[30] Chondrocytes are the predominant intraarticular source of PPi, and it has been shown that chondrocytes and other cell types from older individuals produce it at increased levels.[31] During *in vitro* senescence of fibroblasts, an increase in PPi has also been observed. Among the metabolic pathways leading to PPi formation, the hydrolysis of nucleoside triphosphate (NTP) by the ectoenzyme nucleoside triphosphate pyrophosphohydrolase (NTPPPH) is well characterized and quantitatively important. In this reaction, hydrolysis of NTPs into the monophosphate esters results in the release of PPi. Levels of this enzyme are also elevated in cartilage and synovial fluid from patients with CPPD crystal deposition.[32] Factors responsible for these age-related changes are unknown, and analysis of the regulation of PPi formation has only recently begun. In a study on cartilage organ and chondrocyte monolayer cultures, Rosenthal et al.[33] have shown that TGF-ß is the most potent factor in the stimulation of PPi release. No other growth factor, including EGF, FGF, IGF-I, and IGF-II, was able to directly stimulate PPi, and only EGF and TGF-α synergized with TGF-ß. In these culture systems, TGF-ß induced only a modest increase in NTPPPH levels and this alone could not account for the more marked increase in PPi levels. The identification of the TGF-ß-dependent mechanism for increasing PPi will now allow analysis of the questions as to whether aging chondrocytes secrete increased levels of TGF-ß or have increased sensitivity to this factor and whether this may contribute to increased PPi in aging cartilage. The second regulatory aspect that is still unknown concerns the factor(s) that can increase NTPPPH levels in chondrocytes. The rate of PPi breakdown by pyrophosphatases and alkaline phosphatase also determines overall PPi levels and influences CPPD deposition. Levels of phosphatases in aging cartilage or synovial fluids and the factors that control the expression of these enzymes in chondrocytes have yet to be characterized.

Other types of calcium crystals [such as hydroxyapatite (HA)] are also increased in aging cartilage. Deposition of HA crystals is, in part, controlled by the vitamin K-dependent proteins of bone and cartilage, bone, and matrix gla proteins.[34] Matrix gla protein is found in cartilage and synthesized by chondrocytes *in vitro*.[35,36] Gamma carboxylation of MGP is vitamin K dependent. Inhibition of gamma-carboxylation by the vitamin K antagonist warfarin or by vitamin K deficiency results in the deposition of HA crystals and cartilage calcification.[37] There is evidence for reduced levels of gamma-carboxylation in older individuals, as shown for BGP in plasma; this was corrected by the administration of vitamin K.[38] It is possible that there are impairments in the overall levels of MGP or in the degree of gamma-carboxylation in aging cartilage, but this has not been studied. As suggested by the therapeutic modulation of gamma-carboxylation of plasma BGP, it seems possible to directly regulate chondrocyte function (gamma-carboxlation of MGP) by the administration of vitamin K.

F. CARTILAGE HOMEOSTASIS AND RESPONSES TO INJURY IN AGING

Ageing of cartilage is associated with alterations in extracellular matrix, increased deposition of calcium crystals, and decreased cellularity that have been described above. Since cartilage matrix is continuously being replaced, it is conceivable that the altered composition of matrix may, at least in part, be due to defects in their biosynthesis by chondrocytes. In this section, we will summarize the small number of studies that report on quantitative and qualitative changes in matrix biosynthesis in the aging of chondrocytes.

Chondrocytes in adult cartilage express a low level of replicative and biosynthetic activity and this probably reflects homeostatic tissue maintenance function. The relative low level of this activity is consistent with the rather slow turnover of cartilage extracellular matrix. Changes in chondrocyte adaptive function are more apparent and can be more easily studied in conditions of increased tissue turnover, such as in responses to injury. Age-related aspects in cartilage responses to injury have been investigated in models of inflammatory arthritis. Cartilage damage in antigen-induced arthritis was more severe in old, as compared to young mice. This occurred despite the fact that the severity of inflammation was somewhat lower in older animals. Older tissues appeared to have an impairment in their ability to antagonize and repair loss of cartilage matrix.[39] Upon intraarticular administration of IL-1 to young and old animals, there was a similar degree of suppression of proteoglycan synthesis. A marked difference, however, was observed, in that older chondrocytes showed a slower recovery from the IL-1-induced suppression, and, thus, in the restoration of cartilage matrix.[40] With human articular chondrocytes it has been shown that aging is associated with a decreased sensitivity of the cells to the catabolic effects of IL-1 and TNF.[41] Collectively, these studies indicate that there may be a decreased sensitivity of older chondrocytes to inflammatory stimuli or IL-1. However, the reduced repair capacity of older cells may account for persistent tissue damage.

III. REGULATION OF CHONDROCYTES IN AGING

In discussing regulation of chondrocytes in the aging process, we will review receptors for extracellular matrix antigens. Cytokines, neuropeptides, and growth factors are then reviewed as soluble chondrocyte regulatory factors.

Adult cartilage is avascular, aneural, and it does not contain lymphatics. Communication of chondrocytes is thus limited to homotypic cell-cell interactions, recognition of the extracellular matrix, and soluble regulatory signals generated by chondrocytes themselves or originating from cells in the synovial cavity and synovial membrane. Mechanical stimuli represent a critical signal for chondrocytes and the maintenance of the cartilage extracellular matrix. This is documented by the profound changes in cartilage that occur in response to joint immobilization.[42] At present, however, there is no definitive information on the mechanism of translation of mechanical stimuli into intracellular signals in chondrocytes. Cell surface integrins may be involved with this, but there is as yet no data in support of this notion.

A. MATRIX RECEPTORS

Interaction of cells with extracellular matrix is mediated by cell surface integrins.[43,44] Cartilage matrix components (such as collagen type II, fibronectin, and thrombospondin) are ligands for integrins. Integrins are heterodimers composed of an α and a ß chain. Three major families are defined on the basis of the ß chain present in the complexes.[45] We have recently characterized integrins that are displayed on chondrocytes. The most strongly expressed receptors on human articular chondrocytes from healthy adults are $\alpha5\beta1$ integrins of the very late activation antigen (VLA) family which have been shown in other cell types to serve as adhesion receptors for fibronectin; high levels of $\alpha v\beta3$ were also detected. This receptor binds diverse antigens such as vitronectin, fibronectin, osteopontin, thrombospondin, and van Willebrand factor.[44] In addition, other members of the ß1 family, including $\alpha1\beta1$, $\alpha2\beta1$, and $\alpha3\beta1$ were expressed at lower levels; these receptors bind collagen, laminin, and fibronectin. Integrins can recognize intact extracellular matrix and thus anchor cells to the tissue. These receptors also serve an important role in the response to tissue injury since they recognize degradation products of the extracellular matrix.[43,44] It has been shown in other cell systems that binding of such ligands to their integrin receptors induces functional cellular responses. Thus, collagen binding to cells can induce collagenase release[46] and, in rabbit synoviocytes, the binding of fibronectin induced the expression of both collagenase and stromelysin.[47] A further role of integrins that is now being examined concerns the organization of individual matrix proteins into functional extracellular matrix tissue. Cell surface integrins thus play a critical role in chondrocyte function and cartilage structure. Changes in the levels of their expression, or in their ability to activate cellular responses, can be the basis for a defective chondrocyte function in cartilage homeostasis and repair.

B. REGULATORY FACTORS AND THEIR RECEPTORS

Potential sources or cartilage regulatory factors are synovium, subchondral bone, and cartilage itself. Regulatory signals from synovial tissue and fluid are either generated in processes that originate outside cartilage (such as in arthritis) or in conditions where cartilage extracellular matrix has been destroyed and matrix degradation products act on synoviocytes. Whether or which signals originate from subchondral bone during adult life is not known, although there are multiple levels at which these two structures interact. One hypothesis that has generated widespread interest is that subchondral bone becomes increasingly stiff as a function of aging and that this enhances exposure of cartilage and chondrocytes to mechanical stress.[48] Unclear as to cause or consequence, microfractures are frequently seen in this situation of subchondral bone sclero-sis. In these fracture clefts, ingrowth of blood vessels into cartilage and remodeling of the osteochondroal junction can occur.[49] Fractures and neovascularization provide a potential basis for increased access of bone-derived signals to articular chondrocytes, but this has not yet been examined.

A substantial amount of information on cartilage as the source of regulatory factors has been accumulated in studies that were performed over the course of the past few years. These studies showed that chondrocytes are able to produce most of the factors that are known to modulate their function, as well as additional factors that do not directly act on cartilage but on cells in synovium and on inflammatory cells. The following discussion will address three principal groups of regulatory factors — cytokines, neuropeptides and growth factors. Cellular origin, regulation of expression, receptors, and effects on chondrocyte function will be discussed.

1. Cytokines

Interleukin-1 (IL-1) and tumor necrosis factor — Articular chondrocytes produce IL-1.[50] Expression of this cytokine in chondrocytes can be induced by IL-1 itself (unpublished), but other inducers that may be relevant to cartilage have not been characterized. IL-1 is probably one of the most important factors that cause cartilage degradation (reviewed in Reference 51). This occurs through the induction of proteases in chondrocytes and synoviocytes[52,53] and through the inhibition of glucosaminoglycan synthesis.[54] IL-1 stimulates the production of other cytokines, some of which synergize with IL-1, while others antagonize its effects. IL-1 receptor antagonist (IL-1 ra) inhibits the biological action of IL-1; its gene is related to that of IL-1 and most cell types that produce IL-1 also produce IL-1 ra.[55] Information on IL-1 ra in chondrocytes is very limited. We have been able to detect gene expression by the polymerase chain reaction, but there is no information on the relative levels of IL-1 and IL-1 ra after activation and during aging of chondrocytes. Tumor necrosis factor (TNF) is a cytokine with similar catabolic activities to IL-1 (reviewed in Reference 55a). It induces the production of proteases by chondrocytes, inhibits glucosaminoglycan synthesis, and stimulates the prodution of other cytokines. It is unknown whether chondrocytes produce TNF.

Interleukin-6 (IL-6) — IL-6 is was originally characterized as a differentiation factor for B lymphocytes and a potent inducer of the hepatic acute phase protein synthesis (reviewed in Reference 56). Human articular chondrocytes produce this cytokine in levels that are similar to its other cell sources, such as mononuclear phagocytes or fibroblasts. IL-6 synthesis in chondrocytes is induced by IL-1, TNF, and bacterial and viral products. Interestingly, IL-6 mRNA expression and secretion of the IL-6 proteins are stimulated by TGF-ß but not by PDGF, FGF, or EGF.[57] Receptors for IL-6 are expressed on cartilage cells[58] and IL-6 modulates several chondrocyte functions. In marked contrast to IL-1 and TNF, IL-6 does not induce the production of metalloproteinases, but it is a potent stimulator of the tissue inhibitor of metalloproteinases (TIMP) in chondrocytes, fibroblasts, and synoviocytes.[59] IL-6 can stimulate proliferation of chondrocytes, perichondrial cells, and chondrosarcoma cells (Reference 58 and unpublished results). Thus, IL-6 has qualitatively distinct effects on cartilage as compared to IL-1, and it may serve as a cytokine that balances the catabolic effects of IL-1.

Leukemia inhibitory factor — Leukemia inhibitory factor (LIF) is a more recently identified cytokine which inhibits differentiation of embryonal stem cells and cholinergic neurons. It stimulates the production of a similar set of hepatic acute-phase proteins that is induced by IL-6 and inhibits lipoprotein lipase (reviewed in Reference 60). Effects on connective tissue that have been thus far demonstrated include the stimulation of bone resorption by recombinant LIF *in vitro*.[61] Overexpression of LIF *in vivo* by transfected cells resulted in the development of myelofibrosis, bone resorption, and excess new bone formation.[62] Human articular chondrocytes express the LIF gene, synthesize LIF protein, and secrete the factor in a biologically active form.[63] Expression of LIF is induced by LIF, IL-1, TNF, and bacterial lipopolysaccharides. All three isoforms of TGF-ß increase LIF gene expression. The autoinduction of LIF indicated that chondrocytes are LIF responsive and presumably express LIF receptors. In beginning to characterize effects of LIF on other chondrocyte functions, we have been able to show that it induces the expression of the genes for several proinflammatory cytokines. This raises the possibility that LIF be may a chondrocyte regulator that is qualitatively similar to IL-1. Further studies will be needed in order to define whether LIF can also modulate the production of proteases and extracellular matrix.

Colony stimulating factors — Colony stimulating factors (CSF) do not have demonstrated direct effects on cartilage, but do activate cells in synovial fluid and synovial membrane. There are four genetically distinct human CSFs, G-CSF and GM-CSF, macrophage CSF, and multi-CSF (IL-3). Of these, GM-CSF and G-CSF have been shown to be produced by human articular chondrocytes.[64] IL-1 is their most potent inducer and interferons, PDGF, and FGF do not detectably stimulate them. The biological significance of chondrocyte production of GM-CSF and G-CSF is seen in chronic inflammatory arthritides where these cytokines contribute to the activation of mononoclear phagocytes and thus to cartilage destruction and chronicity of the process.

Intercrine-α/Interleukin-8 (IL-8)— The intercrine cytokine family contains at least 10 distinct cytokines which have a molecular mass of 8 to 10 kDa and are structurally related (reviewed in Reference 65). Two subfamilies have been defined and we will discuss IL-8 and CTAP III as examples of intercrine-α and MCP-1 as a member of the intercrine-ß subfamily.

IL-8 is one of the most potent chemotactic factors for neutrophils.[66] It also promotes enhanced production of oxidative and 5-lipoxygenase products, and increases surface expression of certain leukocyte adhesive proteins. Importantly, IL-8 triggers both neutrophil degranulation and neutrophil degranulation-stimulated cartilage catabolism *in vitro*.[67] In addition, IL-8 stimulates T lymphocyte chemotaxis and it activates mast cells.[66] IL-8 is proinflammatory following local injection *in vivo,* and its biological effects are long lived because of its relative resistance to proteolytic inactivation.[66] We demonstrated that human articular chondrocytes express the IL-8 gene and secrete bioactive IL-8. The best inducers of IL-8 were IL-1 and bacterial LPS. TGF-ß increased IL-8 production and, remarkably, synergized with IL-1.[68] These observations suggest a mechanism whereby chondrocytes may both directly initiate and modulate neutrophil-mediated inflammation and cartilage damage. IL-8 is related to CTAP-III,[69] a factor which has significant effects on the functional properties of chondrocytes. This raised the possibility that chondrocyte secretion of IL-8 may also have autacoid effects. However, we have observed that human recombinant IL-8 does not alter a number of chondrocyte functions (unpublished observations) and it is conceivable that chondrocytes, like fibroblasts,[70] may lack IL-8 receptors.

CTAP III has been detected in monocytes and platelets. It is not known whether it is expressed in chondrocytes, but it does stimulate glucosaminoglycan synthesis in chondrocytes.

Intercrine-β/Monocyte chemoattractant protein-1 (MCP-1) — Members of the intercrine-ß family have been shown to predominantly affect monocyte functions.[65] Human articular chondrocytes in monolayer and cartilage organ culture secreted monocyte chemotactic activity. This was increased after IL-1 stimulation and by antibody neutralization studies shown to be due to the presence of the monocyte chemoattractant protein-1 (MCP-1).[71] IL-1 stimulated chondrocytes *de novo* synthesized and secreted MCP-1 proteins which migrated at 13 and 15 kDa under reducing conditions in SDS PAGE.

The 0.7-kb MCP-1 mRNA was induced in chondrocytes after stimulation with IL-1, TNF, LPS, TGF-ß, and PDGF. Leukemia inhibitory factor (LIF) was identified as a novel inducer of MCP-1. MCP-1 mRNA induction by LIF occurred very rapidly (within 2 h) and declined by 5 h. IL-1 effects were detectable by 2 h and maximal only after 5 to 8 h, while TGF-ß caused a slow induction that was apparent by 5 h. Dexamethasone reduced MCP-1 mRNA levels induced by IL-1 and serum. Retinoic acid (RA) potentiated the induction of MCP-1 mRNA by serum, IL-1, PMA, and LPS.

Induction of chondrocyte MCP-1 gene expression cartilage organ culture was studied by *in situ* hybridization. Stimulation with IL-1 resulted in the expression of MCP-1 mRNA in superficial and in deeper layers of cartilage. LPS induction of MCP-1 mRNA was slower, but also progressed through all cell layers.

Figure 1 demonstrates the simultaneous expression of multiple cytokine genes by activated human articular chondrocytes.

2. Neuropeptides

Cartilage is not innervated. Intraarticular nerve endings are found in synovium, periosteum, joint capsule, and ligaments.[72] Chondrocyte function may be regulated by neuropeptides released from intraarticular nerve endings, but it is not clear whether neuropeptides can remain biologically active in an environment that contains degrading enzymes, whether they can migrate through the cartilage extracellular matrix, or whether neuropeptide receptors are expressed on these cells. Although these fundamental questions have not been answered, there is circumstantial evidence for a role of neural stimuli in joint development and

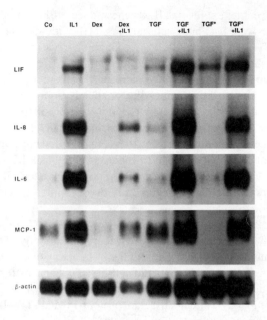

FIGURE 1. Cytokine gene expression by human articular chondrocytes. Human articular chondrocytes were stimulated with the indicated agents for 5 h. RNA was isolated, separated on agarose gels, and transferred onto nylon filters. These were sequentially probed for the presence of mRNA transcripts for LIF, IL-8, IL-6, and MCP-1. mRNA load was documented by hybridization with a ß-actin probe.

homeostasis. Neuropeptides that are present in the unmyelinated type C nerve fibers have been studied as mediators of neuroimmune interactions.[73,74] One of the best-characterized neuropetides in this context is substance P(SP), one of the principal mediators of pain.[75] This peptide is involved with the pathogenesis of experimental arthritis[76] and has been shown to modulate functions of lymphocytes[75] and monocytes.[77] In monocytes, it induces the production of the cytokines IL-1, TNF, and IL-6.[78] Through this effect, SP can contribute indirectly to the regulation of connective tissue metabolism. In addition, SP has also been shown to directly affect fibroblast proliferation[79] and protease production by synoviocytes[80] and chondrocytes.[81] Collectively, these effects of SP are consistent with a role of this neuropeptide in connective tissue development, homeostasis, and disease. The association of SP with limb buds in the newt is an indication for a possible function of the peptide in development.[82] Observations on denervation that is induced experimentally or occurs during human diseases[83] suggest a trophic role for the nervous system in the maintenance of connective tissues. Patients with familial dysautonomia (Riley Day syndrome) have an increased incidence of osteoarthritis. This had previously been interpreted as the result of increased mechanical trauma, but may also be related to the effects of neuropeptides on connective tissue metabolism.

A new perspective on the role of neuropeptides in cell regulation is based on the findings that some neuropeptides can be produced by nonneuronal cells. This was first shown for the prepro-enkephalin gene in lymphocytes[84] and other bone marrow-derived cells.[85,86] In these cells, prepro-enkephalin RNA is expressed and the propeptide is processed in a tissue-specific manner into smaller, biologically active peptides.[86,87] *In vitro* studies with opioid antagonists suggested that enkephalins may not only have opioid function, but may exert regulatory effects on cell proliferation and differentiation in neural tissues.[88-90] We showed that cultured human articular chondrocytes express high levels of 1.4-kb prepro-enkephalin mRNA.[91] Chondrocytes store met-enkephalin intracellularly and secrete this neuropeptide in mature, as well as in precursor form. Gene expression was induced by serum factors. High levels of prepro-enkephalin mRNA were detected in proliferating chondrocytes, but not in confluent, contact-inhibited cells. Phorbol-myristate-acetate and dibutyrate cyclic-AMP, but not dexamethasone, increased prepro-enkephalin mRNA. Furthermore, the two growth factors transforming growth factor-ß (TGF-ß) and platelet-derived growth factor (PDGF) up-regulated gene expression.

To assess the significance of met-enkephalin production by chondrocytes, we analyzed functional characteristics of chondrocytes that are associated with prepro-enkephalin expression. Expression of this gene is not a mere function of chondrocyte activation, but is part to a more specific program in chondrocyte responses. This notion is based on the induction of its expression by PDGF and TGF-ß and its inhibition by IL-1. All three factors stimulate secretory functions of chondrocytes, but TGF-ß and IL-1 induce distinct sets of genes — the former resulting in the formation of extracellular matrix, the latter causing its degradation. Thus, in the context of secretory responses, prepro-enkephalin expression appears to be part of the anabolic program in chondrocytes. To confirm this potential link between expression of this neuropeptide gene and cell proliferation, double-labeling studies for [³H]thymidine incorporation and prepro-enkephalin mRNA expression were performed. These experiments clearly showed that TGF-ß treatment of primary chondrocytes not only induced expression of prepro-enkephalin mRNA, but also resulted in a parallel increase in the number of proliferating cells. On the other hand, IL-1 treatment decreased proliferation and opioid gene expression to the same degree. Interestingly, TGF-ß also led to formation of dense cell clusters which are referred to as "cartilage nodules". Proliferating cells were very rare in these clusters, indicating that these chondrocytes are contact inhibited and differentiated matrix-producing cells. Prepro-enkephalin expression was similarly rare in cartilage nodules. These findings provide insight into the regulation of prepro-enkephalin expression during functionally distinct phases of chondrocyte activation. Under physiological conditions, adult articular cartilage consists of highly differentiated cells which produce extracellular matrix, but rarely divide. Cartilage nodules represent an *in vitro* equivalent for this. Immediately after isolation, chondrocytes are surrounded by pericellular matrix and only a few cells enter S-phase. In primary cultures, cells become adherent and begin to proliferate. This process increases in subculture, and at all stages, the number of prepro-enkephalin-expressing cells is closely linked to the number of proliferating cells. This interpretation is consistent with the findings on met-enkephalin expression *in vivo*. Met-enkephalin immunoreactivity is present in perichondrial cell layers where chondrocytes show a high rate of proliferation, but it is absent in mature cartilage which contains only a very low number of proliferating cells (unpublished results). The concept that prepro-enkephalin expression reflects a particular state of chondrocyte activation is also supported by the findings that retinoic acid (RA) suppresses chondrocyte proliferation (unpublished results), with a simultaneous reduction in prepro-enkephalin expression. RA has profound effects in the regeneration of limb structures,[92] and in many instances has been used to induce differentiation or maturation of cell lines *in vitro*.[93] Consistent with this are reports on prepro-enkephalin expression at certain stages of organogenesis,[94,95] in rapidly proliferating granulosa cells of the ovaries,[96] or in immature spermatocytes of the testes.[97] *In vivo* conditions that can be associated with prepro-enkephalin expression by articular chondrocytes are organogenesis or, in the mature organism, situations of cartilage remodeling such as in arthritis or in responses to trauma.[98]

Age-related changes in the enkephalin system have been observed in humans and mollusks where plasma levels of met-enkephalin profoundly decrease and only a reduced percentage of granulocytes respond to enkephalins.[99] The relationship of preproenkephalin expression to chondrocyte proliferation was also observed in senescent chondrocytes where this mRNA could not be detected.

3. Growth and Differentiation Factors

Transforming growth factor-ß (TGF-ß) — TGF-ßs represent one family of factors that have been detected in bone and cartilage[100-105] and modulate a wide variety of chondrocyte functions. TGF-ß stimulates chondrocyte proliferation, synthesis of extracellular matrix components,[106,107] other regulatory factors, and modulates the effects and receptors of other factors. In mammalian tissues, genes for three TGF-ßs have been characterized and recombinant forms of the proteins produced.[100,101] Human articular chondrocytes release active TGF-ß and larger quantities in latent form.[108] By neutralization with specific antibodies, it was shown that three isoforms TGF-ß1, TGF-ß2, and TGF-ß3, were secreted.[109] The cytokines IL-1 and IL-6 and the growth factors bFGF and PDGF increasd the production of TGF-ß (Figure 2). Analysis of the inducers of TGF-ß gene expression demonstrated that the three isoforms are differentially regulated. IL-6 induced TGF-ß1, while IL-1 preferentially increased TGF-ß3 mRNA levels. bFGF stimulated the release of TGF-ß activity, but had no effect on TGF-ß mRNA levels; PDGF regulated both mRNA and protein. Recombinant and purified forms of TGF-ß1, TGF-ß2, and TGF-ß3 induced similar dose-dependent increases in chondrocyte proliferation which are detectable at low doses and of greater magnitude than the effects of bFGF and PDGF. These results demonstrated that chondrocytes produce TGF-ß and that expression of the three isoforms is

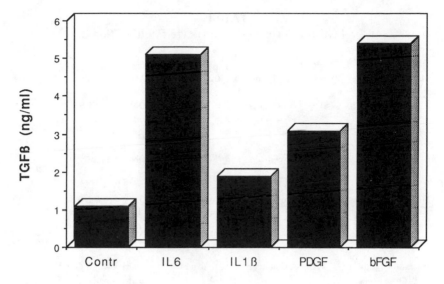

FIGURE 2. Cytokines and growth factors stimulate TGF-ß release from chondrocytes. Primary chondrocytes were cultured for 48 h in serum-free media in the presence of the indicated factors (all at 10 ng/ml) and supernatants were tested in the CCL64 assay for total TGF-ß activity (after transient acidification).

differentially regulated. Induction by growth factors or cytokines can occur at the mRNA level and/or posttranslationally.

A series of previous studies has analyzed the effects of TGF on chondrocyte proliferation.[110-118] In evaluating the findings obtained in these studies, it is important to consider that the effects of a given growth factor on chondrocyte proliferation are modulated by variables such as the species from which chondrocytes were isolated, age of the donor, number of *in vitro* subcultures, primary vs. subculture as monolayer or in soft agar, and the presence or absence of additional factors. The effect of these conditions is likely to be responsible for the divergent results that have been reported for some of the growth factors. Table 1 summarizes the studies on TGF-ß and chondrocyte proliferation. Most of these studies analyzed the effects of TGF-ß on cells from young animals and found stimulation of DNA synthesis. One study on rabbit growth plate chondrocytes found that TGF-ß inhibited proliferation in serum-free cultures, but stimulated it in the presence of serum.[112] All other studies reported stimulation of DNA synthesis, but biphasic dose-response curves were often observed. At high doses, TGF appeared less mitogenic than at lower doses. This may either be related to TGF-ß effects on functional subsets in the different cultures or to a differential effect on matrix synthesis vs. DNA synthesis at high and low doses. Whether different affinities of TGF-ß receptors are involved with this is unknown. In primary human articular chondrocytes from adult donors, we always observed monophasic dose-dependent stimulation of proliferation. However, in articular chondrocytes from young rabbits and in rabbit perichon-

TABLE 1
TGF-ß Effects on Chondrocyte Proliferation

Author	Species	Age (weeks)	Cell type	Culture system	TGF-ß effect	Ref.
Skantze	Rabbit	4	Articular cho	Soft agar	+ DNA with EGF; collagen; –GAG synthesis	110
O'Keefe	Chick	3–4	Growth plate cho	1º Monolayer	+ DNA; biphasic	111
Hiraki	Rabbit	Young	Growth plate cho	1º Monolayer	– DNA serum-free; + DNA with serum; + GAG in confluent culture	112
Iwamoto	Rabbit	4	Articular cho	Soft agar	+colonies with FGF; no effect with EGF, PDGF, Ins	113
Rosier	Chick	3–4	Growth plate cho	1º Monolayer	+[3H]Thy uptake; –collagen synthesis; –alkaline phosphatase	114
Vivien	Rabbit	3	Articular cho	1º Monolayer	Proliferation in 2% FBS; +proliferation in 10% FBS	116
Battegay	Human	?	?Cho, ?subculture	Monolayer	+[3H]Thy biphasic	117

Note: + denotes stimulation; – denotes inhibition.

drium, we also observed biphasic dose-response curves in the presence of serum. TGF-ß effects on chondrocyte proliferation are profoundly affected by the extent of *in vitro* subculture and, thus, of chondrocyte dedifferentiation. We have found with human articular chondrocytes and rabbit perichondrial cells that the magnitude of the TGF-ß effect on chondrocyte proliferation decreases with the number of *in vitro* subcultures. This occurs during presenescent states when, at the same time, the magnitude of FGF-induced proliferation increases or remains unchanged. Thus, proliferation in response to TGF-ß appears to be a function of differentiated chondrocytes and this supports a role of TGF-ß as a principal chondrocyte growth factor.

In contrast to the progress that has been made in cloning the multiple isoforms of TGF-ß and defining biological effects of these proteins, the TGF-ß receptors have only recently been partially characterized.[119-121] There appear to be three distinct membrane molecules that can bind TGF-ß. A proteoglycan, termed "betaglycan", of 200 to 300 kDa exists in soluble and membrane-anchored form and binds TGF-ß1, TGF-ß2, and TGF-ß3 with an affinity of 30 to 300 pM. Since this molecule is expressed on cells that are resistant to the biological effects of TGF-ß, it is thought not to be involved with TGF-ß signal transduction. The two other receptors, type I and type II, are glycoproteins at 53 and 70 to 100 kDa, respectively. Their affinity for TGF-ß1 is 5 to 50 pM, and both bind TGF-ß1, TGF-ß2, and TGF-ß3. Although more than 100 cell types have been analyzed for TGF-ß receptors,[121] only one chondrosarcoma cell line has been shown to express all three receptors. The only data available on chondrocytes is from a study of chick growth plate chondrocytes which were shown to express high and low affinity TGF-ß receptors.[114]

Bone Morphogenetic Proteins (BMP) — A group of regulatory molecules that is part of the TGF-ß family of growth factors are the bone morphogenetic proteins (BMP). Their initial characterization was based on the ability of demineralized bone extracts to promote bone formation when injected subcutaneously.[122] They induce the sequence of events that begins with the infiltration by inflammatory cells, the proliferation of mesenchymal cells which then differentiate into chondrocytes. This is followed by vascularization, the development of osteoblasts, and the formation of new bone. Four distinct genes that encode the human BMPs have been cloned.[123] In chick limb bud mesoderm, the recombinant form of human BMP-2B induced cartilage formation, as indicated by the production of proteoglycans and type II collagen.[124] Interestingly, in this system, TGF-ß inhibited radiosulfate incorporation into proteoglycans, but synergistically stimulated it together with BMP-2B.[125] BMP-2B also increased the expression of alkaline phosphatase, while osteogenin (BMP-3) stimulated proteoglycan formation but not alkaline phosphatase.[125] Information on the expression of BMPs by chondrocytes is very limited. Effects on chondrocytes[126] and the presence of BMP immunoreactivity in chondrosarcoma cells[127] have been demonstrated. Age effects on bone induction by demineralized bone powder include both a decrease in bone formation with increasing age of the recipient as well as with the age of the animal from which bone powder was prepared.[128]

Platelet derived growth factor (PDGF) — PDGF is a growth stimulus for a wide variety of cell types. Three forms of this dimeric growth factor that are either homodimers or heterodimers of the PDGF A and PDGF B chains are known. Two types of PDGF receptors have been identified. They differ in their affinity for the different forms of PDGF. PDGF AA stimulates only the α receptor, while PDGF BB stimulates both α and ß receptors. The ß receptor is usually not expressed on normal cells *in vivo*. It is found on many cultured cells and, *in vivo*, it is probably up-regulated by products of inflammatory cells.[129] PDGF expression is induced by several other growth factors. TGF-ß stimulates PDGF B chain mRNA and secretion in AKR2B cells[130] and the expression of PDGF A chain mRNA and secretion of PDGF AA protein in smooth muscle cells.[131] Similarly, IL-1 induces PDGF secretion in fibroblasts[132] and PDGF increases IL-1 receptor expression.[133] Based on these findings, it has been proposed that PDGF functions as the actual mitogen and that the effects of its inducers in cell proliferation are indirect.[117] Several sudies have examined the regulation of PDGF receptors *in vitro*. TGF-ß induces the expression of PDGF α receptors in fibroblasts,[134] but inhibits the expression of PDGF ß receptor in 3T3 cells.[135]

Recently, rabbit articular chondrocyes were shown to express the gene for the A chain of PDGF and to secrete the PDGF AA protein. In these cells, PDGF AA was induced by IL-1 and this was inhibited by TGF-ß.[136] There appears to be consensus on the function of PDGF as a chondrocyte growth factor. One study which contained a detailed charaterization of the role of PDGF as the principal mitogen for smooth muscle cells also reported that the growth stimulatory effects of TGF-ß are dependent on the induction of PDGF.[117] However, the data presented are difficult to interpret since the effects were minimal and the nature of the chondrocytes was not described. In human articular chondrocytes, we found PDGF consistently less potent than TGF-ß or bFGF and, during chondrocyte aging, responsiveness to PDGF decreased earlier and more profoundly as compared to TGF-ß or bFGF.

Fibroblast growth factors (FGF) — Fibroblast growth factors have been purified from different tissues and stimulate the proliferation and differentiation of mesenchymal, epithelial, and neuroectodermal cells. This family of growth factors includes acidic (aFGF) and basic FGF (bFGF), the oncogenes int-2 and hst, FGF-5, FGF-6, and keratinocyte growth factor.[137] Three receptors for FGF have recently been cloned. They are related proteins with tyrosine kinase activity and bind both aFGF and bFGF.[138] Binding of FGF to its receptors requires complex formation with a proteoglycan (see next section). FGFs are potent chondrocyte growth factors *in vitro* and *in vivo*. Prins et al.[139] showed that FGF purified from bovine pituitary stimualted DNA synthesis in monolayer cultures of rabbit articular chondrocytes. In organ cultures of adult bovine cartilage, bFGF increased cell proliferation. EGF and IGF-I alone had no effect, but synergized with bFGF.[140] In this culture system, all factors enhanced glucosaminoglycan synthesis. The same set of growth factors also stimulated proteoglycan synthesis by isolated chondocytes from rat rib growth plate.[141] In chick growth plate chondrocytes, FGF caused a 13.5-fold increase, TGF-ß a 3.5-fold increase, and both factors dramatically synergized to stimulate a 73-fold increase in [^{3}H]thymidine incorporation. Intracellular levels of cAMP did not increase under any of these conditions.[142] The ability of FGF to regulate chondrocyte function *in vivo* was studied in a rabbit model of cartilage repair. Incisional lesions in articular cartilage usually do not heal. The administration of bFGF via osmotic pumps resulted in chondrocyte proliferation, matrix synthesis, and complete filling of the defects.[98] The effects of FGF on chondrocyte differentiation have been studied on growth plate cells, as well as in mature cells for the regulation of collagen expression. In soft agar culture, FGF enhanced colony formation of chondrocytes which continued to synthesize cartilage-specific proteoglycans.[143] Kato et al.[144] analyzed rabbit growth plate chondrocytes in a culture system where the cells undergo terminal differeniation into hypertrophic chondrocytes which produce high levels of alkaline phosphatase, reduced amounts of proteoglycan, and calcify extracellular matrix. The addition of bFGF at early stages in this system inhibited all features of terminal differntiation and thus maintained chondrocytic phenotype of the cells. In embryonic chicken sternal chondrocytes, FGF and TGF-ß synergistically inhibited type II collagen mRNA and protein expression.[145] This was shown to be related to a decrease in transcription from the promoter of the pro-α-1 chain of type II collagen.[145] Under the same *in vitro* treatment, FGF and TGF-ß synergized in the stimulation of cell proliferation. It is thus possible that the decrease in type II collagen expression represents a decrease in protein synthetic and secretory function of cells that are stimulated to rapidly proliferate, rather than dedifferentiation.

In several chondrocyte responses, bFGF potentiates the effects of IL-1. FGF by itself has very weak or no detectable effects on the production of proteases in chondrocytes. However, it strongly synergizes with IL-1 and induces dose-dependent increases in the presence of low concentrations of IL-1, which are ineffective.[146] This interaction is, in part, related to the up-regulation of IL-1 receptors by bFGF.[147] Hulkower showed in rabbit articular chondrocytes that the synergy between FGF and IL-1 in the induction of gelatinase does not depend on protein kinase C. Inhibitors of protein kinase C even enhanced PGE2 and protease levels induced by low doses of IL-1.[148] Chin et al.[149] also showed that, in rabbit articular chondrocytes, IL-1 and bFGF synergized in the induction of PGE2 production. However, IL-1 completely inhibited FGF-induced proliferation and this was associated with a decrease in high-affinity FGF receptors.[149] In the same cultures, bFGF increased IL-1 receptor mRNA and protein expression.

Insulin-like Growth Factors/Growth Hormone — The insulin-like growth factors (IGFs) are small molecular weight proteins with structural homology to proinsulin. They are bound to carrier proteins and circulate in serum. IGF-II is preferentially expressed in immature cells during fetal growth and development, and decreases as a function of differentiation. Two types of receptors have been identified. Type I receptor is related to the insulin receptor.[150] It binds and mediates the mitogenic effects of both IGF-I and IGF-II. Some very intriguing recent findings showed that the type II receptor is the mannose-6-phosphate receptor which is closely linked to the Tme (T-associated maternal locus effect) locus.[151] The mannose-6-phosphate receptor has a separate binding site for IGF-II, in addition to its binding site for mannose-6 phosphate. This receptor is a genomically imprinted gene and it is expressed only from the maternal chromosome. In contrast, its ligand, IGF-II, is poorly transcribed from the maternal gene[150] so that receptor and ligand are oppositely imprinted. At present, the function of the type II receptor is unclear. It does not appear to mediate mitogenic effects of IGF-II and has been proposed to serve as a "molecular sink" that binds and degrades IGF-II. Initially, IGFs were thought to be produced only in the liver, but it has been shown that other tissues (including cartilage) can produce IGF. The initial analyses of IGF-I/somatomedin C expression in cartilage were based on the controversy of whether the effects of growth hormone on bone growth were a direct action or dependent on the production of somatomedins. Immuno-histochemical analysis of rat growth plate showed strong staining of cells in the proliferating zone, but only weak reactivity in the germinal and hypertrophic areas.[152] IGF immunoreactivity was reduced in hypophysectomized animals and induced after they were treated with growth hormone (GH). A similar distribution and GH-dependent expression of IGF-I mRNA in rat epiphyseal growth plate was demonstrated by *in situ* hybridization.[153] Articular chondrocytes from 2 to 3-month-old rabbits secreted IGF-I and IGF-II *in vitro*. Levels were increased by insulin, but GH had no effect on proliferation or on the levels of IGF.[154] In rabbit articular chondrocytes, bFGF stimulated, but TGF-ß inhibited IGF-I production.[155]

Receptors for GH are expressed on chondrocytes in the hypertrophic zone of neonatal rabbit tibiae. In 20 to 50-day-old rabbits, this was reduced to reserve and proliferative chondrocytes and absent in growth plates from rabbits with closed epiphysis. Receptors were also detected on human infant costal chondrocytes.[156] Gene expression of the GH receptor was shown to be induced by GH and serum in cultured rat epiphyseal chondrocytes.[157] These findings suggest that chondrocytes during growth are responsive to the direct action of GH, but not after cessation of growth; and this is consistent with the apparent lack of direct GH actions on mature cartilage.

IGF-I receptors were detected by radioligand binding studies in all zones of bovine growth plate,[158] suggesting that IGF may not only be involved with cell replication but also with differentiated functions of the chondrocytes. On isolated bovine epiphyseal growth plate chondrocytes, both type I and type II receptros were demonstrated.[159] Cultured costal chondrocytes from 4 to 6-week-old rats specifically bound IGF-I, and it was shown that the ligand-receptor complexes were internalized and degraded in lysosomes, suggesting that under these culture conditions the predominatly expressed receptor was type II.[160]

The major chondrocyte functions that are regulated by IGF are proliferation and glucosaminoglycan production. IGF activity was originally identified as a "sulfation" factor. Recombinant IGF-I, serum, and synovial fluid stimulated proteoglycan synthesis in bovine cartilage organ culture. Neutralizing antibody to IGF reduced the stimulatory effect of serum and completely blocked the activity in synovial fluid.[161] Even in the presence of IL-1 or TNF which cause a depletion of cartilage matrix, IGF-I stimulated proteoglycan synthesis in explants of porcine cartilage.[162] However, cartilage explants from mice with experimentally induced arthritis were unresponsive to the anabolic effects of IGF-I. Since these explants could be stimulated by forskolin, it is possible that the unresponsiveness to IGF-I is related to down-regulation of its receptors by inflammatory stimuli.[163] Vetter et al.[164] showed that IGF-II is a more potent stimulus of clonal growth in human fetal chondrocytes as compared to IGF-I. In adult chondrocytes, IGF-I was more potent and GH had no detectable effect on growth of either cell type. The stimulatory effect of IGF-I on adult cells decreased with increasing culture period, but this was not the case for fetal cells. This finding suggests that responsiveness to IGF-I in adult cells may decrease as a function of *in vitro* senescence.

C. GROWTH FACTORS AS LIGANDS FOR PROTEOGLYCAN RECEPTORS

Several growth factors (such as TGF-ß and BMPs) have originally been isolated from bone and cartilage. It was unclear whether these were growth factor proteins that were extracted from cells within these tissues, or whether they were bound to extracellular matrix. Proteoglycans which were previously recognized only as extracellular matrix structural proteins have now been shown to serve two additional and important functions. Proteoglycans can serve as ligands for cell surface integrin receptors and proteoglycans when anchored in the cytoplasmic membrane or as part of extracellular matrix, can serve as receptors for growth factors.[165]

One of the receptors for TGF-ß is a proteoglycan and therefore termed "betaglycan". It exists in a membrane-bound, as well as in soluble form in plasma. This molecule binds TGF-ß with high affinity, but it is not active in signal transduction.[166] Two proteoglycans that have structural similarity in their core proteins were recently shown to bind and thereby neutralize the biological activity of TGF-ß.[167] Binding of TGF-ß occurred to the core protein and, when present in this complex, TGF-ß is probably unable to bind to its receptor. Since TGF-ß stimulates the production of decorin, this may represent negative regulation of TGF-ß activity.

More detailed charaterization of the functional significance of growth factor binding to proteoglycans has recently been presented for bFGF. FGF was purified as a heparin-binding growth factor, and binding to heparan sulfate protects it from proteolytic degradation. It has now been demonstrated that binding to heparan sulfate is essential for the ability of FGF to bind to its receptor. The proteoglycan that interacts with FGF can either be in free form or bound to the cell membrane (such as syndecan). Binding to heparan sulfate is thought to induce a conformational change in the FGF molecule that allows it to bind to its receptor and activate the receptor-associated tyrosine kinase.

D. INTERACTIONS OF REGULATORY FACTORS

From the information that we have summarized on the individual cytokines, neuropeptides, and growth factors, it is apparent that these agents are integrated in a network of cartilage regulatory factors. Regulators for which receptors are expressed on chondrocytes can affect their own synthesis (autoinduction has been demonstrated in chondrocytes for IL-1, LIF, and TGF-ß1[168] of other regulators) their receptors, or modulate their effect on chondrocyte responses (Table 2). TGF-ß has a particularly broad spectrum of modulatory effects on other regulators. One example that has been characterized in chondrocytes is its interaction with IL-1. TGF-ß inhibits IL-1 effects on protease production,[169] IL-1 production,[170] and the expression of IL-1 receptors.[171] Table 3 lists some of these regulatory factor interactions. Since chondrocytes produce most of these factors, it is often difficult to determine, even with in vitro culture systems, which cellular responses are the direct consequence of a particular agonist and which are dependent on the production of a secondary mediator by the chondrocytes. It is also not clear whether the existance of several stimuli that can induce qualitatively similar responses in chondrocytes indeed represent independent and redundant regulation, or whether they depend on the induction of a principal regulator for a particular chondrocyte response. In some cases, such as with the induction of TGF-ß3 by IL-1, these interactions are seemingly paradoxical. IL-

TABLE 2
Chondrocyte Regulatory Factors

Factor	Major functions in cartilage
IL-1	Induction of proteases, PGE2, other cytokines; inhibition of GAG synthesis
TNF	Similar catabolic effects as IL-1
IL-1 ra	Inhibits IL-1 activity
GM-CSF	Stimulation of HLA DR, induction of cytokines
IL-6	Protease inhibitor production, chondrocyte proliferation
IL-8	Neutrophil chemotactic factor
MCP-1	Monocyte chemotactic factor
LIF	Induction of Cytokines
TGF-ßs	Chondrocyte proliferation, modulates IL-1 effects; promotes formation of ECM
PDGF	Proliferation
bFGF	Proliferation, differentiation, protease production
IGF	Proliferation, glucosaminiglycan synthesis
Enkephalins	Regulation of chondrocyte proliferation

TABLE 3
Interactions of Regulatory Factors

Factor	Effect
IL-1	+ IL-1, IL-6, IL-8, MCP-1, LIF, and TGF-ß3 − Preproenkephalin RNA
IL-6	+ TGF-ß1
LIF	+ MCP-1
TGF-ß	+ IL-6, IL-6 receptor; MCP-1 − IL-1 receptor PDGF R PDGF
PDGF	+ MCP-1, TGF-ß1, IL-1 R
FGF	+ TGF-ß, IL-1 R, MCP-1

Note: + denotes stimulation; − denotes inhibition.

1 and TGF-ß are qualitatively opposing mediators. The former is the most potent cartilage catabolic factor, the latter the prototype anabolic factor for cartilage. One possible interpretation for this interaction is that it represents protective counterregulation to protect the tissue from the detrimental effects of IL-1. This notion is consistent with the role of TGF-ß as an inhibitor of IL-1 receptors and actions. It is also remarkable that all cartilage regulatory factors that have been

examined induce at least one isoform of TGF-ß. Based on this, a hypothesis can be proposed that views TGF-ß as a family of principal cartilage regulatory factors which is part of a protective system to limit tissue damage induced by other agents. With tools to eliminate the endogenous production of regulatory factors or to block their receptors becoming increasingly available, it is now possible to test which agents are the principal inducers of chondrocyte responses and which are accessory factors. These experiments will not only improve the understanding of chondrocyte biology, but will also provide an essential basis to address aging phenomena.

E. CHONDROCYTE GROWTH REGULATION IN AGING

The age-related responsiveness of human chondrocytes to the different growth factors has not been systematically evaluated, but for the limited number of agents that have been tested, a decrease in proliferative potential with aging has been observed. Very important basic information on the relative potency of the different chondrocyte growth factors and whether they act directly or via the induction of a second factor or receptor is presently not available. Chondrocytes are sensitive to growth stimulation by a rather wide variety of agents.[139] This is certainly not a unique feature of this cell type but, in contrast to other cell systems, it has not yet been defined as which are the principal mitogens. Very few studies have examined the effect of recombinant growth factors on chondrocyte proliferation as a function of aging. The relationship of chondrocyte proliferation and aging has been studied in two principal experimental systems. Cells from donors at different ages were used in primary culture or after subculture as an *in vitro* model of chondrocyte senescence.

Evans and Georgescu[172] showed that, similar to the original observations by Hayflick[173] on the senescence of fibroblasts, the *in vitro* proliferative potential of rabbit, canine, and human chondrocytes is proportional to the life-span of the species from which the cells were obtained, that it decreased in rabbit chondrocyte cultures with increasing donor age, and that similar to other cell types, chondrocytes can undergo only a limited number of cell divisions *in vitro*. Limited growth capacity *in vitro* was seen in monolayer and clonal cultures. In a series of studies, Adolphe et al. have studied aging phenomena in rabbit articular chondrocytes. Cell density of old cartilage was approximately one third of that of young animals. There was an age-related decrease in cell number at confluency, reduced cell replication rate, and increased cell volume and cell cycle time.[174] *In vitro* aging of chondrocytes from young (1 to 2 months old) rabbits was associated with a decline in proliferative capacity during subculture, with senescence of the cells at passage 8.[175] Senescent cells were larger, had a higher protein content, and increased cytoskeletal components actin, vimentin, and tubulin, which reflects a highly organized, rigid state of the cytoskeleton. The proportion of cells in S/M phases of the cell cycle decreased and the number of binucleate cells increased.[176] With the same *in vitro* system, it was shown that subcultured cells at passage 4 (which corresponds to 50% of their life span) are still able to respond to FGF with increased proliferation and a decrease in size and protein content.[177] Using mouse mandibular chondrocytes in organ culture from animals at different ages, Livine et al.[178] showed a decrease in proliferation rate with increasing donor age. However, proliferation of senescent cells could be restimulated with PTH, PGE-1, and dexamethasone. The stimulatory effect of dexamethasone on DNA labeling indices was even more pronounced in the condyles from older animals, a rather surprising and intriguing finding. Eilam and Nevo studied intracellular calcium levels as differentiation signals in cultured epiphyseal chondrocytes from chick embryos. These cells undergo typical changes in proliferation, cell structure, and proteoglycan synthesis of *in vitro* senescence. Reduced oxygen tension (from 18 to 8%) in these cultures slowed down the development of these changes.[179] Low oxygen tension does not interfere with the metabolic activites of chondrocytes and may be more similar to the milieu in cartilage, a tissue that is not vascularized and thus also has reduced oxygen tension. An alternative explanation is that high oxygen tension leads to enhanced oxygen radical formation and thus to cell damage and accelerated aging. Cells under low oxygen tension had increased levels of intracellular free calcium and this could be a signal involved with slowing down the *in vitro* aging process. Levels of intracellular free calcium decreased during 34-d *in vitro* culture of chick growth plate chondrocytes to approximately 30% of levels in 5-d cultures.[180] The addition of noncharacterized cartilage-derived growth factors[181] that presumably contained TGF-ß led to a decrease in calcium levels. These investigators thus proposed that decreased levels of calcium are associated with and possibly lead to differentiation and the changes that are seen during *in vitro* aging, while the maintenance of high calcium levels enhances proliferation and delays aging. We analyzed growth factor responsiveness of human articular chondrocytes in aging.[182] For these studies, chondrocytes were isolated from knee cartilage obtained at autopsy or reconstructive surgery. Cells were analyzed in primary culture and stimulated with recombinant preparations of the major chondrocyte growth factors. The *in vitro* proliferative potential of the cells decreased with increasing donor age. Figure 3 shows a representative selection of donors at different ages. First, it is apparent that that the overall levels of [^{3}H]thymidine uptake decreased in older donors (65 years old) to approximately 25% of the levels of young adults (33 years old) whose levels were already lower than those of donors prior to closure of epiphysis (5 years old). In all donors, the order of potency was similar, with TGF-ß being the most effective, followed by bFGF and PDGF. In the older age group (60 to 90 years), TGF-ß was the only factor that stimulated proliferation of all cultures of primary cells. The effects of bFGF and PDGF were reduced to a greater extent and, in many donors, no longer detectable. The effects of IL-6 on chondrocyte proliferation also showed a rather interesting age-dependent pattern. IL-6 stimulated proliferation of cells from donors prior to closure of epiphysis with an efficiency similar to that of PDGF. In young adults, it had only a small effect and, in older donors, no detectable effect. IL-6 synergized in all age groups with TGF-ß in the induction

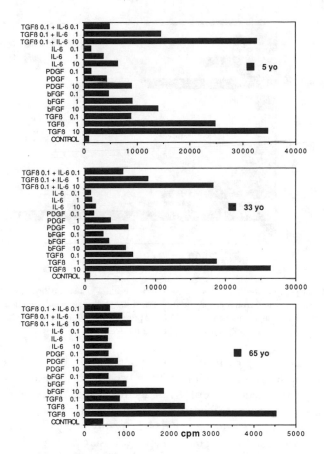

FIGURE 3. Changes in growth factor responsiveness of human articular chondrocytes in aging. Human articular chondrocytes were isolated from normal cartilage of donors at the indicated ages. Cells were cultured under identical conditions for the proliferation studies. Results are shown as [³H]thymidine uptake. Note the difference in the scale of the horizontal axis. The concentrations of the growth factors are shown in nanograms per milliliter (ng/ml).

of chondrocyte proliferation. This synergy was impressive in young adults, but markedly reduced and often absent in the older age group (Figure 3).

These results indicate that *in vitro* proliferation of primary human articular chondrocytes decreases as a function of donor age. The chondrocyte responses to all recombinant growth factors tested showed a similar decline. In all age groups, TGF-ß was the most potent stimulus and, in older cells, often the only factor that could stimulate proliferation of the cells.

Some rather interesting differences were observed with chondrocytes from normal vs. osteoarthritic (OA) old cartilage. Figure 4 shows the comparison of cells from normal articular cartilage of a 67-year-old with osteoarthritic chondrocytes from a 64-year-old subject. Cells from the OA tissue showed a greater response to TGF-ß and expressed the synergy with IL-6 which was absent

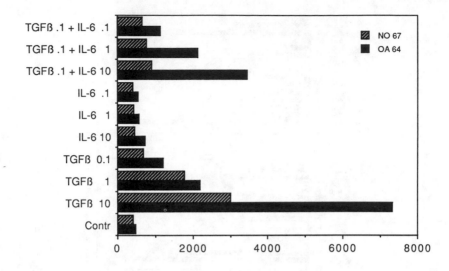

FIGURE 4. Growth factor responsiveness of old normal vs. osteoarthritic chondrocytes. Chondrocytes were isolated from normal and osteoarthritic cartilage, and cultured under identical conditions with the indicated concentrations of TGF-ß1 and IL-6 (ng/ml) for 3 d. Proliferation was measured by [^{3}H]thymidine uptake during the final 12 h of culture.

in the normal aged cartilage. Increased proliferation, TGF-ß responsiveness and reexpression of the synergy with IL-6 have been shown in a larger series of samples to be a consistent feature of OA vs. aged cartilage.[182] These observations raise questions as to whether this enhancement of chondrocyte proliferation in OA specimens from older donors reflects the ability of aged chondrocytes to express increased proliferative potential upon exposure to the particular stimuli that are generated in OA cartilage or whether these are recruited cells that perform attempts at cartilage repair in OA.

F. GROWTH FACTORS, SIGNAL TRANSDUCTION, AND CELLULAR SENESCENCE

The molecular characterization of some of the events that control cell growth has provided the basis to analyze mechanisms potentially involved with cellular senescence and age-related decreases in proliferative capacity.

However, there is as yet no ligand-receptor signal transduction system that has been characterized in detail in chondrocytes or with respect to changes that can occur in aging. In view of the essential role of TGF-ß in the regulation of chondrocyte proliferation and differentiation, we will summarize the information that has been generated on chondrocytes and complement this with aspects that have been studied in other cell systems.

TGF-ß is a bifunctional regulator of cell proliferation. It is a potent growth inhibitor for lymphocytes and epithelial cells, as well as being one of the principal growth factors for cartilage cells. Growth inhibition by TGF-ß has been shown to be, at least in part, related to its effect on the phosphorylation of the retinoblastoma (RB)[183] gene product. RB is a tumor suppressor gene which was discovered on the basis of mutations in retinoblastoma cells that render the protein incapable of inhibiting cell proliferation. RB is known as a 110-kDa nuclear protein and a member of the zinc finger family of transcription factors. In quiescent cells, it is unphosphorylated; upon mitogenic stimulation of cells, it is phosphorylated and this appears to be one essential event in the progression of cells from late G1 to S phase. The antiproliferative effect of RB can be mediated either via binding to regulatory elements in growth-related genes or by binding to other transcription factors that regulate growth. The promoter of the *c-fos* gene, whose product is part of the AP-1 transcription factor complex, is inhibited by RB and this is mediated via RB binding to a specific binding motif.[184] A series of nuclear proteins that is associated with cell proliferation has very recently been shown to bind to the RB protein. These include *c-myc*,[185] cyclin,[186] and several species that are as yet uncharacterized.[187] In epithelial cells, TGF-ß inhibits the serum-induced proliferation. This is associated with an inhibition of RB phosphorylation.[183] These data represent one example where growth inhibition induced by an extracellular factor is associated with an inhibition of RB phosphorylation, but it does not represent a direct relationship of both phenomena. The interaction of TGF-ß and RB has become more complex with the observation that the RB protein can regulate the TGF-ß1 promoter.[188] Depending on the cell type used in the transfection studies, RB either activated of repressed the TGF-ß1 promoter. This effect of RB on TGF-ß1 gene expression is presumably mediated via an element that is similar to the RB control element in the *c-fos* promoter. In cell types such as epithelial cells, where TGF-ß is growth inhibitory, this may represent a positive feedback mechanism to enhance antiproliferative effects. A further level of interactions between TGF-ß and RB may exist since cells that lack functional RB often do not express TGF-ß receptors,[189] thus suggesting that RB may be involved with the activation of TGF-ß receptor genes.

The role of RB phosphorylation in cellular senescence has recently been studied. Human diploid fibroblasts at late passage do not proliferate in response to serum, and under these conditions, they also failed to phosphorylate RB.[190] Similar results were obtained in studies on Syrian hamster embryonal cells;[191] in this system, it was also shown that cells that escape senescence are capable of normal RB phosphorylation.

This relationship between TGF-ß inhibition of RB phosphorylation and antiproliferative effects thus seems to be consistent for cells that are growth inhibited by TGF-ß. However, since TGF-ß is a major stimulus of chondrocyte proliferation, the question of its effect on RB phosphorylation in in chondrocytes therefore raises an interesting paradox. Based on all published studies with other growth factors and RB phosphorylation, one can expect that TGF-ß will induce RB phosphorylation in chondrocytes, which would be the first demonstration that the same factor can inhibit and induce RB phosphorylation. Studies on the regulatory system that involve TGF-ß and RB will generate important new insight into mechanisms of chondrocyte aging.

IV. CONCLUSION

The information summarized in this chapter suggests that joint aging is characterized by reduced cartilage cellularity and that this predisposes it to impaired matrix integrity. During the aging of cartilage, chondrocytes synthesize altered matrix components and their proliferative potential decreases.

These studies have identified interesting phenomena that provide the basis for analysis of molecular mechanisms or defects that are responsible for the age-associated decrease in proliferative potential and biosynthetic functions of chondrocytes.

The limited number of studies that have addressed regulation of chondrocytes in aging indicates that some of these age-related changes may be related to regulatory imbalances. This notion would open up possibilities of targeting these mechanisms to restore some of the cellular functions that decrease with aging. The life span of vascular endothelial cells is reduced by IL-1 and TNF.[192,193] Sencescent endothelial cells contained high levels of IL-1 mRNA transcripts, and the addition of antisense oligonucleotides which inhibited the endogenous production of IL-1 prevented senescence and extended the proliferative life span of the cells.[193] This provides one example for potential applications of research on regulatory factors to modulate age-related defects in cell function.

ACKNOWLEDGMENTS

Drs. Peter M. Villiger and Pierre-André Guerne generated most of the data that is shown in this chapter. The work was supported by Grants AG07996 and AR39799 from the National Institutes of Health.

REFERENCES

1. **Sokoloff, L.,** Ageing and degenerative diseases affecting cartilage, in *Cartilage, Biomedical Aspects,* Hall, B. K., Ed., Academic Press, New York, 1983, 110.
2. **Kirstein, M., Brett, J., Radoff, S., Ogawa S., Stern, D., and Vlassara, H.,** Advanced protein glycosylation induces transendothelial human monocyte chemotaxis and secretion or platelet-derived growth factor: role in vascular disease of diabetes and aging, *Proc. Natl. Acad. Sci. U.S.A.,* 87, 9010, 1990.
3. **Meachim, G.,** Effect of age on the thickness of adult articular cartilage at the shoulder joint, *Ann. Rheum. Dis.,* 38, 43, 1971.
4. **Meachim, G.,** Light microscopy of india ink preparations of fibrillated cartilage, *Ann. Rheum. Dis.,* 31, 457, 1972.
5. **Vetter, U., Helbing, G., Heit W., Pirsig, W., Sterzig, K., and Heinze, E.,** Clonal proliferation and cell density of chondrocyes isolated from human fetal epiphyseal, human adult articular and nasal septal cartilage. Influence of hormones and growth factors, *Growth,* 49, 229, 1985.
6. **Quintero, M., Mitrovic, D. R., Stankovic, A., de Seze, S., Miravet, L., and Ryckewaert, A.,** Aspects cellulaire du vieilissement du cartilage articulaire. I. Cartilage condylien a surface normale, preleve dans les genoux normaux, *Rev. Rheum.,* 51, 375, 1984.

7. **Quintero, M., Mitrovic, D. R., Stankovic, A., de Seze, S., Miravet, L., and Ryckewaert, A.,** Aspects cellulaire du vieilissement du cartilage articulaire. II. Cartilage condylien a surface fissurée préleve dans les genoux normaux at arthrosiques, *Rev. Rheum.,* 51, 445, 1984.

8. **Vignon, E., Arlot, M., Patricot, L. M., and Vignon, G.,** The cell density of human femoral head cartilage, *Clin. Orthop.,* 121, 303, 1976.

9. **Engelmann, G. and Leutert, G.,** The aging alterations of the arytenoid cartilage, *Z. Mikrosk. Anat. Forsch.,* 103, 597, 1989.

10. **Hardingham, T. and Bayliss, M.,** Proteoglycans of articular cartilage: changes in aging and in joint disease, *Semin. Arthritis Rheum.,* 20 (3 Suppl. 1), 12, 1990.

11. **Hardingham, T. E., Venn, G., and Bayliss, M. T.,** Chondrocyte responses in cartilage and in experimental osteoarthritis, *Br. J. Rheumatol.,* 30 (suppl. 1), 37, 1991.

12. **Brand, K. D. and Fife, R.,** Ageing in relation to the pathogenesis of osteoarthritis, *Clin. Rheum. Dis.,* 12, 117, 1986.

13. **Roughley, P. J. and White, R. J.,** Age-related changes in the structure of the proteoglycan subunits forom human articular cartilage, *J. Biol. Chem.,* 255, 217, 1980.

14. **Rapraeger, A.,** Transforming growth factor (type ß) promotes the addition of chondroitin sulfate chains to the cell surface proteoglycan (syndecan) of mouse mammary epithelia, *J. Cell Biol.,* 109, 2509, 1989.

15. **Inoue, H., Kato, Y., Iwamoto, M., Hiraki, Y., Sakuda, M., and Suzuki, F.,** Stimulation of cartilage matrix proteoglycan synthesis by morphologically transformed chondrocytes grown in the presence of fibroblast growth factor and transfroming growth factor-beta, *J. Cell. Physiol.,* 138, 329, 1989.

16. **Holmes, M. W. A., Bayliss, M. T., and Muir, H.,** Hyaluronic acid in human articular cartilage: age-related changes in content and size, *Biochem. J.,* 250, 435, 1988.

17. **Mort, J. S., Poole, A. R., and Roughley, P. J.,** Age-related changes in the structure of proteoglycan link proteins present in normal human articular cartilage, *Biochem. J.,* 214, 269, 1983.

18. **Plaas, A. H. K., Sandy, J. D., and Kimura, J. H.,** Biosynthesis of cartilage proteoglycan and link protein by articular chondrocytes from immature and mature rabbits, *J. Biol. Chem.,* 263, 7560, 1988.

19. **Flannery, C. R., Urbanek, P. J., and Sandy J. D.,** The effect of maturation and aging on the structure and content of link proteins in rabbit articular cartilage, *J. Orthop. Res.,* 8, 78, 1990.

20. **Madsen, K., Moskalewski, S., von der Mark, K., and Friberg, U.,** Synthesis of proteoglycans, collagen and elastin by cultures of rabbit articular chondrocytes — relation to age of the donor, *Dev. Biol.,* 96, 63, 1983.

21. **Thonar, E. J-M. A., Buckwalter, J. A., and Kuettner, K. E.,** Maturation related differences in the structure and composition of proteoglycans synthesized by chondrocytes from bovine articular cartilage, *J. Biol. Chem.,* 261, 2467, 1986.

22. **Plaas, A. H. K. and Sandy, J. D.,** Age-related decrease in the link-stability of proteoglycan aggregates formed by articular chondrocytes, *Biochem. J.,* 220, 337, 1984.

23. **Sandy, J. D., Barrach, H. J., Flannery, C. R., and Plaas, A. H. K.,** The biosynthetic response of mature chondrocytes in early osteoarthritis, *J. Rheumatol.,* S14, 16, 1987.

24. **Wang, S. Y., Merrill, C., and Bell, E.,** Effects of aging and long-term subcultivation on collagen lattice contraction and intra-lattice proliferation in three rat cell types, *Mech. Ageing Dev.,* 44, 127, 1988.

25. **Kumazaki, T., Robetorye, R. S., Robetorye, S. C., and Smith, J.,** Fibronectin expression increases during in vitro cellular senescence: correlation with increased cell area, *Exp. Cell Res.,* 195, 13, 1991.

26. **Magnuson, V. L., Young, M., Schattenberg, D. G., Mancini, M. A., Chen, D. L., Steffensen, B., and Klebe, R, J.,** The alternative splicing of fibronectin pre-mRNA is altered during aging and in response to growth factors, *J. Biol. Chem.,* 266, 14654, 1991.

27. **Borsi, L., Castellani, P., Risso, A. M., Leprini, A., and Zardi, L.,** Transforming growth factor-ß regulates the splicing pattern of fibronectin messenger RNA precursor, *FEBS,* 261, 175, 1990.

28. **Burton-Wuster, N., Leipold, H. R., and Lust, G.,** Dibutyryl cyclic AMP decreases expression of ED-A fibronectin by canine chondrocytes, *Biochem. Biophys. Res. Commun.,* 154, 1088, 1988.

29. **Wilkins, E., Dieppe, P., and Madison, P.,** Osteoarthrosis and articular chondrocalcinosis in the elderly, *Ann. Rheum. Dis.,* 42, 280, 1983.

30. **Russell, R. G. G., Bisaz, S., and Fleisch, H.,** Inorganic pyrophosphate in plasma, urine and synovial fluid of patients with pyrophosphate arthropathy (chondrocalcinosis and pseudogout), *Lancet,* 2, 899, 1970.

31. **Lust, G., Nuki, G., and Seegmiller, J. E.,** Inorganic pyrophosphate and proteoglycan metabolism in cultured human articular chondrocytes and fibroblasts, *Arthritis Rheum.,* 19, 479, 1976.

32. **Rachow, J. W. and Ryan, L. M.,** Adenosine triphosphate pyrophosphohydrolase and neutral inorganic pyrophosphatase in pathologic joint fluids: elevated pyrophosphohydrolase in calcium pyrophosphate dihydrate deposition disease, *Arthritis Rheum.,* 28, 1283, 1985.

33. **Rosenthal, A. K., Cheung, H. S., and Ryan, L. M.,** Transforming growth factor beta 1 stimulates inorganic pyrophosphate elaboration by porcine cartilage, *Arthritis Rheum.,* 34, 904, 1991.

34. **Poser, J. W. and Price, P. A.,** A method for decarboxylation of γ-carboxyglutamic acid-containing protein from bovine bone, *J. Biol. Chem.,* 254, 431, 1979.

35. **Hale, J. E., Fraser, J. D., and Price, P. A.,** The identification of matrix Gla protein in cartilage, *J. Biol. Chem.,* 263, 5820, 1988.

36. **Cancela, L. M. and Price, P. A.,** Retinoic acid induces matrix Gla protein gene expression in human cells, *Endocrinology,* in press, 1992.

37. **Price, P. A., Williamson. M. K., Haba, T., Dell, R. B., and Lee, W. S. S.,** Excessive mineralization with growth plate closure in rats on chronic Warfarin treatment, *Proc. Natl. Acad. Sci. U.S.A.,* 79, 7734, 1982.

38. **Knapen, M. H., Hamuly, K., and Vermeer, C.,** The effect of Vitamin K supplementation on circulating osteocalcin (bone Gla protein) and urinary calcium secretion, *Ann. Intern. Med.,* 111, 1001, 1989.

39. **Van Beuningen, H. M., Van den Berg, W. B., Schalkwijk, J., Arntz, O. J., and Van de Putte, L. B. A.,** Age- and sex-related differences in the antigen-induced arthritis in C57Bl/10 mice, *Arthritis Rheum.,* 32, 789, 1989.

40. **Van Beuningen, H. M., Arntz, O, J., and Van den Berg, W. B.,** In vivo effects of interleukin-1 on articular cartilage. Prolongation of proteoglycan metabolic disturbances in old mice, *Arthritis Rheum.,* 34, 606, 1991.

41. **Hickery, M. S., Vilim, V., Bayliss, M. T., and Hardingham, T. E.,** Effect of interleukin-1 and tumor necrosis factor-α on the turnover of proteoglycans in human articular cartilage, *Biochem. Soc. Trans.,* 18, 953, 1990.

42. **Saamanen, A. M., Tammi, M., Kiviranta, I., Jurvelin, J., and Helminen, H.,** Levels of chondroitin-6 sulfate and nonaggregating proteoglycans at articular cartilage contact sites in the knees of young dogs subjected to moderate running exercise, *Arthritis Rheum.,* 32, 1282, 1989.

43. **Ruoslahti, E.,** Integrins, *J. Clin. Invest.,* 87, 1, 1991.

44. **Yamada, K. M.,** Adhesive recognition sequences, *J. Biol. Chem.,* 266, 12809, 1991.

45. **Woods, V. L., Gesink, D. S., Pacheco, H. O., Amiel, D. A., Akeson, W. H., and Lotz, M.,** Integrin expression by human articular chondrocytes, *Arthritis Rheum.,* 34, 540, 1991.

46. **Biswas, C. and Dayer, J. M.,** Stimulation of collagenase production by collagen in mammalian cell cultures, *Cell,* 18, 1035, 1979.

47. **Werb, Z., Tremble, P. M., Behrendtsen, O., Crowley, E., and Damsky, C. H.,** Signal transduction through the fibronectin receptor induces collagenase and stromelysin gene expression, *J. Cell Biol.,* 109, 877, 1989.

48. **Radin E. L. and Paul, I. L.,** Does cartilage compliance reduce skeletal impact loads? The relative force attenuatinng properties of articular cartilage, synovial fluid, periarticular tissues, and bone, *Arthritis Rheum.,* 13, 139, 1970.

49. **Sokoloff, L.,** Osteoarthritis as a remodeling process, *J. Rheumatol.,* 14 (Suppl. 14), 7, 1987.

50. **Dingle, J. T., Davies, M. E., Mativi, B. Y., and Middleton, H. F.,** Immunohistochemical identification of interleukin-1 in activated chondrocytes, *Ann. Rheum. Dis.,* 49, 889, 1990.

51. **Pujol, J. P. and Loyau, G.,** Interleukin-1 and osteoarthritis, *Life Sci.,* 41, 1187, 1987.

52. **Mizel, S. B., Dayer, J. M., Krasne, S. M., and Mergenhagen, S. E.,** Stimulation of rheumatoid synovial cell collagenase and prostaglandin production by partially purified lymphocyte activating factor (interleukin-1), *Proc. Natl. Acad. Sci. U.S.A.,* 78, 2474, 1981.

53. **Saklatvala, J., Pilsworth, L. M. C., Sarsfield, S. J., Gavrilovlc, J., and Heath, J. K.,** Pig catabolin is a form of interleukin-1, *Biochem. J.,* 224, 461, 1984.

54. **Tyler, J. A.,** Articular cartilage cultured with catabolin (pig interleukin-1) synthesizes a decreased number of normal proteoglycan molecules, *Biochem. J.,* 227, 869, 1985.

55. **Eisenberg, S. P., Evans, R. J., Arend, W. P., Verderber, E., Brewer, M. T., Hannum, C. H., and Thompson, R. C.,** Primary structure and functional expression from complementary DNA of a human interleukin-1 receptor antagonist, *Nature (London),* 343, 341, 1990.

56. **Kishimoto, T.,** Interleukin-6, *Blood,* 74, 1, 1989.

57. **Guerne, P. A., Carson, D. A., and Lotz, M.,** IL-6 production by human articular chondrocytes: modulation of its synthesis by cytokines, growth factors, and hormones in vitro, *J. Immunol.,* 144, 499, 1990.

58. **Guerne, P. A., Vaughan, J. H., Carson, D. A., Terkeltaub, R., and Lotz, M.,** Interleukin-6 and joint tissues, *Ann. N.Y. Acad. Sci.,* 557, 558, 1989.

59. **Lotz, M. and Guerne, P. A.,** Interleukin-6 induces the synthesis of tissue inhibitor of metalloproteinases-1/erythroid potentiating activity (TIMP-1/EPA), *J. Biol. Chem.,* 266, 2017, 1991.

60. **Gearing, D. P.,** Leukemia inhibitory factor: does the cap fit?, *Ann. N.Y. Acad. Sci.,* 628, 9, 1991.

61. **Reid, I. R., Lowe, C., Cornish, J., Skinner, S. J. M., Hilton, D. J., Wilson, T. A., Gearing, D. P., and Martin, T. J.,** Leukemia inhibitory factor: a novel-bone active cytokine, *Endocrinology,* 126, 1416, 1990.

62. **Metcalf, D. and Gearing, D. P.,** A fatal syndrome in mice engrafted with cells producing high levels of leukemia inhibitory factor (LIF), *Proc. Natl. Acad. Sci. U.S.A.,* 86, 5948, 1989.

63. **Lotz, M., Moats, T., and Villiger, P. M.,** Leukemia inhibitory factor expression by human articular chondrocytes, *Arthritis Rheum.,* 34, 5141, 1991.

64. **Campbell, I. K., Novak, U., Cebon, J., Layton, J. E., and Hamilton, J. A.,** Human articular cartilage and chondrocytes produce hemopoietic colony stimulating factors in culture in response to IL-1, *J. Immunol.,* 147, 1238, 1991.

65. **Oppenheim, J. J., Zachariae, C. O. C., Mukaida, N., and Matsushima, K.,** Properties of the novel proinflammatory supergene "intercrine" cytokine family, *Ann. Rev. Immunol.,* 9, 617, 1991.

66. **Baggiolini, M., Walz, A., and Kunkel, S. L.,** Neutrophil-activating peptide-1/interleukin 8, a novel cytokine that activates neutrophils, *J. Clin. Invest.,* 84, l045, 1989.

67. **Elford, P. R. and Cooper, P. H.,** Induction of neutrophil-mediated cartilage degradation by interleukin-8, *Arthritis Rheum.,* 34, 325, 1991.

68. **Lotz, M., Terkeltaub, R., and Villiger, P. M.,** Cartilage and joint inflammation: regulation of IL-8 expression by human articular chondrocytes, *J. Immunol.,* 148, 466, 1992.

69. **Walz, A. and Baggiolini, M.,** Generation of the neutrophil-activating peptide NAP-2 from platelet basic protein or connective tissue-activating peptide III through monocyte proteases, *J. Exp. Med.,* 171, 449, 1990.

70. **Grob, P. M., David, E., Warren, T. C., DeLeon, R. P., Fartina, P. R., and Homon, C. A.,** Characterization of a receptor for human monocyte-derived neutrophil chemotactic factor/interleukin-8, *J. Biol. Chem.,* 265, 8311, 1990.

71. **Villiger, P. M., Terkeltaub, R., and Lotz, M.,** Monocyte chemoattractant protein-1 is expressed in human articular chondrocytes and differentially regulated by dexamethasone and retinoic acid, *J. Clin. Invest.,* in press, 1992.

72. **Wyke, B.,** The neurology of joints: a review of general principles, *Clin. Rheum. Dis.,* 7, 223, 1981.

73. **Payan, D. C.,** Neuropeptides and inflammation. The role of Substance P, *Ann. Rev. Med.,* 40, 341, 1989.
74. **Morley, J. E., Kay, N. E., Solomon, G. F., and Plotnikoff, N. P.,** Neuropeptides: conductors of the immune orchestra, *Life Sci.,* 41, 527, 1987.
75. **Pernow, B.,** Substance P, *Pharmacol. Rev.,* 35, 85, 1983.
76. **Levine, J. D., Clark, R., Devor, M., Helms, C., Mosjkowitz, M. A., and Basbaum, A. J.,** Intraneuronal Substance P contributes to the severity of experimental arthritis, *Science,* 266, 547, 1984.
77. **Hartung, H. P., Wolters, K., and Toyka, K. V.,** Substance P: binding properties and studies on cellular responses in guinea pig macrophages, *J. Immunol.,* 136, 3856, 1986.
78. **Lotz, M., Vaughan, J. H., and Carson, D. A.,** Effect of neuropeptides on production of inflammatory cytokines by human monocytes, *Science,* 241, 1218, 1988.
79. **Nilsson, J., von Euler, A. M., and Dalsgaard, C. J.,** Stimulation of connective tissue cell growth by Substance P and Substance K, *Nature (London),* 315, 61, 1985.
80. **Lotz, M., Carson, D. A., and Vaughan, J. H.,** Substance P activation of rheumatoid synoviocytes: neural pathway in pathogenesis of arthritis, *Science,* 235, 893, 1987.
81. **Matsubara, T., Saura, R., Uno, K., and Hirohata, K.,** Modulation of articular chondrocyte metabolism by substance P: possible involvement of neural pathway in cartilage degradation, *Arthritis Rheum.,* 33, 1134, 1990.
82. **Globus, M., Vethamany-Globus, S., Kesik, A., and Milton, G.,** Roles of neural peptide Substance P and calcium in blastoma cell proliferation in the newt notophthalmus viridescens, in *Limb Development and Regeneration,* Alan R. Liss, New York, 1983, 513.
83. **Thompson, M. and Bywaters, E. G. L.,** Unilateral rheumatoid arthritis following hemiplegia, *Ann. Rheum. Dis.,* 21, 370, 1962.
84. **Zurawski, G., Benedik, M., Kamb, B. J., Abrams, J. S., Zurawski, S. M., and Lee, F. D.,** Activation of mouse T-helper cellsinduces abundant preproenkephalin mRNA synthesis, *Science,* 232, 772, 1986.
85. **Martin, J., Prystowsky, M. B., and Angeletti, R. H.,** Preproenkepahlin mRNA in T-cells, macrophages, and mast cells, *Neurosci. Res.,* 18, 82, 1987.
86. **Vindrola, O., Padros, M. R., Sterin-Prync, A., Ase, A., Finkilman, S., and Nahmod, V.,** Proenkephalin system in human polymorphonuclear cells. Production and release of a novel 1. 0-kD peptide derived from synenkephalin, *J. Clin. Invest.,* 86, 531, 1990.
87. **Kuis, W., Villiger, P. M., Leser, H. G., and Lotz, M.,** Differential processing of proenkephalin-A by human peripheral blood monocytes and T lymphocytes, *J. Clin. Invest.,* 88, 817, 1991.
88. **Zagon, I. S., McLaughlin, P. J., Goodman, S. R., and Rhodes, R. E.,** Opioid receptors and endogenous opioids in diverse human and animal cancers, *J. Natl. Cancer Inst.,* 79, 1059, 1987.
89. **Hauser, K. F., McLaughlin, P. J., and Zagon, I. S.,** Endogenous opioids regulate dendritic growth and spine formation in developing rat brain, *Brain Res.,* 416, 157, 1990.
90. **Verbeeck, M. A., Draaijer, M., and Burbach, J. P.,** Selective downregulation of the preproenkephalin gene during differentiation of a multiple neuropeptide-co-expressing cell line, *J. Biol. Chem.,* 265, 18087, 1990.
91. **Villiger, P. M. and Lotz, M.,** Expression of prepco-enkephalin by human articular chondrocytes is linked to cell proliferation, *EMBO J.,* 11, 135, 1992.
92. **Giguere, V., Ong, E. S., Evans, R. M., and Tabin, C. J.,** Spacial and temporal expression of the retinoic acid receptor in the regenerating amphibian limb, *Nature (London),* 337, 566, 1989.
93. **Amos, B. and Lotan, R.,** Retinoid-sensitive cells and cell lines, *Meth. Enzymol.,* 190, 217, 1990
94. **Keshet, E., Polakiewicz, R. D., Itin, A., Ornoy, A., and Rosen, H.,** Proenkephalin A is expressed in mesodermal lineages during organogenesis, *EMBO J.,* 8, 2917, 1989.

95. **Polakiewicz, R. D. and Rosen, H.,** Regulated expression of proenkephalin A during ontogenic development of mesenchymal derivative tissues, *Mol. Cell. Biol.,* 10, 736, 1990.

96. **Rosen, H., Polakiewicz, R. D., and Simantov, R.,** Expression of proenkephalin A mRNA and enkephalin-containing peptides in cultured fibroblasts, *Biochem. Biophys. Res. Commun.,* 171, 722, 1990.

97. **Kilpatrick, D. L. and Millette, C. F.,** Expression of proenkephalin messenger RNA by mouse spermatogenic cells, *Proc. Natl. Acad. Sci. U.S.A.,* 83, 5015, 1986.

98. **Cuevas, P., Burgos, J., and Baird, A.,** Basic fibroblast growth factor (FGF) promotes cartilage repair in vivo, *Biochem. Biophys. Res. Commun.,* 156, 611, 1988.

99. **Stefano, G. B., Kimura, T., Stefano, J. M., Finn, J. P., Leung, M. K., Smith, E. M., and Hughes, T. K.,** Age-related differences in met-enkephalin levels and numbers of opioid responsive cells in vertebrate and invertebrate immune systems, *Prog. Neuroendocrin-immunol.,* 4, 92, 1991.

100. **Massagué, J.,** The transforming growth factor-ß family, *Ann. Rev. Cell Biol.,* 6, 597, 1990.

101. **Roberts, A. B. and Sporn, M. B.,** Transforming growth factor-ß, *Adv. Cancer Res.,* 51, 107, 1988.

102. **Seyedin, S. M., Thompson, A. Y., Bentz, H., Rosen, D. M., McPherson, J. M., Conti, A., Siegel, N. R., Galluppi, G. R., and Piez, K. A.,** Cartilage-inducing factor-A, *J. Biol. Chem.,* 261, 5693, 1986.

103. **Ellingsworth, L. R., Breenan, J. E., Fok, K., Rosen, D. M., Bentz, H., Piez, K. A., and Seyedin, S. M.,** Antibodies to the N-terminal portion of cartilage-inducing factor A and transforming growth factor ß, *J. Biol. Chem.,* 261, 12362, 1986.

104. **Seyedin, S. M., Segarini, P. R., Rosen, D. M., Thompson, A. H., Bentz, H., and Graycar, J.,** Cartilage-inducing factor-ß is a unique protein structurally and functionally related to transforming growth factor-ß, *J. Biol. Chem.,* 262, 1946, 1987.

105. **Thompson, N. L., Flanders, K. C., Smith, J. M., Ellingsworth, L. R., Roberts, A. B., and Sporn M. B.,** Expression of transforming growth factor-ß1 in specific cells and tissues of adult and neonatal mice, *J. Cell. Biol.,* 108, 661, 1989.

106. **Redini, F., Galera, P., Mauviel, A., Loyau, G., and Pujol, J. P.,** Transforming growth factor ß stimulates collagen and glycosaminoglycan biosynthesis in cultured rabbit articular chondrocytes, *FEBS,* 234, 172, 1988.

107. **O'Keefe, R. O., Puzas, J. E., Brand, S., and Rosier, R. N.,** Effects of transforming growth factor-ß on matrix synthesis by chick growth plate chondrocytes, *Endocrinology,* 122, 2953, 1988.

108. **Villiger, P. M., ten Dijke, P., and Lotz, M.,** IL-1ß and IL-6 differentially regulate expression of TGFß isoforms in human articular chondrocytes, *J. Biol. Chem.,* submitted.

109. **Villiger, P. M. and Lotz, M.,** Differential expression of TGFß isoforms by human articular chondrocytes in response to growth factors, *J. Cell. Physiol.,* in press, 1992.

110. **Skantze, K., Brinckerhoff, C. E., and Collier, J. P.,** Use of agarose culture to measure the effect of transforming growth factor-ß and epidermal growth factor on rabbit articular chondrocytes, *Cancer Res.,* 45, 4416, 1985.

111. **O'Keefe, R. J., Puzas, J. E., Brand, S., and Rosier, R. N.,** Effect of transforming growth factor-ß on DNA synthesis by growth plate chondrocytes: modulation by factors present in serum, *Calcif. Tissue Int.,* 43, 352, 1988.

112. **Hiraki, Y., Inoue, H., Hirai, R., Kata, Y., and Suzuki, F.,** Effect of transforming growth factor ß on cell proliferation and glycosaminoglycan synthesis by rabbit growth-plate chondrocytes in culture, *Biochim. Biophys. Acta,* 969, 91, 1988.

113. **Iwamoto, M., Sato, K., Nakashima, K., Fuchihata, H., Suzuki, F., and Kato, Y.,** Regulation of colony formation of differentiated chondrocytes in soft agar by transforming growth factor-beta, *Biochem. Biophys. Res. Commun.,* 159, 1006, 1989.

114. **Rosier, R. N., O'Keefe, R. J., Crabb, I. D., and Puzas, J. E.,** Transforming growth factor beta: an autocrine regulator of chondrocytes, *Connect. Tissue Res.,* 20, 295, 1989.

115. **Osborn, K., Trippel, S. B., and Mankin., H. J.,** Growth factor stimulation of adult articular cartilage, *J. Orthop. Res.,* 7, 35, 1989.

116. **Vivien, D., Galera, P., Lebrun, E., Loyau, G., and Pujol, J. P.,** Differential effects of transforming growth factor-ß and epidermal growth factor on the cell cycle of cultured rabbit articular chondrocytes, *J. Cell. Physiol.,* 143, 534, 1990.

117. **Battegay, E. J., Raines, E. W., Seifert, R. A., Bowen-Pope, D. F., and Ross, R.,** TGFß induces bimodal proliferation of connective tissue cells via complex control of an autocrine PDGF loop, *Cell,* 63, 515, 1990.

118. **Guerne, P. A. and Lotz, M.,** Interleukin-6 and transforming growth factor-ß synergistically stimulate chondrosarcoma cell proliferation, *J. Cell. Physiol,* 149, 117, 1991.

119. **Cheifetz, S., Bassols, A., Stanley, K., Ohta, M., Greenberger, J., and Massagué, J.,** Heterodimeric transforming growth factor-ß. Biological properties and interaction with three types of surface receptors, *J. Biol. Chem.,* 263, 10783, 1988.

120. **Cheifetz, S., Hernandez, H., Laiho, M., ten Dijke, P., Iwata, K. K., and Massague, J.,** Distinct transforming growth factor-ß (TGFß) receptor subsets as determinants of cellular responsiveness to three TGFß isoforms, *J. Biol. Chem.,* 265, 20533, 1990.

121. **Massagué, J., Cheifetz, S., Boyd, F. T., and Andres, J.,** TGFß receptors and TGFß binding proteoglycans: recent progress in identifying their functional properties, *Ann. N.Y. Acad. Sci.,* 593, 59, 1990.

122. **Wozney, J. M., Rosen, V., Celeste, A. J., Mitsock, L. M., Whitters, M. J., Kriz, R. W., Hewick, R. M., and Wang, E. A.,** Novel regulators of bone formation: molecular clones and activity, *Science,* 242, 1528, 1989.

123. **Urist, M. R., Iwata, H., Cecotti, P. L., Dorfman, R. L., Boyd, S. D., McDowell, R. M., and Chien, C.,** Bone morphogenesis in implants of insoluble gelatin, *Proc. Natl. Acad. Sci. U.S.A.,* 70, 3511, 1973.

124. **Chen, P., Carrington, J. L., Hammonds, R. G., and Reddi, A. H.,** Stimulation of chondrogenesis in limb bud mesoderm cells by recombinant human bone morphogenetic protein 2B (BMP-2B) and modulation by transforming growth factor ß1 and ß2, *Exp. Cell Res.,* 195, 509, 1991.

125. **Carrington, J. L., Chen, P., Yanagishita, M., and Reddi, A, H.,** Osteogenin (bone morphogenetic protein-3) stimulates cartilage formation by chick limb bud cells in vitro, *Dev. Biol.,* 146, 406, 1991.

126. **Nogami, H., Ono, Y., and Oohira, A.,** Bioassay of chondrocyte differentiation by bone morphogenetic protein, *Clin. Orthop.,* 258, 295, 1990.

127. **Yang, L. J. and Jin, Y.,** Immunohistochemical observations on bone morphogenetic protein in normal and abnormal conditions, *Clin. Orthop.,* 257, 249, 1990.

128. **Jergesen, H. E., Chua, J., Kao, R. T., and Kaban, L. B.,** Age effects on bone induction by demineralized bone powder, *Clin. Orthop.,* 268, 253, 1991.

129. **Heldin, C. H. and Westermark, B.,** Platelet-derived growth factor: mechanism of action and possible in vivo function, *Cell Reg.,* 1, 555, 1990.

130. **Shipley, G. D., Tucker, R. F., and Moses, H. L.,** Type ß transforming growth factor/growth inhibitor stimulates entry of monolayer AKR-2B cells into S-phase after a prolonged prereplicative interval, *Proc. Natl. Acad. Sci. U.S.A.,* 82, 4147, 1985.

131. **Majack, R. A., Majewski, M. W., and Goodman, L. V.,** Role of PDGF-A expression in the control of vascular smooth muscle cell growth by TGFß, *J. Cell Biol.,* 111, 239, 1990.

132. **Raines, E. W., Dower, S. K., and Ross, R.,** Interleukin-1 mitogenic activity for fibroblasts and smooth muscle cells is due to PDGF-AA, *Science,* 243, 393, 1989.

133. **Bonin, P. D. and Singh, J. S.,** Modulation of interleukin-1 receptor expression and interleukin-1 response in fibroblasts by platelet-derived growth factor, *J. Biol. Chem.,* 263, 11052, 1988.

134. **Ishikawa, O., LeRoy, E. C., and Trojanowska, M.,** Mitogenic effect of transforming growth factor ß1 on human fibroblasts involves the induction of platelet-derived growth factor α receptors, *J. Cell. Physiol.,* 145, 181, 1990.

135. **Gronwald, R. G. K., Seifert, R. A., and Bowen-Pope, D. F.,** Differential regulation of two platelet-derived growth factor receptor subunits by transforming growth factor-ß, *J. Biol. Chem.,* 264, 8120, 1989.

136. **Peracchia, F., Ferrari, G., Poggi, A., and Rotilio, D.,** IL-1ß-induced expression of PDGF-AA isoform in rabbit articular chondrocytes is modulated by TGFß1, *Exp. Cell Res.*, 193, 208, 1991.

137. **Burgess, W. H. and Maciag, T.,** The heparin-binding (fibroblast) growth factor family of growth factors, *Ann. Rev. Bioch.*, 58, 575, 1989.

138. **Keegan, K., Johnson, D. E., Williams, L. T., and Hayman, M. J.,** Isolation of an additional member of the fibroblast growth factor receptor family, FGFR-3, *Proc. Natl. Acad. Sci. U.S.A.*, 88 1095, 1991.

139. **Prins, P., Lipman, J., and Sololoff, L.,** Effect of purified growth factors on rabbit articular chondrocytes in monolayer culture, *Arthritis Rheum.*, 25, 1217, 1982.

140. **Osborn, K. D., Trippel, S. B., and Mankin, H. J.,** Growth factor stimulation of adult articular cartilage, *J. Orthop. Res.*, 7, 35, 1989.

141. **Makower, A., Wroblewski, J., and Pawlowski, A.,** Effects of IGF-1, EGF, and FGF on proteoglycans synthesized by fractionated chondrocytes of rat rib growth plate, *Exp. Cell Res.*, 179, 498, 1988.

142. **Crabb, I. D., O'Keefe, R. J., Puzas, J. E., and Rosier, R. N.,** Synergistic effect of transforming growth factor beta and fibroblast growth factor on DNA synthesis in chick growth plate chondrocytes, *J. Bone Miner. Res.*, 5, 1105, 1990.

143. **Kato, Y., Iwamoto, M., and Koike, T.,** Fibroblast growth factor simulates colony formation of differentiated chondrocytes in soft agar, *J. Cell. Physiol.*, 133, 491, 1987.

144. **Kato, Y. and Iwamoto, M.,** Fibroblast growth factor is an inhibitor of chondrocyte terminal differentiation, *J. Biol. Chem.*, 265, 5903, 1990.

145. **Horton, W. E., Jr., Higginbotham, J. D., and Chandrasekhar, S.,** Transforming growth factor-beta and fibroblast growth factor act synergistically to inhibit collagen II synthesis through a mechanism involving regulatory DNA sequences, *J. Cell. Physiol.*, 141, 8, 1989.

146. **Phadke, K.,** Fibroblast growth factor enhances the interleukin-1-mediated chondrocytic protease release, *Biochem. Biophys. Res. Commun.*, 142, 448, 1987.

147. **Chandrasekhar, S. and Harvey, A. D.,** Induction of interleukin-1 receptors on chondrocytes by fibroblast growth factor: a possible mechanism for modulation of interleukin-1 activity, *J. Cell. Physiol.*, 138, 236, 1989.

148. **Hulkower, K. I., Georgescu, H. I., and Evans, C. H.,** Evidence that responses of articular chondrocytes to interleukin-1 and basic fibroblast growth factor are not mediated by protein kinase C, *Biochem. J.*, 276, 157, 1991.

149. **Chin, J. E., Hatfield, C. A., Krzesicki, R. F., and Herblin, W. F.,** Interactions between interleukin-1 and basic fibroblast growth factor on articular chondrocytes. Effects on cell growth, prostanoid production, and receptor modulation, *Arthritis Rheum.*, 34, 314, 1991.

150. **Haig, D. and Graham, C.,** Genomic imprinting and the strange case of the insulin-like growth factor II receptor, *Cell*, 64, 1045, 1991.

151. **Barlow, D. P., Stroger, R., Herrmann, B. G., Saito. K., and Schweifer, N.,** The mouse insulin-like growth factor type-2 receptor is imprinted and closely linked to the Tme locus, *Nature (London)*, 349, 84, 1991.

152. **Nilsson, A., Isgaard, J., Lindahl, A., Dahlstrom, A., Skottner, A., and Isaksson, O. G.,** Regulation by growth hormone of number of chondrocytes containing IGF-I in rat growth plate, *Science*, 233, 571, 1986.

153. **Nilsson, A., Carlsson, B., Isgaard, J., Isaksson, O. G., and Rymo, L.,** Regulation by GH of insulin-like growth factor-I mRNA expression in rat epiphyseal growth plate as studied with in-situ hybridization, *Endocrinology*, 125, 67, 1990.

154. **Froger, B., Gaillard B., Hossenlopp, P., Adolphe, M., and Binoux, M.,** Production of insulin-like growth factors and their binding proteins by rabbit articular chondrocytes: relationships with cell multiplication, *Endocrinology*, 124, 2365, 1989.

155. **Elford, P. R. and Lamberts, W. J.,** Contrasting modulation by transforming growth factro-ß1 of insulin-like growth factor-I production in osteoblasts and chondrocytes, *Endocrinology*, 127, 1635, 1990.

156. **Barnard, R., Haynes, K. M., Werther, G. A., and Waters, M. J.,** The ontogeny of growth hormone receptros in the rabbit tibia, *Endocrinology*, 122, 2562, 1988.

157. **Nilsson, A., Carlsson, B., Mathews, L., and Isaksson, O. G.,** Growth hormone regulation of the growth hormone receptor mRNA in cultured rat epiphyseal chondrocytes, *Mol. Cell. Endocrinol.,* 70 237, 1990.

158. **Lindahl, A., Nilsson, A., and Isaksson, O. G.,** Effects of growth hormone and insulin-like growth factor-I on colony formation of rabbit epiphyseal chondrocytes at different stages of maturation, *Endocrinology,* 115, 263, 1987.

159. **Trippel, S. B., Chernausek, S. D., Van Wyk, J. J., Moses, A. C., and Mankin, H. J.,** Demonstration of type I and type II somatomedin receptors on bovine growth plate chondrocytes, *J. Orthop. Res.,* 6, 817, 1988.

160. **Schalch, D. S., Sessions, C. M., Farley, A. C., Masakawa, A., Emler, C. A., and Dills, D. G.,** Interaction of insulin-like growth factor I/somatomedin-C with cultured rat chondrocytes: receptor binding and internalization, *Endocrinology,* 118, 1590, 1986.

161. **Schalkwijk, J., Joosten, L. A., Van den Berg, W. B., van Wyk, J. J., and Van de Putte, L. B.,** Insulin-like growth factor stimulation of chondrocyte proteoglycan synthesis by human synovial fluid, *Arthritis Rheum.,* 32, 66, 1989.

162. **Tyler, J. A.,** Insulin-like growth factor I can decrease degradation and promote synthesis of proteoglycan in cartilage exposed to cytokines, *Biochem. J.,* 260, 543, 1989.

163. **Schalkwijk, J., Joosten, L. A., Van den Berg, W. B., and Van de Putte, L. B.,** Chondrocyte nonresponsiveness to insulin-like growth factor I in experimental arthritis, *Arthritis Rheum.,* 32, 894, 1989.

164. **Vetter, U., Zapf, J., Heit, W., Halbing, G., Heinze, E., Froesch, E. R., and Teller, W. M.,** Human fetal and adult chondrocytes. Effect of insulin-like growth factors I and II, insulin and growth hormone on clonal growth, *J. Clin. Invest.,* 77, 1903, 1986.

165. **Ruoslahti, E. and Yamaguchi, Y.,** Proteoglycans as modulators of growth factor activities, *Cell,* 64, 867, 1991.

166. **Andres, J. L., Stanley, K., Cheifez, S., and Massague, J.,** Membrane-anchored and soluble forms of betaglycan, a polymorphic proteoglycan that binds transforming growth factor-ß, *J. Cell Biol.,* 109, 3173, 1989.

167. **Yamaguchi, Y., Mann, D. M., and Ruoslahti, E.,** Negative regulation of transforming growth factor-ß by the proteoglycan decorin, *Nature (London),* 346, 281, 1990.

168. **Van Obberghen-Schilling, E., Roche, N. S., Flanders, K. D., Sporn, M. B., and Roberts, M.,** Transforming growth factor ß1 positively regulates its own expression in normal and transformed cells, *J. Biol. Chem.,* 263, 7741, 1988.

169. **Chandrasekhar, S. and Harvey, A. K.,** Transforming growth factor-ß is a potent inhibitor of IL-1 induced protease activity and cartilage proteoglycan degradation, *Biochem. Biophys. Res. Commun.,* 157, 1352, 1988.

170. **Leser, H. G., Villiger, P., and Lotz, M.,** TGFß and monocyte differentiation. Effects on surface marker expression and cytokine production, *Arthritis Rheum.,* 34, 5153, 1991.

171. **Harvey, A. K., Hrubey, P. S., and Chandrasekhar, S.,** Transforming growth factor-beta inhibition of interleukin-1 activity involves down-regulation of interleukin-1 receptors on chondrocytes, *Exp. Cell Res.,* 195, 376, 1991.

172. **Evans, C. H. and Georgescu, H. I.,** Observations on the senescence of cells derived from articular cartilage, *Mech. Ageing Dev.,* 22, 179, 1983.

173. **Hayflick, L. and Moorhead, P. S.,** The serial cultivation of human diploid cell strains, *Exp. Cell. Res.,* 25, 585, 1961.

174. **Adolphe, M., Ronot, X., Jaffray, P., Hecquet C., Fontagne, J., and Lechat, P.,** Effects of donor's age on growth kinetics of rabbit articular chondrocytes in culture, *Mech. Ageing Dev.,* 23, 191, 1983.

175. **Dominice, J., Levasseur, C., Larno, S., Ronot, X., and Adolphe, M.,** Age-related changes in rabbit articular chondrocytes, *Mech. Ageing Dev.,* 37, 231, 1986.

176. **Ronot, X., Gaillard, B., Froger, B., Hainque, B., and Adolphe, M.,** In vitro aging of articular chondrocytes identified by analysis of DNA and tubulin content and relationship to cell size and protein content, *Cytometry,* 9, 436, 1988.

177. **Froger, B., Gaillard B., Charrier, A. M., Thenet, S., Ronot, X., and Adolphe, M.,** Growth-promoting effects of acidic and basic fibroblast growth factor on rabbit articular chondrocytes aging in culture, *Exp. Cell Res.,* 183, 388, 1989.

178. **Livne, E., Weiss, A., and Silbermann, M.,** Articular chondrocytes lose their proliferative activity with aging yet can be restimulated by PTH-(1-84), PGE1, and dexamethasone, *J. Bone Miner. Res.,* 4, 539, 1989.

179. **Nevo, Z., Beit-Or, A., and Eilam, Y.,** Slowing down aging of cultured embryonal chick chondrocytes by maintainance under lowered oxygen tension, *Mech. Ageing Dev.,* 45, 157, 1988.

180. **Beit-Or, A., Nevo, Z., Kalina, M., and Eilam, Y.,** Decrease in the basal levels of cytosolic free calcium in chondrocytes during aging in culture: possible role as differentiation-signal, *J. Cell. Physiol.,* 144, 197, 1990.

181. **Eilam, Y., Beit-Or, A., and Nevo, Z.,** Decrease in cytosolic free Ca^{++} and enhanced proteoglycan synthesis induced by cartialge derived growth factors in cultured chondrocyes, *Biochem. Biophys. Res. Commun.,* 132, 770, 1985.

182. **Guerne, P. A. and Lotz, M.,** Chondrocyte growth factor responsiveness in aging and osteoarthritis, *Arthritis Rheum.,* 34, 562, 1991.

183. **Laiho, M., DeCaprio, J. A., Ludlow, J. W., Livingston, D. M., and Massague, J.,** Growth inhibition by TGFß linked to suppression of retinoblastoma protein phosphorylation, *Cell,* 62, 175, 1990.

184. **Robbins, P. D., Horowitz, J. M., and Mulligan, R. C.,** Negative regulation of human *c-fos* expression by the retinoblastoma gene product, *Nature (London),* 346, 668, 1990.

185. **Rustig, A. K., Dyson, N., and Bernards, R.,** Amino-terminal domains of *c-myc* and *N-myc* proteins mediate binding of the retinoblastoma gene product, *Nature (London),* 352, 541, 1991.

186. **Bandara, L. R., Adamczewski, J. P., Hunt, T., and La Thangue, N. B.,** Cyclin A and the retinoblastoma gene product complex with a common transcription factor, *Nature (London),* 352, 249,1991.

187. **Kaeplin, W. G., Pallas, D. C., DeCaprio, J. A., Haye, F. J., and Livingston, D. M.,** Identification of cellular proteins that can interact specifically with the T1/E1A binding region of the retinoblastoma gene product, *Cell,* 64, 512, 1991.

188. **Kim, S. J., Lee, H. D., Robbins, P. D., Busam, K., Sporn, M. B., and Roberts, A. B.,** Regulation of transforming growth factor beta 1 gene expression by the product of the retinoblastoma-susceptibility gene, *Proc. Natl. Acad. Sci. U.S.A.,* 88, 3052, 1991.

189. **Kimchi, A., Wang, X. F., Weinberg, R. A., Cheifetz, S., and Massague, J.,** Absence of TGFß receptors and growth inhibitroy responses in retinoblastoma cells, *Science,* 245, 196, 1988.

190. **Stein, G. H., Beeson, M., and Gordon, L.,** Failure to phosphorylate retinoblastoma gene product in senescent human fibroblasts, *Science,* 249, 666, 1990.

191. **Futreal, P. A. and Barrett, J. C.,** Failure of senescent cells to phosphorylate the RB protein, *Oncogene,* 6, 1109, 1991.

192. **Shimada, Y., Ito, H., Kaji, K., and Fukuda, M.,** Tumor necrosis factor reduces lifespan of human endothelial cells in vitro, *Mech. Ageing Dev.,* 55, 245, 1990.

193. **Maier, J. A., Voulalas, P. , Roeder, D., and Maciag, T.,** Extension of the life-span of human endothelial cells by an interleukin-1 alpha antisense oligomer, *Science,* 249, 1570, 1990.

Chapter 11

REGULATION OF CHONDROCYTES
IN RHEUMATOID ARTHRITIS

Gust Verbruggen and Eric M. Veys

TABLE OF CONTENTS

I. INTRODUCTION

The chondrocyte in articular cartilage is responsible for the homeostasis of the surrounding intercellular matrix. The cartilage cells respond to subtle changes in their environment while maintaining the composition of this tissue. These environmental changes include mechanical as well as biological factors. In synovial joints, articular cartilage is in close contact with the synovial membrane.

Articular involvement in rheumatoid arthritis is characterized by a chronic inflammation of the synovial membrane. Microscopic studies are suggestive of ongoing immunologic processes in this tissue. The synovial pannus is invaded by lymphomyeloid cells. These cells cooperate through the secretion of various cytokines. The biological action of these mediators is not restricted to immunocompetent cells alone. Cytokines affect a wide variety of cells. Hence, connective tissue cells (i.e., chondrocytes in an inflamed synovial joint) will be influenced by a variety of these immune cell products.

In order to understand the changes in the bioactivity of the cartilage cell in rheumatoid arthritis, we will discuss: (1) the immune response and the immune processes — and cytokines involved — in the synovial membrane in this disease; (2) the effects of these cytokines on the chondrocyte; (3) the role of the chondrocyte in the production of cytokines; (4) the effects of cytokines on cartilage explants and in *in vivo* models of joint inflammation; and (5) the observations made in articular cartilage from patients with rheumatoid arthritis.

II. THE IMMUNE RESPONSE

A. THE GENERATION OF A SPECIFIC IMMUNE RESPONSE

Microscopic studies of the synovial membrane in rheumatoid joints show that the synovial tissue is invaded by mononuclear cells. The organization of these cells in typical lymphoid nodules, where macrophage-like or dendritic cells are surrounded by T-lymphocytes and plasma cells, is suggestive of on-going immunologic or immunization processes in these areas. As a rule, the immune response starts with the ingestion of an antigen by an antigen-presenting cell. The antigens will be processed inside the cell and small parts of it will be exposed on the outer surface of the cell on HLA [class II MHC (DR) or class I] molecules. The HLA-DR-"antigen" complex can be recognized by a T-(helper/inducer) lymphocyte. The antigen-presenting cell will then liberate interleukin(IL)-1. IL-1 will force the T-(helper/ inducer)lymphocyte to secrete IL-2. IL-2 will stimulate the propagation of a clone of antigen-specific T-helper cells. Clones of antigen-specific T-suppressor/cytotoxic cells are generated by a similar (though not identical) mechanism. Antigen-specific T-helper cells will further stimulate specific cellular and humoral responses through the secretion of different cytokines. Among other activities, interferon-γ will enhance DR expression by antigen-presenting cells, resulting in an amplification of the immune cascade. IL-4 and IL-6, secreted by T-helper cells, will increase B-cell activity.

As a rule, specific cellular and humoral immune responses will eliminate the targeted antigens and, when these antigens are no longer present or presented, all immune activity will cease.

B. CYTOKINES IN THE SYNOVIAL MEMBRANE IN RHEUMATOID ARTHRITIS

It can be argued that somehow, in rheumatoid arthritis, the antigen(s) is (are) not eliminated. Persistent antigens cause persistent immune activity. Hence, the cells required for this activity, as well as the secreted cytokines, will be constantly present in the area involved. Synovial fluids and tissues in rheumatoid arthritis thus contain high levels of both activated macrophage- and fibroblast-like cell lymphokines, such as IL-1-ß,[1-7] IL-6,[6,8-12] tumor necrosis factor(TNF)-α,[6,13-16] granulocyte-macrophage colony stimulating factor (GM-CSF),[6,17-19] macrophage colony stimulating factor (MCSF),[20,21] transforming growth factor (TGF)-ß,[8-9,17,22] platelet derived growth factor (PDGF),[17] and fibroblast growth factor (FGF).[23]

Considering the role of activated T-lymphocytes as the cells primarily responsible for the perpetuation of immunological and inflammatory processes, one would expect high levels of T-cell cytokines in inflamed synovial tissues in rheumatoid arthritis. Activated T-cells produce IL-2, IL-3, IL-4, TNF-ß, and interferon-γ. However, these cytokines are hardly ever detected in synovial fluids and in synovial membranes in rheumatoid arthritis.[6,13,20,24] Consequently, the role of the activated T-cell in maintaining the chronic immune and inflammatory events in the rheumatoid synovium is not fully established.[25,26] Chondrocytes in articular cartilage in rheumatoid arthritis will thus be influenced by the macrophage/fibroblast-derived cytokines.

III. MACROPHAGE/FIBROBLAST-DERIVED CYTOKINES: SYNERGISTIC AND ANTAGONISTIC EFFECTS ON THE IMMUNOLOGICAL CASCADE AND ON CONNECTIVE TISSUE CELL METABOLISM

Secretion and function of cytokines is a complex matter. Most cytokines are produced by highly different cells. Cytokines promote or inhibit the secretion and function of other cytokines or enhance their own secretion and function. They promote or inhibit the expression of receptors for other cytokines. Most of these cytokines thus synergize or antagonize each other's effects.

Cytokines do not exclusively act on lymphomyeloid cells. Most of them profoundly affect the metabolism of connective tissue cells. Although progressive destruction is prominent in rheumatoid joints, catabolic processes are continuously counteracted by repair processes. Destructive as well as repair activities result from the effects of various cytokines released during the immune cascade and inflammation.

A. CYTOKINES THAT PROMOTE CONNECTIVE
TISSUE DESTRUCTION

Most of the degradation of joint tissues (cartilage, bone, and tendons) in rheumatoid arthritis is accounted for by the effects of IL-1 and TNF. Secretion and biological effects of these cytokines are promoted by GM-CSF, FGF, and PDGF.

1. IL-1

IL-1 is primarily produced by monocytes and macrophages.[27,28] Two forms (IL-1-α and -ß) of this 17-kDa protein have been described. IL-1-α and -ß have the same biological activities. However, IL-1-ß is the predominant form of IL-1. IL-1 has pronounced effects on articular cartilage chondrocytes. The factor that was originally described as "catabolin" and induced cartilage depletion *in vitro*,[29,30] was found to be identical to IL-1.[31]

IL-1 enhances the transcription of mRNA for collagenase, stromelisin, and other metalloproteinases by chondrocytes. Increased amounts of these enzymes are then secreted.[32-35] IL-1-activated chondrocytes are thus capable of degrading their own extracellular matrix.[28-30,35-39]

IL-1 strongly inhibits the production of cartilage-specific proteoglycan and it changes the production of collagen.[38-41] The synthesis of cartilage-specific types II and IX collagen is blocked, whereas the production of types I, V, and III collagen is promoted.[42,43] IL-1 promotes the synthesis of fibroblast types of collagen by chondrocytes that have been dedifferentiating to fibroblast-like cells.[43] A similar observation has been made regarding the synthesis of proteoglycan and hyaluronic acid: IL-1 promotes the production of hyaluronic acid by dedifferentiated chondrocytes.[39] Under the influence of IL-1, chondrocytes destroy their original surrounding matrix and, at the best, replace it with a fibrocartilaginous tissue.

IL-1 enhances the transcription of mRNA for and the production of phospholipase A[44,45] and causes an increase in the production of PGE_2,[32,37,46,47] a known inhibitor of chondrocyte mitotic activity.[48] IL-1-activated chondrocytes show a decrease in proliferation rates,[49,50] which probably is mediated by PGE_2.[50] Adequate tissue repair through the generation of a new cartilage cell population is thus hampered in rheumatoid arthritis (Figure 1).

2. TNF-α

TNF-α is also synthesized in the rheumatoid synovial membrane. This 17-kDa cytokine is primarily produced by monocytes and macrophages. Both IL-1 and TNF-α are simultaneously secreted by monocytes. Both IL-1 and TNF-α stimulate their own, as well as each other's transcription by the macrophage. Furthermore, they synergize each other's effects on the chondrocyte. Both IL-1 and TNF depress chondrocyte DNA synthesis rates, inhibit proteoglycan[36,39,51] and chondrocyte-specific collagen synthesis,[43] and enhance proteoglycanase and collagenase secretion.[43,52-53] IL-1 and TNF-α thus provoke connective tis-

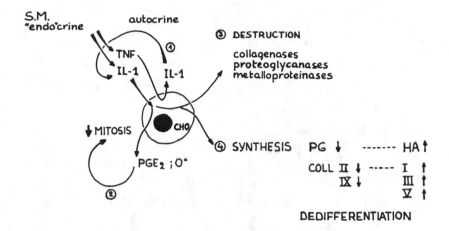

FIGURE 1. Metabolism of extracellular matrix macromolecules by the chondrocyte (CHO) under the influence of IL-1/TNF. TNF enhances the secretion of IL-1 (1) and both these cytokines mediate (2) the production of PGE_2 (mitosis), (3) the secretion of enzymes (catabolism), and (4) the synthesis of nonspecific matrix molecules (dedifferentiation). S.M.: synovial membrane; PG: proteoglycan; and COLL: collagen.

sue catabolism. TNF is much less potent than IL-1 in stimulating proteoglycan breakdown and inhibiting proteoglycan synthesis.[36,39]

Degradation in response to TNF, but not to IL-1, could be blocked by a polyclonal antibody to TNF,[36] and a polyclonal antibody to IL-1 could block proteoglycan breakdown in response to both cytokines, thereby suggesting that TNF may be mediating proteoglycan degradation by inducing the production of IL-1.[36]

Both cytokines provoke prostaglandin and hydrogen peroxide production.[47,54,55] TNF is less effective here than IL-1. The production of reactive oxygen intermediates by chondrocytes may be an important mechanism by which chondrocytes induce structural alterations in the extracellular cartilage matrix. Effects of IL-1 and of TNF on chondrocyte metabolism are summarized in Figure 1.

3. GM-CSF

GM-CSF belongs to a heterogeneous group of growth factors that act on bone progenitor cells. It has been shown that other mature cells are under the influence of this pleiotropic cytokine. In rheumatoid arthritis, GM-CSF is probably produced and secreted by macrophages and by fibroblasts in the inflamed synovium[18] and significant levels of GM-CSF have been reported in rheumatoid synovial fluids.[18,19,56]

It is not known whether GM-CSF acts directly on articular chondrocytes, or whether chondrocytes, as fibroblasts, produce GM-CSF. However, this cytokine is important because of its synergy with IL-1 and TNF-α. IL-1 and TNF-α increase GM-CSF production by fibroblasts.[57,58] On the other hand, GM-CSF

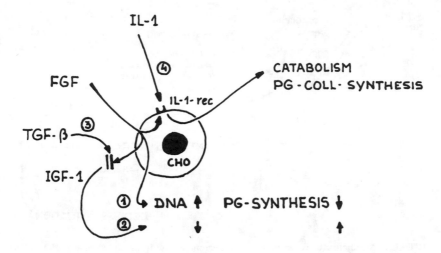

FIGURE 2. Metabolism of extracellular matrix macromolecules by the chondrocyte (CHO) under the influence of FGF/TGF-ß. FGF induces proliferation and decreases the synthesis of matrix macromolecules (1). This effect is counterbalanced by the simultaneous release of IGF-I (2). In the presence of TGF-ß, the secretion of IGF-I is blocked and FGF can exert its original effects (3). These effects are enhanced by IL-1 since FGF induces the expression of IL-1-receptors (4).

enhances the function (IL-1 production, HLA class II expression) of antigen-presenting cells, that is, macrophages.[56,59] TNF-α possibly synergizes with GM-CSF when this increase in macrophage function is considered.[56]

4. FGF

Heparin binding growth factor (HBGF)-1 is the precursor of acidic fibroblast growth factor (FGF). HBGF-1 mRNA and its product were detected intracellularly in RA synovium. HBGF-1 staining was intense in the synovium of rheumatoid arthritic (RA) patients and correlated with the extent and intensity of synovial mononuclear cell infiltration.[60] Basic FGF is produced by the same inflammatory cells.[61] Basic(b)FGF stimulates chondrocyte DNA[62-64] and inhibits sulfated proteoglycan synthesis and type II collagen mRNA.[62] It is difficult to predict how FGF will affect chondrocyte metabolism *in vivo*. bFGF stimulates IGF-I release by chondrocytes[65] and the effects of IGF-I on cartilage cells are opposed to those of FGF. IGF-I blocks chondrocyte DNA synthesis and enhances the production of both sulfated-proteoglycan and collagen (Figure 2).[62] Furthermore, important quantities of TGF-ß are released in rheumatoid synovitis. So, when possible effects of FGF on chondrocytes in rheumatoid cartilage are to be considered, the synergy between both basic and acidic FGFs and TGF-ß must be taken into consideration.[66] TGF-ß-1 inhibits IGF-I release and weakens basic FGF-induced IGF-I release by chondrocytes.[65]

FGF synergizes with IL-1. Although FGF alone does not induce neutral metalloproteinases or increase PGE_2 synthesis, this cytokine progressively enhances the synthesis of PGE_2, collagenase, and gelatinase by chondrocytes responding to IL-1.[67] The potentiation of interleukin-1 catabolic effects by FGF may be related to its ability to induce additional interleukin-1 receptors on the chondrocyte cell surface (Figure 2).[68]

5. PDGF

PDGF has been reported in rheumatoid synovial fluids. It originates from the inflamed synovial membrane,[17] where it is produced by macrophages.[61] PDGF is mitogenic for adult articular chondrocytes in monolayer cultures.[69]

B. CYTOKINES THAT MAY PROMOTE CONNECTIVE TISSUE REPAIR

TGF-ß is considered to be a differentiation factor for cartilage. There is a possible synergy between this cytokine and IL-6. Specific inhibitors of IL-1 and TNF are also discussed below since their activities will decrease chondrocyte catabolic activity and allow connective tissue repair.

1. IL-6

IL-6 is a 26-kDa cytokine. In the rheumatoid synovium, it is produced by monocytes and fibroblasts. High levels of this cytokine are detected in synovial fluid. Articular chondrocytes also secrete IL-6 and this production appears to be stimulated by TNF-α and by IL-1.[70,71] Among some growth factors known to act on chondrocytes, and relevant to this subject, only TGF-ß, FGF, and PDGF were able to significantly increase the synthesis of IL-6.[71]

IL-6 does not stimulate fibroblasts and chondrocytes to secrete collagen or proteoglycan-degrading enzymes. Nor does this cytokine provoke production of PGE_2 by these cells. Therefore, the effects of TNF and IL-1 on proteoglycan metabolism by human chondrocytes are probably not mediated by IL-6. DNA synthesis in human articular cartilage was reported to be increased by IL-6 and TGF-ß. The effects of both cytokines were synergistic.[72] Studies on the biosynthesis of proteoglycan revealed similar synergy between IL-6 and TGF-ß on chondrocyte function.[72] It was found that IL-6 stimulates TGF-ß secretion and mRNA expression,[72] and that, in turn, TGF-ß induces the secretion of IL-6[71] and the expression of IL-6 receptors (Figure 3).[72]

2. TGF-ß

TGF-ß is a 25-kDa cytokine that is produced by platelets, macrophages and synovial fibroblasts,[73] bone cells, and chondrocytes.[72,73] It is produced by the rheumatoid synovial membrane *in vivo*.[8,9] Aside from its proinflammatory influences on lymphomyeloid cells, TGF-ß primarily affects connective tissue cells. Human TGF-ß stimulates the synthesis of matrix proteoglycan and type II collagen, and inhibits matrix degradation.[73-75] Human TGF-ß also partially inhibits IL-1-stimulated loss of proteoglycan from cartilage *in vitro*.[74,75] These observations suggest a possible role for TGF-ß in limiting cartilage proteoglycan loss in inflammatory conditions such as rheumatoid arthritis. However, the effect of TGF-ß on connective tissue cells in inflammatory conditions is not that clear-cut.

TGF-ß-1 inhibits IGF-I release and also blocks basic FGF-induced IGF-I release by chondrocytes.[65] In some circumstances, TGF-ß could thus hinder

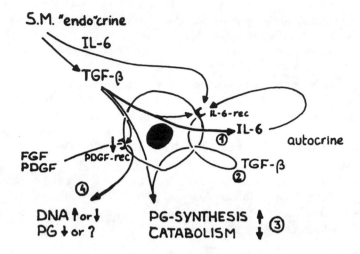

FIGURE 3. Metabolism of extracellular matrix macromolecules by the chondrocyte (CHO) under the influence of IL-6/TGF-ß. TGF-ß induces the expression of receptors for IL-6 (1) and increases the effects and the secretion (1) of IL-6. IL-6 induces the secretion of TGF-ß (2). TGF-ß/IL-6 synergy results in "tissue repair" (3). TGF-ß can increase or decrease the effects of PDGF/FGF on cartilage cell proliferation.

IGF-induced differentiation and repair by articular chondrocytes and exert both agonist and antagonist effects in cartilage (Figure 2). Conflicting reports were further published on the effects of TGF-ß on DNA synthesis by chondrocytes.[76-78] If TGF-ß increases DNA synthesis in chondrocytes, some observations suggest that this cytokine is involved in the control of cartilage growth, possibly by increasing the responsiveness of chondrocytes to FGF[76,79] or PDGF.[78]

TGF-ß has been shown to induce proliferation of connective tissue cells at low concentrations by stimulating autocrine PDGF secretion. However, at higher concentrations of TGF-ß, proliferation was decreased by down-regulation of receptors for PDGF.[78]

3. The Natural Cytokine Inhibitors
a. IL-1 Receptor Antagonist

Various IL-1 inhibiting activities have been described.[80] Most of them have not been purified and their mechanism of action has yet to be established. One of these IL-1 inhibitors specifically binds to the IL-1-receptors of immune and inflammatory cells, but has no IL-1-like activity. Therefore, this factor is termed the IL-1 receptor antagonist (IL-1-ra).[81,82] IL-1-ra is a 22-kDa protein and is produced by monocytes originating from various tissues.[83-85] IL-1-ra is produced when monocytes mature to macrophages.[86] The production of the receptor antagonist may be mediated by cytokines present in inflammatory foci, such as GM-CSF.[87] When IL-1-ra binds to the IL-1 receptor on chondrocytes, the induction of PGE_2 and tissue-degrading enzymes by these cells is prevented.[88-90]

IL-1 inhibitors, other than the 22-kDa IL-1-ra, have been shown to exist. Some of these are produced *in vitro* by cells involved in synovial inflammation.[91,92] Monocytes are able to secrete IL-1 or their inhibitors in response to different signals.[93,94] IL-1 inhibiting activities were detected in synovial fluids from patients with rheumatoid arthritis,[7] and it is plausible that IL-1-ra and IL-1 inhibitors can regulate the biosynthesis and turnover of extracellular matrix molecules by chondrocytes in this disease.

b. TNF Inhibitors

A specific 31 to 33-kDa inhibitor of TNF has been described.[95] This protein is produced *in vitro* by cells cultured from synovial fluids obtained from patients having rheumatoid arthritis.[92] The TNF-α inhibitor may be a soluble form of the cell surface receptor for TNF-α.[96] The TNF inhibitor binds to TNF-α and prevents this cytokine from reaching its target and exerting its biological activity.[95,97,98]

IV. CHONDROCYTES AND THE PRODUCTION OF CYTOKINES: THE GENERATION OF AN AUTOCRINE CYTOKINE PATHWAY

Under inflammatory conditions, the chondrocytes are able to produce different cytokines, thereby controlling extracellular matrix metabolism via autocrine routes. Articular cartilage chondrocytes themselves were shown to produce IL-1-α[52,99,100] and TNF.[52,101] It can be assumed that, by this autocrine pathway, these cells are capable of enhancing catabolic events in articular cartilage of rheumatoid patients. Other cytokines (such as FGF) can amplify IL-1-induced catabolic activity through an upregulation of the expression of cell-surface receptors for IL-1.[102] The same chondrocytes are also capable of producing IL-6[52,71,76,101,103,104] and TGF-ß.[105] Considering the synergy between IL-6 and TGF-ß on chondrocyte function,[72] it can be anticipated that chondrocytes themselves are able to amplify their repair function in certain conditions. The fact that the spontaneous production and secretion of IL-6 by human cartilage cells is enhanced by IL-1-ß makes it acceptable that these cells can down-regulate catabolic events induced by "pro-inflammatory" cytokines.[104]

These observations illustrate that articular chondrocytes in rheumatoid arthritis are strongly intricated in the degradation, as well as in the preservation of the surrounding extracellular matrix. Their function is orchestrated by the balancing effects of various cytokine systems.

V. CYTOKINE EFFECTS ON ARTICULAR CHONDROCYTES IN EXPLANT CULTURE

There is ample evidence that cartilage cells, encased in their extracellular matrix, are affected by cytokines liberated in the surrounding tissues. Human articular cartilage, cultured *in vitro* in the presence of IL-1 and/or TNF, is resorbed by enzymatic activity. Living chondrocytes secrete the enzymes. IL-1, for instance, has no capacity to stimulate glycosaminoglycan (GAG) release from or inhibit GAG synthesis by dead cartilage and inhibitors of protein synthesis, or suppress cytokine-stimulated cartilage matrix resorption.[29,30,35,37,38,53]

In addition to inducing cartilage matrix resorption, these cytokines also inhibited the incorporation of radiosulfate into cultured cartilage, which is a reflection of the suppression of GAG synthesis.[38-41] Also, there is no doubt that the exogenous differentiation factors are operational with explanted cartilage. TGF-ß causes partial inhibition of IL-1-induced cartilage degradation *in vitro*.[74,75]

Co-culture of living articular cartilage explants with activated peripheral blood monocytes from patients with rheumatoid arthritis strongly enhanced the proteoglycan depletion of living cartilage.[106] These observations suggest a possible role for cytokines and growth factors in causing, as well as limiting cartilage damage in rheumatoid arthritis.

If these cytokine effects occur *in vivo*, they must lead to cartilage destruction. Intraarticular injections of IL-1 cause depletion of proteoglycans in cartilage[107-110] and suppress matrix proteoglycan synthesis.[111] In established *in vivo* models of synovitis, the intraarticular injection of IL-1 is followed by an accentuation of the inflammatory response and by destruction of articular cartilage.[112-114] Likewise, the injection of recombinant human tumor necrosis factor-alpha (TNF-α) induced depletion of cartilage proteoglycan in some *in vivo* models.[109] The effects of intraarticular injections of TNF-α are not as prominent as those of IL-1. However, injection of both TNF-α and IL-1 into joints elicits a more severe inflammatory response and articular damage than the intraarticular injection of TNF or IL-1 alone. TNF-α thus augments the inflammatory effects of IL-1.[110] However, these experimental situations are not exactly a model for the chronic inflammation in rheumatoid arthritis in humans.

VI. THE CHONDROCYTE IN ARTICULAR CARTILAGE IN RHEUMATOID ARTHRITIS

As can be expected from cytokine availability in the rheumatoid joint, the chondrocyte in rheumatoid arthritis undoubtedly shows overt catabolic activity. Immunochemical methods that permit the identification and analysis of type II collagen degradation *in situ*, showed areas of pronounced pericellular and territorial staining for collagen degradation in human rheumatoid arthritis cartilage as compared with nondiseased tissues, indicating that chondrocytes are responsible, in part, for tissue degradation.[35] It is thus conceivable that these degradation products are found in the rheumatoid synovial cavity. Collagen type II fragments were found in synovial fluid phagocytes.[115] Urinary concentrations of articular cartilage collagen cross-links, pyridinoline, were higher in patients with rheumatoid arthritis as compared with normal controls.[116] This index correlated with clinical measures of joint involvement and biochemical variables of inflammatory activity.[116]

Extraction yields for proteoglycans were higher for cartilage of rheumatoid joints, when compared with normal cartilage.[117] Synovial fluids obtained from patients with rheumatoid arthritis showed high concentrations of glycosamino-glycans or proteoglycan fragments.[118-120] However, no significant correlation was detected between the concentration of proteoglycan degradation products and IL-1 levels, or the numbers or types of inflammatory cells in these fluids.[118] Concentrations of proteoglycan turnover products were lower in synovial fluids from patients with rheumatoid arthritis than in fluids from patients with reactive or gouty arthritis.[119,120] These observations illustrate the differences between chronic and acute synovial inflammatory situations. In chronic and protracted inflammatory conditions such as rheumatoid arthritis, articular cartilage may have suffered sustained proteoglycan loss during the initial stages of the disease. When the proteoglycan reserves are decreased, the amounts of degradation products recovered from the joint will decrease. Last, but not least, the synovial membrane will contain other and different cell types. Agonist and antagonist actions of cytokines on chondrocyte metabolism will be more evident. Proteoglycan synthesis can be decreased as well as increased in rheumatoid cartilage. One would expect decreased synthesis rates under an IL-1/TNF regimen. However, TGF-ß and IGF-I, shown to be present in these fluids, may counteract IL-1 and TNF activity. Rheumatoid synovial fluid contains some IGF-like activity that stimulates proteoglycan synthesis in cartilage explant culture.[121]

Proteoglycan synthesis is altered *in vivo* in rheumatoid articular cartilage. Synthesis rates of cartilage-specific proteoglycans may decrease. Chondrocytes under the influence of cytokines may produce other than the normal extracellular macromolecules.[39,42,43] There appears to be some evidence for the fact that chondrocytes in rheumatoid cartilage *in vivo* produce noncartilage-specific proteoglycans.[117] Chondrocytes in cartilage specimens from rheumatoid joints were found to synthesize increased proportions of small-size proteoglycans, enriched in dermatansulfate and lacking the ability to interact with hyaluronic acid.[117] When proteoglycan and collagen metabolism are involved, it is plausible that cartilage cells in rheumatoid arthritis will exhibit very different activities during the course of the disease. The possibility that the original chondrocytes may be replaced is certainly to be considered. High PGE_2 levels inside the articular cartilage may down-modulate proliferation rates. On the other hand, various growth factors that are produced during inflammation will initiate mitotic activity. A new population of chondrocytes, less differentiated, may be generated. It is known that chondrocytes grown in different environments dedifferentiate to fibloblast-like cells.[122,123] Dedifferentiation of the cells would explain the finding of noncartilage-specific extracellular matrix molecules in rheumatoid cartilage.[117]

VII. CONCLUSION: THE ULTIMATE STATE OF ARTICULAR CARTILAGE IN A RHEUMATOID JOINT RESULTS FROM A BALANCED EFFECT OF POSITIVE AND NEGATIVE MEDIATORS ON THE CHONDROCYTE

Both the lymphomyeloid cells responsible for immune responses and for inflammation, and the connective tissue cells responsible for the integrity of the joint tissues, are under the influence of the same cytokine systems. Although the effects of the individual cytokines are diverse, it is possible to classify individual factors based on their effects on specific aspects of cartilage resorption or repair. The principal cytokines and growth factors that have been shown to negatively affect articular cartilage include IL-1-α and ß, TNF-α, the colony-stimulating factors, PDGF and FGF. Cytokine inhibitors and the so-called growth and differentiation factors (including TGF-α and TGF-ß, IGF-I, and IL-6) may counterbalance the effects of the first group.

Chondrocytes also are capable of producing these cytokines. This generation of an autocrine loop constitutes a second way to enhance or attenuate the above-mentioned effects. Cellular replication can be hampered, as well as stimulated, by cytokines released in the course of an inflammatory joint disease. How far chondrocytes in rheumatoid cartilage will replace themselves will probably depend on the local situation. However, mitotic activity in an altered environment can lead to loss of specific function or dedifferentiation. Influenced by these factors, the chondrocytes in rheumatoid arthritis are able to destroy their pericellular environment. With changing conditions, original or newly generated cells can restore this environment by a new — though not necessarily better one (Figure 4).

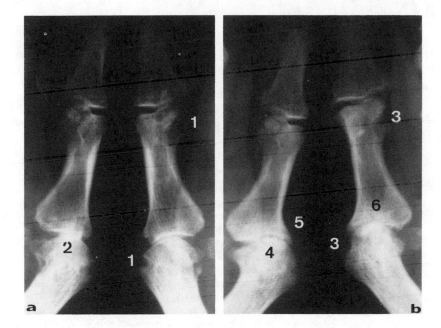

FIGURE 4. The chondrocyte in rheumatoid arthritis. a: Catabolic action by chondrocytes under the influence of cytokines — X-Ray of the thumbs of a patient having active rheumatoid arthritis. (1): Pocked lesions in cartilage and bone (catabolism) are filled with invasive pannus tissue. (2): Joint space narrowing (loss of cartilage as a consequence of enzymatic activity). b: Tissue repair by cartilage and bone cells influenced by cytokines during another stage of the disease (patient in remission for approximately 3 years). (3): Healing of erosions supposes proliferation and in-growth of new cells and the deposition of a newly synthesized extracellular matrix. (4): Regeneration of a new joint space (proliferation and differentiation of chondrocytes and synthesis of new extracellular tissue). (5): Formation of osteophytes (bone cells also respond to cytokines) preventing, or failing to prevent, (6): dislocation.

REFERENCES

1. **Fontana, A., Hengartner, H., Weber, E., Fehr, K., Grob, P. J., and Cohen, G.,** Interleukin 1 activity in the synovial fluid of patients with rheumatoid arthritis, *Rheumatol. Int.,* 2, 49, 1982.
2. **Wood, D. D., Ihrie, E. J., Dinarello C. A., and Cohen, P. L.,** Isolation of an interleukin-1-like factor from human joint effusions, *Arthritis Rheum.,* 26, 975, 1983.
3. **Nouri, A. M. E., Panayi, G. S., and Goodman, S. M.,** Cytokines and the chronic inflammation of rheumatoid disease. I. The presence of interleukin-1 in synovial fluids, *Clin. Exp. Immunol.,* 55, 295, 1984.
4. **Miossec, P., Dinarello, C. A., and Ziff, M.,** Interleukin-1 lymphocyte chemotactic activity in rheumatoid arthritis synovial fluid, *Arthritis Rheum.,* 29, 461, 1986.
5. **Miyasaka, N., Sato, K., Goto, M., Sasano, M., Natsuyama, M., Inoue, K., and Nishioka, K.,** Augmented interleukin-1 production and HLA-DR expression in the synovium of rheumatoid arthritis patients: possible involvement in joint destruction, *Arthritis Rheum.,* 31, 480, 1988.
6. **Firestein, G. S., Alvaro-Garcia, J. M., Maki, R., and Alvaro-Garcia, J. M.,** Quantitative analysis of cytokine gene expression in rheumatoid arthritis, *J. Immunol.,* 144, 3347, 1990.

7. **Smith, J. B., Bocchieri, M. H., Sherbin-Allen, L., Borofsky, M., and Abruzzo, J. L.,** Occurrence of interleukin-1 in human synovial fluid: detection by RIA, bioassay and presence of bioassay-inhibiting factors, *Rheumatol. Int.,* 9, 53, 1989.

8. **Fava, R., Olsen, N., Keski-Oja, J., Moses, H., and Pincus, T.,** Active and latent forms of transforming growth factor-beta activity in synovial effusions, *J. Exp. Med.,* 169, 291, 1989.

9. **Lafyatis, R., Thompson, N. L., Remmers, E. F., Flanders, K. C., Roche, N. S., Kim, S.-J., Case, J. P., Sporn, M. B., Roberts, A. B., and Wilder, R. L.,** Transforming growth factor-beta production by synovial tissues from rheumatoid patients and streptococcal cell wall arthritic rats: studies on secretion by synovial fibroblast-like cells and immunohistologic localization, *J. Immunol.,* 143, 1142, 1989.

10. **Guerne, P.-A., Zuraw, B. L., Vaughan, J. H., Carson, D. A., and Lotz, M.,** Synovium as a source of interleukin 6 in vitro: contribution to local and systemic manifestations of arthritis, *J. Clin. Invest.,* 83, 585, 1989.

11. **Houssiau, F. A., Devogelaer, J.-P., Van Damme, J., Nagant de Deuxchaisnes, C., and Van Snick, J.,** Interleukin-6 in synovial fluid and serum of patients with rheumatoid arthritis and other inflammatory arthritides, *Arthritis Rheum.,* 31, 784, 1988.

12. **Hirano, T., Matsuda, T., Turner, M., Miyasaka, N., Buchan, G., Tang, B., Sato, K., Shimizu, M., Maini, R., Feldmann, M., Kishimoto, T.,** Excessive production of interleukin 6/B cell stimulatory factor-2 in rheumatoid arthritis, *Eur. J. Immunol.,* 18, 1797, 1988.

13. **Saxne, T., Palladino, M. A., Jr., Heinegard, D., Talal, N., and Wollheim, F. A.,** Detection of tumor necrosis factor alpha but not tumor necrosis factor beta in rheumatoid arthritis synovial fluid and serum, *Arthritis Rheum.,* 31, 1041, 1988.

14. **Yocum, D. E., Esparza, L., Dubry, S., Benjamin, J. B., Volz, R., and Scuderi, P.,** Characteritics of tumor necrosis factor production in rheumatoid arthritis, *Cell. Immunol.,* 122, 131, 1989.

15. **Hopkins, S. J. and Meager, A.,** Cytokines in synovial fluid. II. The presence of tumor necrosis factor and interferon, *Clin. Exp. Immunol.,* 73, 88, 1988.

16. **Husby, G., and Williams, R. C., Jr.,** Synovial localization of tumor necrosis factor in patients with rheumatoid arthritis, *J. Autoimmunol.,* 1, 363, 1988.

17. **Lafyatis, R., Thompson, N., Remmers, E., Flanders, K., Roberts, A., Sporn, M., and Wilder, R. L.,** Demonstration of local production of PDGF and TGF-beta by synovial tissue from patients with rheumatoid arthritis (abstract), *Arthritis Rheum.,* 31(Suppl. 4), S62, 1988.

18. **Xu, W. D., Firestein, G. S., Taetle, R., Kaushansky, K., and Zvaifler, N. J.,** Cytokines in chronic inflammatory arthritis. II. Granulocyte-macrophage colony stimulating factor in rheumatoid synovial effusions, *J. Clin. Invest.,* 83, 876, 1989.

19. **Williamson, D. J., Begley, C. G., Vadas, M. A., and Metcalf, D.,** The detection and initial characterisation of colony-stimulating factors in synovial fluid, *Clin. Exp. Immunol.,* 72, 67, 1988.

20. **Firestein, G. S., Xu, W. D., Townsend, K., Broide, D., Alvaro-Garcia, J., Glasebrook, A., and Zvaifler, N. J.,** Cytokines in chronic inflammatory arthritis. I. Failure to detect T cell lymphokines (IL-2 and IL-3) and presence of macrophage colony-stimulating factor (CSF-1) and a novel mast cell growth factor in rheumatoid synovitis, *J. Exp. Med.,* 168, 1573, 1988.

21. **Smith, J. B., Bocchieri, M. H., Smith, J. B., Jr, Sherbin-Allen, L., and Abruzzo, J. L.,** Colony stimulating factor occurs in both inflammatory and noninflammatory synovial fluids, *Rheumatol. Int.,* 10, 131, 1990.

22. **Lotz, M., Kekow, J., and Carson, D. A.,** Transforming growth factor-beta and cellular immune responses in synovial fluids, *J. Immunol.,* 144, 4189, 1990.

23. **Melnyk, V. O., Shipley, G. D., Sternfeld, M. D., Sherman, L., and Rosenbaum, J. T.,** Synoviocytes synthesize, bind, and respond to basic fibroblast growth factor, *Arthritis Rheum.,* 33, 493, 1990.

24. **Firestein, G. S. and Zvaifler, N. J.,** Peripheral blood and synovial fluid monocyte activation in inflammatory arthritis. II. Low levels of synovial fluid and synovial tissue interferon suggest that gamma-interferon is not the primary macrophage activating factor, *Arthritis Rheum.,* 30, 864, 1987.

25. **Firestein, G. S. and Zvaifler, N. J.,** How important are T cells in chronic rheumatoid synovitis?, *Arthritis Rheum.,* 33, 768, 1990.

26. **Hovdenes, J., Gaudernack, G., Kvien, T. K., Hovdenes, A. B., and Egeland, T.,** Mitogen-induced interleukin 2 and gamma interferon production by CD4+ and CD8+ cells of patients with inflammatory arthritides. A comparison between cells from synovial fluid and peripheral blood, *Scand. J. Immunol.,* 30, 597, 1989.

27. **Dinarello, C. A.,** Interleukin 1 and its biologically related cytokines, *Adv. Immunol.,* 44, 153, 1989.

28. **Saklatvala, J., Curry, V. A., and Sarsfield, S. J.,** Purification to homogeneity of pig leucocyte catabolin, a protein that causes cartilage resorption in vitro, *Biochem. J.,* 215, 385, 1983.

29. **Dingle, J. T., Saklatvala, J., Hembry, R., Tyler, J., Fell, H. B., and Jubb, R.,** A cartilage catabolic factor from synovium, *Biochem. J.,* 184, 177, 1979.

30. **Fell, H. B. and Jubb, R. W.,** The effect of synovial tissue on the breakdown of articular cartilage in organ culture, *Arthritis Rheum.,* 20, 1359, 1977.

31. **Saklatvala, J., Pilsworth, L. M. C., Sarsfield, S. J., Gravilovic, J., and Heath, J. K.,** Pig catabolin is a form of interleukin 1, *Biochem. J.,* 224, 461, 1984.

32. **Evéquoz, V., Bettens, F., Kristensen, F., Trechsel, U., Stadler, B. M., Dayer, J.-M., de Weck, A. L., and Fleisch, H.,** Interleukin 2-independent stimulation of rabbit chondrocyte collagenase and prostaglandin E_2 production by an interleukin 1-like factor, *Eur. J. Immunol.,* 14, 490, 1984.

33. **Campbell, I. K., Golds, E. E., Mort, J. S., and Roughley, P. J.,** Human articular cartilage secretes characteristic metal dependent proteinases upon stimulation by mononuclear cell factor, *J. Rheumatol.,* 13, 20, 1986.

34. **Pasternak, R. D., Hubbs, S. J., Caccese, R. G., Marks, R. L., Camaty, J. M., and Di Pasquale, G.,** Interleukin-1 stimulates the secretion of proteoglycan — and collagen degrading proteases by rabbit articular chondrocytes, *Clin. Immunol. Immunopathol.,* 41, 351, 1986.

35. **Dodge, G. R. and Poole, A. R.,** Immunohistochemical detection and immunochemical analysis of type II collagen degradation in human normal, rheumatoid, and osteoarthritic articular cartilages and in explants of bovine articular cartilage cultured with interleukin 1, *J. Clin. Invest.,* 83, 647, 1989.

36. **Pratta, M. A., Di Meo, T. M., Ruhl, D. M., and Arner, E. C.,** Effect of interleukin-1-beta and tumor necrosis factor-alpha on cartilage proteoglycan metabolism in vitro, *Agents Actions,* 27, 250, 1989.

37. **Tyler, J. A.,** Insulin-like growth factor 1 can decrease degradation and promote synthesis of proteoglycan in cartilage exposed to cytokines, *Biochem. J.,* 260, 543, 1989.

38. **Smith, R. J., Rohloff, N. A., Sam, L. M., Justen, J. M., Deibel, M. R., and Cornette, J. C.,** Recombinant human interleukin-1 alpha and recombinant human interleukin-1 beta stimulate cartilage matrix degradation and inhibit glycosaminoglycan synthesis, *Inflammation,* 13, 367, 1989.

39. **Yaron, I., Meyer, F. A., Dayer, J. M., Bleiberg, I., and Yaron, M.,** Some recombinant human cytokines stimulate glycosaminoglycan synthesis in human synovial fibroblast cultures and inhibit it in human articular cartilage cultures, *Arthritis Rheum.,* 32, 173, 1989.

40. **Dingle, J. T.,** The effect of synovial catabolin on cartilage synthetic activity, *Connect. Tissue Res.,* 12, 277, 1984.

41. **Jubb, R. W. and Fell, H. B.,** The effect of synovial tissue on the synthesis of proteoglycan by the articular cartilage of young pigs, *Arthritis Rheum.,* 23, 545, 1980.

42. **Goldring, M. B., Birkhead, J., Sandell, L. J., Kimura, T., and Krane, S. M.,** Interleukin 1 suppresses expression of cartilage-specific types II and IX collagens and increases type I and III collagens in human chondrocytes, *J. Clin. Invest.,* 82, 2026, 1988.

43. **Lefebvre, V., Peeters-Joris, C., and Vaes, G.,** Modulation by interleukin 1 and tumor necrosis factor alpha of production of collagenase, tissue inhibitor of metalloproteinases and collagen types in differentiated and dedifferentiated articular chondrocytes, *Biochim. Biophys. Acta,* 1052, 366, 1990.

44. **Stevens, T. M., Chin, J. E., McGowan, M., Giannaras, J., and Kerr, J. S.,** Phospholipase A2 (PLA2) activity in rabbit chondrocytes, *Agents Actions,* 27, 385, 1989.

45. **Kerr, J. S., Stevens, T. M., Davis, G. L., McLaughlin, J. A., and Harris, R. R.,** Effects of recombinant interleukin-1 beta on phospholipase A2 activity, phospholipase A2 mRNA levels, and eicosanoid formation in rabbit chondrocytes, *Biochem. Biophys. Res. Commun.,* 165, 1079, 1989.

46. **Meats, J. E., McGuire, M. K. B., and Russell, R. G. G.,** Human synovium releases a factor which stimulates chondrocyte production of PGE and plasminogen activator, *Nature (London),* 286, 891, 1980.

47. **Campbell, I. K., Piccoli, D. S., and Hamilton, J. A.,** Stimulation of human chondrocyte prostaglandin E$_2$ production by recombinant human interleukin-1 and tumour necrosis factor, *Biochim. Biophys. Acta,* 1051, 310, 1990.

48. **Goodwin, J. S., Bankhurst, A. D., and Messner, R. P.,** Suppression of T-cell mitogenesis by prostaglandin, *J. Exp. Med.,* 146, 1719, 1977.

49. **Iwamoto, M., Koike, T., Nakashima, K., Sato, K., and Kato, Y.,** Interleukin 1: a regulator of chondrocyte proliferation, *Immunol. Lett.,* 21, 153, 1989.

50. **Verbruggen, G., Veys, E. M., Malfait, A. M., Declercq, L., Van den Bosch, F., and De Vlam, K.,** Influences of human recombinant interleukin-1 on human articular cartilage chondrocytes. Mitotic activity and proteoglycan metabolism, *Clin. Exp. Rheum.,* 9, 1, 1991.

51. **Kolibas, L. M. and Goldberg, R. L.,** Effect of cytokines and antiarthritic drugs on glycosaminoglycan synthesis by bovine articular chondrocytes, *Agents Actions,* 27, 245, 1989.

52. **Shinmei, M., Masuda, K., Kikuchi, T., and Shimomura, Y.,** The role of cytokines in chondrocyte mediated cartilage degradation, *J. Rheumatol.,* 16 (Suppl. 18), 32, 1989.

53. **Campbell, I. K., Piccoli, D. S., Roberts, M. J., Muirden, K. D., and Hamilton, J. A.,** Effects of tumor necrosis factor alpha and beta on resorption of human articular cartilage and production of plasminogen activator by human articular chondrocytes, *Arthritis Rheum.,* 33, 542, 1990.

54. **Tiku, M. L., Liesch, J. B., and Robertson, F. M.,** Production of hydrogen peroxide by rabbit articular chondrocytes. Enhancement by cytokines, *J. Immunol.,* 145, 690, 1990.

55. **Campbell, I. K., Piccoli, D. S., and Hamilton, J. A.,** Stimulation of human chondrocyte prostaglandin E2 production by recombinant human interleukin-1 and tumour necrosis factor, *Biochim. Biophys. Acta,* 1051, 310, 1990.

56. **Alvaro-Garcia, J. M., Zvaifler, N. J., and Firestein, G. S.,** Cytokines in chronic inflammatory arthritis. IV. GM-CSF mediated induction of class II MHC antigen on human monocytes: a possible role in rheumatoid arthritis, *J. Exp. Med.,* 170, 865, 1989.

57. **Kaushanski, K., Lin, N., and Adamson, J. W.,** Interleukin 1 stimulates fibroblasts to synthesize granulocyte-macrophage and granulocyte colony-stimulating factors, *J. Clin. Invest.,* 81, 92, 1988.

58. **Alvaro-Garcia, J. M., Zvaifler, N. J., and Firestein, G. S.,** GM-CSF gene expression and production by rheumatoid arthritis (RA) synovial tissue cells stimulated by IL-1-beta and TNF-alpha, (abstract), *Clin. Rheumatol.,* 9, 569, 1990.

59. **Morrissey, P. J., Bressler, L., Park, L. S., Alpert, A., and Gillis, S.,** Granulocyte-macrophage colony stimulating factor augments the primary antibody response by enhancing the function of antigen-presenting cells, *J. Immunol.,* 139, 1113, 1987.

60. **Sano, H., Forough, R., Maier, J. A., Case, J. P., Jackson, A., Engleka, K., Maciag, T., and Wilder, R. L.,** Detection of high levels of heparin binding growth factor-1 (acidic fibroblast growth factor) in inflammatory arthritic joints, *J. Cell. Biol.,* 110, 1417, 1990.

61. **Butler, D. M., Leizer, T., and Hamilton, J. A.,** Stimulation of human synovial fibroblast DNA synthesis by platelet-derived growth factor and fibroblast growth factor, *J. Immunol.,* 142, 3098, 1989.

62. **Demarquay, D., Dumontier, M. F., Tsagris, L., Bourguignon, J., Nataf, V., and Corvol, M. T.,** In vitro insulin-like growth factor I interaction with cartilage cells derived from postnatal animals, *Horm. Res.,* 33, 111, 1990.

63. **Koike, T., Iwamoto, M., Shimazu, A., Nakashima, K., Suzuki, F., and Kato, Y.,** Potent mitogenic effects of parathyroid hormone (PTH) on embryonic chick and rabbit chondrocytes. Differential effects of age on growth, proteoglycan, and cyclic AMP responses of chondrocytes to PTH, *J. Clin. Invest.,* 85, 626, 1990.

64. **Froger-Gaillard, B., Charrier, A. M., Thenet, S., Ronot, X., and Adolphe, M.,** Growth-promoting effects of acidic and basic fibroblast growth factor on rabbit articular chondrocytes ageing in culture, *Exp. Cell Res.,* 183, 388, 1989.

65. **Elford, P. R. and Lamberts, S. W.,** Contrasting modulation by transforming growth factor-beta-1 of insulin-like growth factor-I production in osteoblasts and chondrocytes, *Endocrinology,* 127, 1635, 1990.

66. **Horton, W. E., Jr., Higginbotham, J. D., and Chandrasekhar, S.,** Transforming growth factor-beta and fibroblast growth factor act synergistically to inhibit collagen II synthesis through a mechanism involving regulatory DNA sequences, *J. Cell. Physiol.,* 141, 8, 1989.

67. **Bandara, G., Lin, C. W., Georgescu, H. I., Mendelow, D., and Evans, C. H.,** Chondrocyte activation by interleukin-1: analysis of the synergistic properties of fibroblast growth factor and phorbolmyristate acetate, *Arch. Biochem. Biophys.,* 274, 539, 1989.

68. **Chandrasekhar, S. and Harvey, A. K.,** Induction of interleukin-1 receptors on chondrocytes by fibroblast growth factor: a possible mechanism for modulation of interleukin-1 activity, *J. Cell. Physiol.,* 138, 236, 1989.

69. **Rosenthal, A. K., Cheung, H. S., and Ryan, L. M.,** Stimulation of pyrophos phate production in articular cartilage by a platelet-derived factor is independent of mitogenesis, *J. Lab. Clin. Med.,* 115, 352, 1990.

70. **Bunning, R. A., Russell, R. G., and Van Damme, J.,** Independent induction of interleukin 6 and prostaglandin E_2 by interleukin 1 in human articular chondrocytes, *Biochem. Biophys. Res. Commun.,* 166, 1163, 1990.

71. **Guerne, P. A., Carson, D. A., and Lotz, M.,** IL-6 production by human articular chondrocytes. Modulation of its synthesis by cytokines, growth factors, and hormones in vitro, *J. Immunol.,* 144, 499, 1990.

72. **Guerne, P. A., Villiger, P., and Lotz, M.,** Unique synergy and interactions between IL-6 and TGF-beta in the regulation of chondrocyte function (abstract), *Clin. Rheumatol.,* 9, 572, 1990.

73. **Sporn, M. B., Roberts, A. B., Wakefield, L. M., and de Crombrugghe, B.,** Some recent advances in the chemistry and biology of transforming growth factor-beta, *J. Cell Biol.,* 105, 1039, 1987.

74. **Morales, T. I. and Hascall, V. C.,** Factors involved in the regulation of proteoglycan metabolism in articular cartilage, *Arthritis Rheum.,* 32, 1197, 1989.

75. **Andrews, H. J., Edwards, T. A., Cawston, T. E., and Hazleman, B. L.,** Transforming growth factor-beta causes partial inhibition of interleukin 1-stimulated cartilage degradation in vitro, *Biochem. Biophys. Res. Commun.,* 162, 144, 1989.

76. **Iwamoto, M., Sato, K., Nakashima, K., Fuchihata, H., Suzuki, F., and Kato, Y.,** Regulation of colony formation of differentiated chondrocytes in soft agar by transforming growth factor-beta, *Biochem. Biophys. Res. Commun.,* 159, 1006, 1989.

77. **Vivien, D., Galera, P., Lebrun, E., Loyau, G., and Pujol, J. P.,** Differential effects of transforming growth factor-beta and epidermal growth factor on the cell cycle of cultured rabbit articular chondrocytes, *J. Cell. Physiol.,* 143, 534, 1990.

78. **Battegay, E. J., Raines, E. W., Seifert, R. A., Bowen-Pope, D. F., and Ross, R.,** TGF-beta induces bimodal proliferation of connective tissue cells via complex control of an autocrine PDGF loop, *Cell,* 63, 515, 1990.

79. **Schofield, J. N. and Wolpert, L.,** Effect of TGF-beta 1, TGF-beta 2, and bFGF on chick cartilage and muscle cell differentiation, *Exp. Cell Res.,* 191, 144, 1990.

80. **Larrick, J. W.,** Native interleukin 1 inhibitors, *Immunol. Today,* 10, 61, 1989.

81. **Hannum, C. H., Wilcox, C. J., Arend, W. P., Joslin, F. G., Dripps, D. J., Heimdal, P. L., Armes, L. G., Sommer, A., Eisenberg, S. P., and Thompson, R. C.,** Interleukin-1 receptor antagonist activity of a human interleukin-1 inhibitor, *Nature (London),* 343, 336, 1990.

82. **Mazzei, G. J., Seckinger, P. L., Dayer, J. M., and Shaw, A. R.,** Purification and characterization of a 26-kDa competitive inhibitor of interleukin 1, *Eur. J. Immunol.,* 20, 683, 1990.

83. **Arend, W. P., Joslin, F. G., Thompson, R. C., and Hannum, C. H.,** An interleukin 1 inhibitor from human monocytes: production and characterization of biological properties, *J. Immunol.,* 143, 1851, 1989.

84. **Arend, W. P., Joslin, F. G., and Massoni, R. J.,** Effects of immune complexes on production by human monocytes of interleukin 1 inhibitor, *J. Immunol.,* 134, 3868, 1985.

85. **Galve de Rochemonteix, B., Nicod, L. P., Junod, A. F., and Dayer, J. M.,** Characterization of a specific 20- to 25-kD interleukin-1 inhibitor from cultured human lung macrophages, *Am. J. Respir. Cell Mol. Biol.,* 3, 355, 1990.

86. **Roux-Lombard, P., Modoux, C., and Dayer, J.-M.,** Production of interleukin 1 (IL-1) and a specific IL-1 inhibitor during human monocyte-macrophage differentiation: influence of GM-CSF, *Cytokine,* 1, 45, 1989.

87. **Mazzei, G. J., Bernasconi, L. M., Lewis, C., Mermod, J. J., Kindler, V., and Shaw, A. R.,** Human granulocyte-macrophage colony-stimulating factor plus phorbol myristate ace-tate stimulate a promyelocytic cell line to produce an IL-1 inhibitor, *J. Immunol.,* 145, 585, 1990.

88. **Balavoine, J.-F., de Rochemonteix, B., Williamson, K., Seckinger, P., Cruchaud, A., and Dayer, J.-M.,** Prostaglandin E$_2$ and collagenase production by fibroblasts and synovial cells is regulated by urine-derived human interleukin 1 and inhibitor(s), *J. Clin. Invest.,* 78, 1120, 1986.

89. **Seckinger, P., Williamson, K., Balavoine, J.-F., Mach, B., Mazzei, G., Shaw, A., and Dayer, J.-M.,** A urine inhibitor of interleukin 1 activity affects both interleukin 1-alpha and 1-beta but not tumor necrosis factor alpha, *J. Immunol.,* 139, 1541, 1987.

90. **Seckinger, P., Lowenthal, J. W., Williamson, K., Dayer, J.-M., and Mac-Donald, H. R.,** A urine inhibitor of interleukin 1 activity that blocks ligand binding, *J. Immunol.,* 139, 1546, 1987.

91. **Lotz, M., Tsoukas, C. D., Robinson, C. A., Dinarello, C. A., Carson, D. A., and Vaughan, J. H.,** Basis for defective responses of rheumatoid arthritis synovial fluid lymphocytes to anti-CD3(T3) antibodies, *J. Clin. Invest.,* 78, 713, 1986.

92. **Roux-Lombard, P., Modoux, C., and Dayer, J.-M.,** Inhibitors of IL-1 and TNF-alpha activities in synovial fluids and cultured synovial fluid cell supernatants (abstract), *Calcif. Tissue Int.,* 42, S(A47), 1988.

93. **Schur, P. H., Chang, D. M., Baptiste, P., Uhteg, L. C., and Hanson, D. C.,** Human monocytes produce IL-1 and an inhibitor of IL-1 in response to two different signals, *Clin. Immunol. Immunopathol.,* 57, 45, 1990.

94. **Bories, P. N., Feger, J., Benbernou, N., Rouzeau, J. D., Agneray, J., and Durand, G.,** Prevalence of tri- and tetraantennary glycans of human alpha 1-acid glycoprotein in release of macrophage inhibitor of interleukin-1 activity, *Inflammation,* 14, 315, 1990.

95. **Seckinger, P., Isaaz, S., and Dayer, J.-M.,** Purification and biologic characterization of a specific tumor necrosis factor-alpha inhibitor, *J. Biol. Chem.,* 264, 11966, 1989.

96. **Novick, D., Engelmann, H., Wallach, D., and Rubenstein, M.,** Soluble cytokine receptors are present in normal urine, *J. Exp. Med.,* 170, 1409, 1989.

97. **Engelmann, H., Aderka, D., Rubinstein, M., Rotman, D., and Wallach, D.,** A tumor necrosis factor-binding protein purified to homogeneity from human urine protects cells from tumor necrosis factor toxicity, *J. Biol. Chem.,* 264, 11974, 1989.

98. **Olsson, I., Lantz, M., Nilsson, E., Peetre, C., Thysell, H., Grubb, A., and Adolf, G.,** Isolation and characterization of a tumor necrosis factor binding protein from urine, *Eur. J. Haematol.,* 42, 270, 1989.

99. **Ollivierre, F., Gubler, U., Towle, C. A., Laurencin, C., and Treadwell, B. V.,** Expression of IL-1 genes in human and bovine chondrocytes: a mechanism for autocrine control of cartilage matrix degradation, *Biochem. Biophys. Res. Commun.,* 141, 904, 1986.

100. **Rath, N. C., Oronsky, A. L., and Kerwar, S. S.,** Synthesis of interleukin-1-like activity by normal rat chondrocytes in culture, *Clin. Immunol. Immunopathol.,* 47, 39, 1988.

101. **Shinmei, M., Masuda, K., Kikuchi, T., and Shimomura, Y.,** Interleukin 1, tumor necrosis factor, and interleukin 6 as mediators of cartilage destruction, *Semin. Arthritis Rheum.,* 18 (Suppl. 1), 27, 1989.

102. **Chandrasekhar, S. and Harvey, A. K.,** Induction of interleukin-1 receptors on chondrocytes by fibroblast growth factor: a possible mechanism for modulation of interleukin-1 activity, *J. Cell. Physiol.,* 138, 236, 1989.

103. **Bunning, R. A. D., Russell, R. G. G., and Van Damme, J. V.,** Independent induction of interleukin-6 and prostaglandin E_2 by interleukin 1 in human articular chondrocyte, *Biochem. Biophys. Res. Commun.,* 166, 1163, 1990.

104. **Bender, S., Haudeck, H.-D., Van de Leur, E. V., Dufhues, G., Schiel, X., Lauwerijns, J., Greiling, H., and Heinrich, P. C.,** Interleukin-1 induces synthesis and secretion of interleukin-6 in human chondrocytes, *FEBS Lett.,* 263, 321, 1990.

105. **Guerne, P. A., Villiger, P., and Lotz, M.,** Unique synergy and interactions between IL-6 and TGF-beta in the regulation of chondrocyte function (abstract), *Clin. Rheumatol.,* 9, 572, 1991.

106. **Wilbrink, B., Bijlsma, J. W., Huber-Bruning, O., Van Roy, J. L., Den Otter, W., and Van Eden, W.,** Mycobacterial antigens stimulate rheumatoid mononuclear cells to cartilage proteoglycan depletion, *J. Rheumatol.,* 17, 532, 1990.

107. **Arner, E. C., Di Meo, T. M., Ruhl, D. M., and Pratta, M. A.,** In vivo studies on the effects of human recombinant interleukin-1 beta on articular cartilage, *Agents Actions,* 27, 254, 1989.

108. **Pettipher, E. R., Henderson, B., Hardingham, T., and Ratcliffe, A.,** Cartilage proteoglycan depletion in acute and chronic antigen-induced arthritis, *Arthritis Rheum.,* 32, 601, 1989.

109. **O'Byrne, E. M., Blancuzzi, V., Wilson, D. E., Wong, M., and Jeng, A. Y.,** Elevated substance P and accelerated cartilage degradation in rabbit knees injected with interleukin-1 and tumor necrosis factor, *Arthritis Rheum.,* 33, 1023, 1990.

110. **Henderson, B. and Pettipher, E. R.,** Arthritogenic actions of recombinant IL-1 and tumour necrosis factor alpha in the rabbit: evidence for synergistic interactions between cytokines in vivo, *Clin. Exp. Immunol.,* 75, 306, 1989.

111. **Van De Loo, A. A., and Van Den Berg, W. B.,** Effects of murine recombinant interleukin-1 on synovial joints in mice: measurement of patellar cartilage metabolism and joint inflammation, *Ann. Rheum. Dis.,* 49, 238, 1990.

112. **Stimpson, S. A., Dalldorf, F. G., Otterness, I. G., and Schwab, J. H.,** Exacerbation of arthritis by IL-1 in rat joints previously injured by peptidoglycan-polysaccharide, *J. Immunol.,* 140, 2964, 1988.

113. **Hom, J. T., Bendele, A. M., and Carlson, D. G.,** In vivo administration with IL-1 accelerates the development of collagen-induced arthritis in mice, *J. Immunol.,* 141, 834, 1988.

114. **Staite, N. D., Richard, K. A., Aspar, D. G., Franz, K. A., Galinet, L. A., and Dunn, C. J.,** Induction of an acute erosive monarticular arthritis in mice by interleukin-1 and methylated bovine serum albumin, *Arthritis Rheum.,* 33, 253, 1990.

115. **Moreland, L. W., Stewart, T., Gay, R. E., Huang, G. Q., McGee, N., and Gay, S.,** Immunohistologic demonstration of type II collagen in synovial fluid phagocytes of osteoarthritis and rheumatoid arthritis patients, *Arthritis Rheum.,* 32, 1458, 1989.

116. **Seibel, M. J., Duncan, A., and Robins, S. P.,** Urinary hydroxy-pyridinium crosslinks provide indices of cartilage and bone involvement in arthritic diseases, *J. Rheumatol.,* 16, 964, 1989.

117. **Mitrovic, D. R. and Darmon, N.,** Structural and biochemical abnormalities of articular cartilage in rheumatoid arthritis, *Rheumatol. Int.,* 10, 31, 1990.

118. **Bensouyad, A., Hollander, A. P., Dularay, B., Bedwell, A. E., Cooper, R. A., Hutton, C. W., Dieppe, P. A., and Elson, C. J.,** Concentrations of glycosaminoglycans in synovial fluids and their relation with immunological and inflammatory mediators in rheumatoid arthritis, *Ann. Rheum. Dis.,* 49, 301, 1990.

119. **Pavelka, K. and Seibel, M. J.,** Quantitative detection of keratan sulfate specific epitopes in synovial fluid in inflammatory and degenerative joint diseases, *Z. Rheumatol.,* 48, 294, 1989.

120. **Carroll, G.,** Measurement of sulphated glycosaminoglycans and proteoglycan fragments in arthritic synovial fluid, *Ann. Rheum. Dis.,* 48, 17, 1989.

121. **Schalkwijk, J., Joosten, L. A., Van den Berg, W. B., Van Wyk, J. J., and Van De Putte, L. B.,** Insulin-like growth factor stimulation of chondrocyte proteoglycan synthesis by human synovial fluid, *Arthritis Rheum.,* 32, 66, 1989.

122. **Verbruggen, G. and Veys, E. M.,** Proteoglycan metabolism of connective tissue cells. An in vitro technique and its relevance to in vivo conditions, in *Degenerative Joints,* Verbruggen, G. and Veys, E. M., Eds., Excerpta Medica, Amsterdam, 1982, 113.

123. **Verbruggen, G., Veys, E. M., and Luyten, F. P.,** Is a decrease in proteoglycan content in degenerative cartilage exclusively caused by enzymatical degradation?, in *Degenerative Joints,* Vol. 2, Verbruggen, G. and Veys, E. M., Eds., Excerpta Medica, Amsterdam, 1984, 55.

Chapter 12

REGULATION OF CHONDROCYTES IN OSTEOARTHROSIS

Charles J. Malemud and Thomas M. Hering

TABLE OF CONTENTS

I. INTRODUCTION

Osteoarthrosis (osteoarthritis; OA) ensues as a result of perturbations in chondrocyte metabolism. Whether the initial inciting event occurs in articular cartilage directly, or as a result of mechanical dysfunction of the underlying subchondral bone,[1-3] has been debated. One thing, however, is of certainty in the disease process; the repertoire of chondrocyte-driven metabolic events is altered. In addition to changes in the activity of chondrocytes involved in extracellular matrix degradation, attempts at repair are often seen throughout the course of the OA process. These reparative events often overlap with normal synovial joint remodeling that occur throughout the lifetime of the individual.[1]

The regulation of chondrocytes in OA is under the control of events occurring in cartilage itself, but may be also amplified by the interaction of exogenous factors (i.e., cytokines) with cartilage that potentially alter the phenotypic expression of the cells. Ultimately, the constellation of cellular, molecular, and biochemical events occurring in OA require elucidation in order to address the question as to why injured cartilage does not successfully repair itself and how the OA process may alter inherent attempts at repair.

II. REGULATION OF INTEGRATION OF EXTRACELLULAR MATRIX COMPONENTS: POTENTIAL ALTERATIONS IN OA

A. CARTILAGE COLLAGENS
1. Collagen Isotypes

Collagens are homopolymers or heteropolymers of specific polypeptides representing the products of at least 20 different genes. At least 11 different types described in vertebrate tissues and five types (types II, V, VI, IX, and XI) have been isolated from adult bovine articular cartilage.[4] cDNA clones have been isolated for all of these collagen types, mostly from human and chick cartilage.[5-9] Type II collagen is the most abundant cartilage collagen and is the main component of collagen fibrils, comprising 90 to 95% of total collagenous protein. Type II collagen is a specific product of chondrocytes and vitreous cells. Type V collagen comprises 1 to 2% of total cartilage collagen. It is nearly ubiquitous in all connective tissue, including bone. Type VI collagen comprises 1 to 2% of total cartilage collagen and is a short-helix, microfibrillar collagen of unknown function. It has been immunolocalized to the pericellular region of the cartilage extracellular matrix. Type IX collagen is a component of cartilage collagen fibrils and comprises 1 to 2% of total collagenous protein. Its function is unknown, but it has been found to be associated with type II collagen fibrils with a D-periodic distribution at the fibril surface and is, in fact, covalently cross-linked to type II collagen.[10] It possesses three chains with a helical region interrupted by nonhelical regions and one chain has a bound GAG chain. It may be involved in mediating interactions of collagen with proteoglycans. Type X "short-chain" collagen is a specific product of growth plate hypertrophic chondrocytes and is found in regions undergoing endochondral bone formation. Type XI collagen comprises 2 to 3% of total cartilage collagenous protein. Like type IX collagen, it is known to be associated with type II fibrils but, unlike type IX collagen, may be buried in the interior of the fibrils.[11] Only type XI to type XI cross-links have been detected, with the possibility of some heterotypic cross-linking to type II collagen.[12] The detailed biosynthesis of type XI collagen has been investigated.[13] The function of type XI collagen in cartilage is not known, but the synthesis and extremely slow proteolytic processing of this molecule may affect the formation of heterotypic cartilage collagen fibrils. Type XIII collagen is a "novel" short-chain collagen, the sequence of which has been entirely determined from cDNA clones. It possesses three triple-helical and four noncollagenous domains. The corresponding protein has not yet been isolated, but steady-state mRNA levels, as determined by Northern blot and *in situ* hybridization, are high in cartilage, as well as in skin, intestine, bone, and striated muscle.

2. Changes in OA

One of the earliest changes reported to occur in OA cartilage is an increase in collagen synthesis.[14] Poole et al.[15] have shown evidence for increased synthesis of type II collagen in human OA cartilage. Evidence has been sought for collagen "isotype switching" (i.e., appearance of type I and type III collagen), but studies in animal models of OA[16] have failed to provide justification for conclusions drawn from earlier studies[17] that OA cartilage contains type I collagen. The collagen network may be altered in OA cartilage, as evidenced by the increased hydration of OA cartilage in human OA and animal models of the disease.[16]

Genetic linkage has been demonstrated between the type II collagen gene and primary OA in a number of families showing a predisposition to primary OA.[18-20] Furthermore, there is direct evidence that a mutation in the type II procollagen gene can cause OA.[21] Ala-Kokko et al.[21] have cloned the allele for type II procollagen shown to co-inherit with primary generalized OA and a mild chondrodysplasia.[19] In this family, the coding sequences for type II collagen were normal except for a single base mutation that converts the codon for arginine at position 519 to a codon for cysteine, an amino acid not found in the triple helical domain of type II collagen of any species. Analysis of the type II collagen purified from the articular cartilage of one affected family member revealed that approximately one quarter of the $\alpha 1(II)$ chains present in the polymeric extracellular collagen of the cartilage contained the expected substitution.[22] The protein exhibited signs of posttranslational overmodification and formed disulfide-bonded $\alpha 1(II)$ dimers. These abnormal properties apparently contribute to reduced durability of the articular cartilage and manifest as severe primary generalized OA.

B. AGGREGATING CARTILAGE PROTEOGLYCAN AND LINK PROTEIN

An increased synthesis of proteoglycan is reported to occur early in OA cartilage.[23] Proteoglycans are a diverse family of macromolecules variable in size (core protein molecular mass, 11,000 to about 220,000 Da) and are grouped together because all are substituted with glycosaminoglycan (GAG) chains, sulfated polymers of disaccharides. The presence of GAG chain(s) is the common feature that justifies the broad classification. The GAG chains, which are variable in number from 1 to 100, may bind other molecules (fibronectin, interstitial collagens, laminin, vitronectin, thrombospondin, and growth factors) by virtue of their high negative charge and, in general, proteoglycans occupy a great deal of space in the extracellular matrix. The large aggregating chondroitin sulfate proteoglycan of cartilage (aggrecan) is responsible for tissue resiliency, load distribution, and providing a low frictional surface in the joint. Aggrecan molecules noncovalently bind filaments of hyaluronic acid to form large aggregates, an interaction which is stabilized by a related glycoprotein named "link protein" whose biosynthesis has been extensively investigated.[24]

Interestingly, changes in proteoglycans in OA cartilage do not seem to be directly related to those that occur as a consequence of ageing. In ageing, cartilage proteoglycan molecules undergo significant transformations, including a decrease in the size and number of chondroitin sulfate chains, a decrease in the chondroitin sulfate 4/6 ratio, and an increase in the size and number of keratan sulfate chains.[25] There is greater heterogeneity in proteoglycan monomers purified from adult compared to fetal or juvenile cartilage, in terms of keratan sulfate content, chondroitin sulfate content, and core protein size.[26] It appears that these variations are the result of differences in posttranslational and proteolytic modification of the core protein with increasing age. In OA, there have been reports of an increased chondroitin sulfate 4/6 ratio, and a decrease in keratan sulfate content.[27]

Cartilage chondrocytes are actively involved in maintenance of appropriate tissue concentrations of proteoglycan.[28] In OA, a number of factors may alter this steady state, resulting in changes in proteoglycan concentration. Fetal bovine serum is required for the maintenance of proteoglycan steady-state metabolism in bovine cartilage explant cultures.[29] Insulin-like growth factors (IGFs) have been shown to be the major component of fetal calf serum (FCS) responsible for up-regulating proteoglycan synthesis.[30,31] IGF-I is a component of freshly isolated bovine articular cartilage and likely plays a similar role *in vivo*. TGF-ß is also capable of maintaining proteoglycan homeostasis in cartilage explants.[32] However, its effect on proteoglycan biosynthesis may have been due to a general increase in protein synthesis. In OA, proteoglycans are lost from the tissue at a rate exceeding deposition of newly synthesized proteoglycan. Interleukin-1 (IL-1) can decrease synthesis[33] and increase catabolism of proteoglycans, even in the presence of concentrations of FCS that would normally maintain homeostasis.[34] IL-6 may also be required for the IL-1-induced inhibition of proteoglycan synthesis since IL-1-induced inhibition of proteoglycan synthesis can be partially reversed by antibodies against recombinant human IL-6.[35]

Regulation of proteoglycan aggregation may not be mediated only by alterations in chondrocyte biosynthetic activity, but may be directly influenced by microenvironmental changes in the cartilage. Kimura et al.[36] has demonstrated that proteoglycan monomers acquire the capacity to bind hyaluronic acid early in biosynthesis. In human cartilage explants, however, the affinity of the monomer for hyaluronic acid may increase after secretion.[37,38] Furthermore, it was shown that the affinity of newly synthesized monomer for hyaluronic acid can be markedly enhanced by exposure to alkaline pH.[39] Conversely, acidic medium pH or static tissue compression can slow the acquisition of high-binding affinity for hyaluronic acid by newly secreted proteoglycans in cartilage explants.[40] Compression of cartilage increases the tissue charge density and results in a decrease in intratissue pH. The lowered pH appears to cause a change in the structure of the G1 domain of the proteoglycan. The significance of these findings to alterations in OA cartilage require further study.

Fragmentation of the link protein is observed in extracts of cartilage from aged individuals,[41] possibly related to proteolytic cleavage *in vivo*. An age-related decrease in the link stability of proteoglycan aggregates formed by articular chondrocytes has been reported by Plaas et al.[42] and link protein was reported to be absent in OA cartilage.[43] This finding was disputed by Bayliss and Venn.[44] In an animal model for OA (the OA mouse strain STR/IN), link protein was reported to be present.[45] There has thus far been no evidence that the OA process reduces chondrocyte link protein synthesis. The content of extractable link protein is similar in nonarthritic and OA human cartilage.[46] Although stromelysin can degrade link protein,[47] it is not known whether this degradation alters link protein function. An antigenic determinant in link proteins that is altered in OA cartilage from different individuals has been identified,[48] the nature of which has not yet been detemined.

C. SMALL, NONAGGREGATING CARTILAGE PROTEOGLYCANS

Small, nonaggregating proteoglycans have been identified in bovine[49] and human[50] articular cartilage where they are present in molar amounts similar to that of the large chondroitin sulfate proteoglycan. Three such molecules found in cartilage, as well as other connective tissues, have been named decorin, biglycan, and fibromodulin, and have been shown to be members of a family of related proteins. While decorin and biglycan are substituted with chondroitin sulfate or dermatan sulfate chains,[51] fibromodulin is substituted with keratan sulfate.[52] With age, a decrease in the amount of newly synthesized biglycan and an increase in the synthesis of decorin was found.[53] Biglycan is more abundant in load-bearing cartilage.[54] *In vitro*, mechanical loading of cartilage explants can increase the synthesis of both biglycan and decorin.[55] Apparent alterations in the biosynthesis of small proteoglycans have been observed in OA cartilage. Transforming growth factor-ß (TGF-ß) has been shown to stimulate synthesis of increased amounts of small proteoglycans in normal bovine cartilage[32] and in rabbit chondrocyte cultures.[56,57] OA canine cartilage synthesized greater amounts of small proteoglycan than did control cartilage, and synthesis was further augmented by TGF-ß.[58] In an effort to mimic pathological changes observed in degenerative and inflammatory arthritis due to increased enzymatic activity,

Bartholomew et al.[59] have shown that, following trypsin treatment, there was an increase in the synthesis of small proteoglycans and a decrease in the synthesis of the large proteoglycan.

Decorin or PGII is a small proteoglycan of cartilage and other connective tissues. It is synthesized as a 40-kDa precursor which is processed to a 36-kDa core protein.[60] It possesses a single GAG chain at serine 4,[61] and up to three N-linked oligosaccharides.[62,63] Following a cysteine-rich region at the amino terminus, the majority of the core protein sequence consists of repeats of a 24-amino acid, leucine-rich sequence. A number of functions have been determined for this small proteoglycan. Decorin influences adhesion of fibroblasts to extracellular matrices.[64] The core protein binds to collagen and affects collagen fibrillogenesis.[65] Decorin core protein has been recently shown to bind the cytokine TGF-ß and may therefore modulate the activity of this growth factor.[66] TGF-ß, in turn, stimulates the synthesis of decorin. In OA cartilage, decorin could conceivably play an important role in cartilage repair, as there is evidence for a role for decorin in control of cell proliferation. Schmidt et al.[67] have observed that although most of the decorin produced by cells in culture is secreted into the medium, a small proportion does appear to be associated with the cell surface. In this location, the molecule could serve to bind to growth factors (e.g., TGF-ß can bind the decorin core protein specifically). It could also provide an additional site for interaction of the cell with the extracellular matrix since decorin binds collagen. The expression of recombinant decorin by Chinese hamster ovary cells converts these morphologically altered cells into a more orderly monolayer which grow to a lower saturation density than cells lacking decorin.[68] A similar role for decorin in chondrocytes has yet to be demonstrated. An alteration in the biosynthesis of decorin has been shown to be responsible for a multisystem connective tissue defect in one well-documented case.[69] The affected individual is described as a progeroid variant with signs of the Ehlers-Danlos syndrome. The disorder was traced to a deficiency in galactosyltransferase I which catalyzes the second glycosyl transfer reaction in the assembly of the dermatan sulfate chain, resulting in the secretion of a glycosaminoglycan-free core protein.[70] Scott and Dodd[63] have recently demonstrated that removal of the N-linked oligosaccharides from the C-terminal domain of the proteodermatan sulfate core protein isolated from bovine skin resulted in a tendency of the proteoglycan to self aggregate in solution. Therefore, the N-linked oligosaccharides may play a role in maintaining this proteoglycan in monomeric form in the tissue and promoting its interaction with collagen. Since posttranslational modification of this small proteoglycan is not inconsequential, abnormalities in glycosaminoglycan chains or N-linked glycosylation could result in atypical and functionally altered extracellular matrices. In human OA cartilage, large and small proteoglycans were found to be present in cartilagenous osteophytes. In partially eroded cartilage, the small proteoglycans could not be detected.[71,72] Chondrocytes from human OA articular cartilage have been shown to synthesize larger amounts of small proteoglycans than cells from nonarthritic cartilage.[73] In an immunohistochemical study , the distribution of decorin was shown to be different in normal and OA cartilage.[74]

Biglycan (or PGI) is another abundant, small proteoglycan of cartilage, bone, and other connective tissues whose function is not yet understood. Although its amino acid sequence is homologous to decorin[75] and its tissue distribution is similar, there are indications that each of these proteoglycans is functionally distinct inasmuch as biglycan does not bind collagen. Biglycan is translated as a 42.5-kDa prepro core protein and is processed to a 38-kDa secreted form.[75] It contains two GAG chains at the amino terminus, a small cysteine-rich region, and the majority of the remainder of the core protein sequence consists of repeats of the 24-residue amino acid consensus sequence also found in decorin.[60,76]

Fibromodulin is 59-kDa collagen binding protein which is homologous to decorin and biglycan, suggesting that all three have evolved from a common ancestral gene. It was originally isolated from cartilage [77] and it is present in many types of connective tissues, including cartilage, tendon, skin, sclera, and cornea. The core protein is 42 kDa in size, containing 23-amino acid residue leucine-rich repeats also found in decorin and biglycan. Like decorin, fibromodulin can inhibit collagen fibril formation *in vitro*.[78] The fibromodulin binding site on the collagen molecule may be different from that of decorin since decorin and fibromodulin have additive effects in a collagen fibrillogenesis assay.[78]

D. NONCOLLAGENOUS PROTEINS

Cartilage collagen interacts with chondrocytes through a number of matrix glycoproteins. Anchorin II[79] and chondronectin[80] are associated with the chondrocyte membrane and bind type II collagen. Although the source of fibronectin in the cartilage extracellular matrix is controversial, fibronectin can be detected in the cartilage extracellular matrix using immunofluorescence techniques if the proteoglycans are enzymatically depleted.[81-83] More convincingly, fibronectin has been isolated from canine articular cartilage[84] and fibronectin was found to be more abundant in OA canine cartilage than in normal cartilage. Alterations in the collagenous component of cartilage may promote metabolic changes in chondrocytes through these transducing glycoproteins. A specific set of proteins is produced by a number of different types of cells following stress, including heat shock. Human chondrocytes isolated from OA cartilage were shown to synthesize at least one of these stress proteins (70 kDa) at physiologic temperatures, whereas chondrocytes obtained from nonpathologic cartilage synthesize the same protein only in response to heat shock at temperatures above 39°C.[85] A number of other cartilage-specific noncollagenous proteins, as well as some with a wider distribution, have been described;[54] their relevance to the pathogenesis of OA will become more apparent with further study.

E. HYALURONIC ACID

Hyaluronic acid (HA) synthesis by chondrocytes has been demonstrated in cartilage explant cultures[86,87] and by chondrocytes.[88,89] From cell culture studies, there appear to be two forms of HA; one form contains proteoglycans complexed to it and another form separated in the ultracentrifuge appears to be devoid of proteoglycan.[89] The latter appears to subserve the form already containing proteoglycan and may be the source for new aggregation of proteoglycan to HA.

The relationship between changes in HA during ageing and those reported to occur in OA have been studied. The concentration of HA increases severalfold from birth to 90 years of age,[90] but this increase does not appear to be due to increased HA synthesis.[91] The polymer size of HA also undergoes age-related changes leading to shorter HA chains.[91] Taken together, the studies indicated that, in all liklihood, an accumulation of degraded HA results in smaller proteoglycan aggregates in adult cartilage.

Only a few studies have addressed the issue of HA changes in OA. In a series of investigations, Manicourt et al.[92,93] reported that in experimentally-induced OA in rabbits (partial medial meniscectomy),[94] the proteoglycan aggregate size decreased and was spread more broadly over the intermediate and large size ranges. The "free" HA from the normal controls had the same aggregating capacity for proteoglycan as highly purified commercial HA. By contrast, the aggregating capacity of HA from OA cartilage was dramatically decreased. Furthermore, the HA size was decreased in the OA cartilage from 7.2×10^5 to 1.6×10^5 Da and suggests that the nature of the HA molecules in OA may substantially contribute to the pathogenesis of the disease in this experimental model. Whether or not the reduced HA size is a function of HA synthesis changes in OA, the result of depolymerization, or a combination of both remains to be established.

III. REGULATIONAL CONTROL POINTS (TRANSCRIPTIONAL AND TRANSLATIONAL) IN BIOSYNTHESIS OF EXTRACELLULAR MATRIX COMPONENTS: POTENTIAL ALTERATIONS IN OA

A. COLLAGENS

A comprehensive review has recently been published[95] concerning regulation of expression of collagen genes. Although type I collagen mRNA can be detected in cartilage,[96] synthesis of type I collagen is not believed to be a significant feature of the early pathogenesis of OA,[97,98] although it may be a component of fibrocartilage which can overgrow exposed bone.[99] Chick cartilage is known to contain type I collagen mRNA, but does not normally synthesize type I collagen. The 5′ end of the mRNA for the $\alpha2(I)$ chain in cartilage is different from the 5′ end of the mRNA in other tissues and in cells that actively synthesize $\alpha2(I)$ collagen. This difference is due to the use of a cartilage-specific promoter within intron 2 of the $\alpha2(I)$ collagen gene. This mRNA contains several small open reading frames which are out of frame with the remainder of the $\alpha2(I)$ chain sequence; thus, no $\alpha2(I)$ collagen chains are transcribed.[100]

Potential regulatory sequences in the promoter and the first intron of the type II collagen gene have been described.[101] Evidence is accumulating that regulation of expression of the type II collagen gene involves both negative and positive regulatory elements. The rat COLII promoter has been shown to contain "silencer" elements that inhibit expression of the gene in fibroblasts,[102] but not in chondrocytes, thus suggesting that these sequences are important in tissue-specific regulation of type II collagen expression. The first intron of the human type II procollagen gene contains elements known to be involved in transcription of other genes.[101] Factors including IL-1, TGF-ß, IGF-I, and interferon-γ may be present in the OA joint which can negatively or positively affect transcriptional regulation of type II collagen synthesis. For example, inhibition of collagen type II synthesis by IL-1 is due to a reduction in the transcription of the type II collagen gene. This reduction in gene transcription involves DNA regulatory sequences that determine type II collagen gene expression.[103] Another level of control of type II collagen gene expression is at the level of mRNA splicing. Alternative splicing of type II collagen pre-mRNA results in two possible isoforms of type II procollagen which are differentially expressed, either containing (type IIA) or lacking (type IIB) a 69-amino acid, cysteine-rich domain in the NH_2 propeptide.[104] The alternatively spliced forms of type II collagen were shown to define distinct populations of cells during vertebral development.[105] Although the role of the alternatively expressed isoforms of type II collagen in OA has yet to be investigated, the expression of type IIB appears to correlate with abundant synthesis and accumulation of cartilagenous extracellular matrix[105] and, as such, may be relevant to cartilage repair.

Type X collagen is a transient, developmentally regulated collagen in vertebrates. It appears to be a product of hypertrophic chondrocytes and is present in the extracellular matrix of presumptive ossification zones of cartilage. Isolation and characterization of cDNA and genomic clones have revealed that the type X collagen gene contains a long, open reading frame without introns, the significance of which is not clear.[106] There is evidence that type X collagen expression may be altered in OA. Immunohistochemical staining of OA tissue revealed a change in the staining pattern for type X collagen as compared to normal cartilage.[107] It was observed that type X collagen expression was increased in and just above the zone of calcified cartilage and also in cells exhibiting cloning.

In the rabbit partial meniscectomy model of OA, there may occur a stimulated synthesis and deposition of collagen type XI along with collagen type II. A coordinate increase in collagen type IX was not evident and, in fact, type IX appeared to be depleted.[108] As occurs with type I collagen, the type IX collagen pre-mRNA undergoes chondrocyte-specific alternative splicing. The α1(IX) collagen gene has two promoters for transcription that will result in an α1(IX) chain either possessing or lacking the NC4 domain, which forms a prominent "knob" at the amino terminal end of the type IX collagen molecule.[109] Expression of these two promoters has tissue specificity, with the end result being that the type IX collagen molecule in cartilage possesses the NC4 domain and the type IX collagen in vitreous lacks this domain. The result of the alternative splicing is the synthesis of a type IX collagen molecule in cartilage which may be functionally different from that which is found in vitreous.[110]

B. PROTEOGLYCANS

There is evidence that mRNA for the human, large, cartilage-specific proteoglycan (aggrecan) is alternatively spliced to yield multiple forms.[111,112] cDNA clones have been isolated from libraries of embryonic human cartilage, either containing or lacking a domain at the carboxyl terminus, which bears homology to the epidermal growth factor (EGF).[111] The presence of this domain could potentially provide a mechanism of feedback regulation affecting chondrocyte biosynthetic activity and proliferation. More recently, the complete coding sequence for the human, large aggregating chondroitin sulfate proteoglycan of cartilage was determined, revealing that there exist at least three forms of aggrecan transcripts generated by alternative exon usage.[112] The alternatively spliced domains include a C-reactive protein-like region, as well as an EGF-like region. There is evidence that the splice variant lacking both domains is the predominant mRNA species in fetal/juvenile cartilage.[112] The regulation of the splicing events and the function of the alternatively spliced domains is not presently known. It is becoming apparent, however, that analysis of aggrecan gene expression in pathologic cartilage will require not only a determination of steady-state mRNA levels for the aggregating proteoglycan, but also an additional analysis of the ratio of the possible spliced variants.

C. NONCOLLAGENOUS PROTEINS

At present, little is known concerning interactions between cartilage fibronectin and other unique components of the cartilage extracellular matrix, such as types II, IX, and XI collagens and aggrecan. Bennett et al.[113] have demonstrated that a unique exon designated exon IIIB is expressed in cartilage fibronectin mRNA that is not present in fibronectin mRNA from other cell or tissue types. This finding suggests that this exon may confer upon the fibronectin molecule the ability to bind cartilage-specific collagen types. It will be of interest to examine reparative or OA cartilage for the presence of this cartilage-specific splice variant which may be critical for chondrocyte-matrix adhesive interactions.

IV. CATABOLIC ENZYMES PRODUCED BY CHONDROCYTES AND ALTERATIONS IN THEIR REGULATION IN OA

A. MATRIX METALLOPROTEINASES

Controlled production of extracellular matrix metalloproteases ("matrixins") is probably required for normal development and turnover of cartilage components.[114] Chondrocytes synthesize and secrete these matrix metalloproteinases as proenzymes.[115] Although the mechanism for activation of these enzymes *in vivo* has yet to be elucidated, it has been shown that these enzymes can be activated by trypsin and plasmin *in vitro*.[116] The so-called "interstitial collagenase", also called matrix metalloprotease I (MMP-1) and produced by connective tissue cells, can degrade collagen types I, II, III,[117] and X.[118] Another enzyme called gelatinase or MMP-2 can degrade denatured collagen (gelatin) as well as native type IV and V collagen.[119,120] A third matrix metalloprotease called stromelysin or MMP-3 was found to have a broad range of substrates,[121] including fibronectin, laminin, collagen IV, cartilage proteoglycans, and link protein.[47] Sites of cleavage by stromelysin of collagen types II, IX, X, and XI of cartilage have recently been determined.[122] Furthermore, stromelysin may indirectly effect matrix degradation by activating latent collagenase.[123] There is recent evidence that plasmin is responsible for activation of collagenase in OA cartilage. Martel-Pelletier et al.[124] have demonstrated that the increased level of plasmin in OA cartilage is due to increased plasminogen activator activity.

These enzymes have been implicated in the joint destruction characteristic of OA, indicating that alterations in their normal regulation may occur in disease. Neutral metalloproteinases are significantly elevated in OA human cartilage as compared to normal cartilage,[125,126] and it has been reported that human OA chondrocytes in culture synthesize increased neutral metalloproteinase as well,[127] suggesting a stable phenotypic alteration in OA chondrocytes. Stromelysin activity has been shown to be elevated in extracts of OA cartilage.[128] Collagenase is produced by human OA cartilage in culture in increasing amount with disease progression.[129] However, chondrocytes in culture synthesize mammalian-type collagenase and its activity towards type II collagen is lower than comparable activity against type I and type III collagen.[130] These results suggest that increases in chondrocyte collagenase in OA cartilage may play a less critical role in the final dissolution of the matrix than the antecedent removal of proteoglycans. Transcription via the stromelysin promoter can be induced by IL-1 and is repressed by dexamethasone.[131] Furthermore, Frisch et al.[131] have shown that stromelysin and collagenase genes appear to be coordinately regulated. Rabbit articular chondrocytes produce collagenase and stromelysin in response to IL-1, but not to FGF. FGF, however, potentiates the action of IL-1. It has recently been shown by Hruby et al.[132] that IL-1 and FGF regulate gene expression for these metalloproteases through different mechanisms. The recent finding that stromelysin degrades cartilage collagens[122] and may play a key role in activating the latent collagenase[123] must be considered as important in regulating chondrocyte responses in OA.

IL-1 has been found in significant levels in OA synovial fluid.[133,134] IL-1 receptors are not diminished in number in OA cartilage-derived chondrocytes, nor is the chondrocyte affinity for IL-1 significantly altered.[135] Rabbit[136] and human[133] chondrocytes show evidence for both the IL-1α and IL-1ß form of the IL-1 receptor. Human chondrocytes *in vitro* respond to recombinant IL-1α and IL-1ß by up-regulating the transcription of the stromelysin-1 (i.e., MMP-3) gene,[137] and chondrocytes derived from human OA cartilage synthesize more stromelysin-1 than chondrocytes derived from nonarthritic cartilage.[138] Evidence for the transcription of the stromelysin-1 gene, but not the stromelysin-2 gene, in normal human articular cartilage was obtained by polymerase chain reaction amplification.[96] Since glucocorticoids can repress metalloprotease synthesis, maintenance or loss of the glucocorticoid receptor activity in chondrocytes may be responsible for efficacy of corticosteroids in the treatment of OA.[139]

B. CYSTEINE PROTEASES

The lysosomal cysteine proteinases cathepsins B and L could potentially be involved in extracellular matrix destruction in OA. Cathepsin B is found in the matrix in rheumatoid synovium.[140] Human OA cartilage was shown to have increased cathepsin B activity when compared to normal cartilage.[141] Cathepsin B is capable of degrading proteoglycans,[142] collagen,[143,144] elastin,[145] and fibronectin.[146] Whether or not changes in cathepsin B levels in cartilage actually alter the proteolytically susceptible matrix components, thus playing an important role in OA, remains contentious despite many studies that have addressed the issue. Cathepsin L was shown to possess several-fold greater activity toward collagen and elastin[145] and was recently shown to be more effective than cathepsin B in degrading proteoglycan aggregates.[147] Cathepsin L produced degraded link protein fragments similar in size to those fragments observed to accumulate in adult human cartilage.[41] Since these enzymes are irreversibly denatured at neutral and alkaline pH,[148] their activity in OA cartilage would seem to require a focal acidic microenvironment.

V. INTRACELLULAR EVENTS

Sometime during the early stages of OA and well before any clinical symptoms of disease are apparent, an attempt at chondrocyte proliferation is aborted. The reason for this important event is unknown, but may represent a loss of the ability of chondrocytes to respond to autocrine or paracrine growth factors. Prior to finding altered surface changes (i.e., roughening or fibrillation), a "halo" surrounding chondrocytes which fails to stain histochemically with metachromatic dyes such as toluidine blue has been found.[149,150] This may represent the first sign of altered extracellular matrix, and the fact that it occurs in the pericellular matrix suggests that this event is regulated by chondrocytes, perhaps in response to exogenous signals, mechanical or otherwise. In mid and deep zones of cartilage, topographically restricted pericellular matrix macromolecules such as type VI collagen[151,152] and thrombospondin[153] could play significant roles if, in the absence of proteoglycan, their functional capacity to bind growth factors or enzyme inhibitors is reduced. The fact that at the surface of cartilage type VI collagen is localized around the cells and also in the matrix[154,155] must also be considered in this regard.

Attempts at restoration of the matrix have been defined by Mankin and colleagues as the "hypermetabolic" state of OA[156] and appears to be a function of disease severity.[157] Eventually, those metabolic processes required to restore the integrity of articular cartilage no longer keep up with degradative events and cartilage loss occurs.[158]

A. PROLIFERATION

Abortive attempts at repair of cartilage accompanying the fibrillated clefts in the matrix are well recognized by the so-called "brood" capsules. They represent the proliferation of isogenous chondrocytes, as shown by the incorporation of tritiated thymidine into DNA.[159] Their capacity to synthesize proteoglycan, as evidenced by autoradiography, suggests that in no way does chondrocyte "activation" suppress the synthesis of matrix components;[160] thus, stimulation of chondrocytes in cell culture with growth factors accompanied by marked suppression of proteoglycan synthesis[161] appears related more to aberrant cell cycle responses caused by growth factors than an inherent inability of chondrocytes to proliferate and also synthesize matrix. This attempt to expand the chondrocyte population is, however, not sustained. The appearance of "halos" around the chondrocyte clusters suggests that early proliferative events in the OA process are also accompanied by dissolution of the pericellular matrix.[149,150] Since cell proliferation requires cell movement, the focal rather than diffuse matrix changes seen at that time may be responsible, in part, for the lack of sustained chondrocyte proliferation in OA. The evidence that more diffuse loss of matrix occurs later, and that the pericellular matrix reappears[149] indicates that chondrocyte proliferation in OA is governed by signals conveyed to the chondrocytes by their microenvironment. Studies showing incorporation of tritiated thymidine *in vitro* by chondrocytes derived from cartilage at "end-stage" OA[162] indicates that the cells retain the capacity to proliferate throughout their life span.

B. NUTRITION

Since articular cartilage is an avascular and aneural tissue, it has often been implied that its principle mode of deriving nutrition is via the synovial fluid.[163] When osteophytes form centrally in OA, defects in the subchondral plate occur and a soft tissue containing blood vessels fills the bone marrow cavities. Thus, assuming bone marrow provides an alternative to articular cartilage nutrition coming from synovial fluid, a reduction in the size of the bone marrow spaces, as was seen during the ageing of the femoral head,[164] may be of significance in affecting chondrocyte metabolism. In addition, changes in synovial fluid content could be of significance in OA considering studies which showed a limitation to which articular cartilage nutrition could be derived from synovial fluid.[165] This is likely to be important in chondrocyte metabolism in the deep zone of articular cartilage.[163] How nutritional deficiencies alter the mechanical properties of cartilage or the metabolism of chondrocytes in OA has not been rigorously established, although alterations in subchondral vascularity at sites of earliest cartilage erosions has been considered.

VI. CARTILAGE REPAIR

It seems reasonable to speculate that normal compensatory matrix biosynthesis requires regulation of the individual matrix components as well as their integration. All of the available evidence to date indicates that OA chondrocytes retain the capacity to synthesize high-density aggregating proteoglycan[73] even when the cartilage is obtained from patients undergoing joint replacement for end-stage disease.[166] Recent evidence, however, indicates that a continuum of changes in the proteoglycans occurs during the various stages of OA which correlates with histological matrix changes.[166] Together with studies showing that significant intracellular processing of the glycosylated, newly synthesized proteoglycan does not occur in either nonarthritic or OA chondrocytes in culture,[73] the evidence points to significant postsynthesis modifications in newly synthesized proteoglycans that probably reduce their ability to integrate into the cartilage matrix.[167] Studies performed in organ explant culture, which show that cleavage of proteoglycans occur predominantly in or around the HABR,[168] have been sustained when cartilage proteoglycan aggregates have been systematically examined during the early stages of experimentally induced canine OA[169] and in human OA.[170] Additional cleavage sites that reduce the keratan sulfate content of canine OA cartilage[171] or render the monomeric form of proteoglycan smaller by multiple cleavages in the CS-rich domain have also been implied to occur.[91]

Although experimental models of cartilage healing have long been used to explore the chondrocytes' capacity to undergo intrinsic repair,[99,172-177] many studies do not address the regulation of chondrocyte repair mechanisms in an unstable joint which is likely to be impacting on chondrocyte repair in the OA process. The characteristics of the repair tissue have varied from study to study[178] and may, in part, reflect the source of repair cells. In one study, the source of cells appeared to be subchondral bone which produced a matrix rich in type II collagen,[177] but uncertainty as to the maintenance of this matrix over long periods or to its biomechanical stability over time has made chondrocyte therapy an alternative attractive proposition. Such regimens will require that chondrocytes be expanded *in vitro* with the maintenance of the chondrogenic phenotype, and that a precise understanding of the impact of the preexisting matrix on the fate of the transplanted chondrocytes be obtained.

VII. MECHANICAL PRESSURES

Alterations in the mechanical forces applied to synovial joints affect the metabolism of articular cartilage. For example, mechanical instability plays an important part in the genesis of osteophytic spurs in an experimental model of OA in rabbits.[179,180] The force changes applied by altering the periarticular ligaments and menisci appear to cause the proliferation and migration of a pluripotential cell type to occur.

Articular cartilage extracellular matrix provides the physicochemical mechanism for resistance to compressive, tensile, and shearing forces. Little is known about how mechanical forces transduce signals to chondrocytes, but it must be envisioned that such forces provide stimuli which regulate chondrocyte metabolism. Stresses applied to articular cartilage appear to be related to tissue thickness, cell density, and the supramolecular structure of the matrix. For example, the amount of matrix per cell appears related to the level of stress applied,[181] and the composition of the matrix, with particular regard to the capacity of proteoglycans to form aggregates with HA, appears to differ between weight-bearing and reduced weight-bearing regions of a synovial joint.[182] Whether or not chondrocytes within these topographical regions differ with respect to their capacity to synthesize proteoglycans that can form proteoglycan aggregates or whether mechanical forces applied to an equipotent chondrocyte generates these differences is not known.

Immobilization causes significant alterations in the composition of articular cartilage, most notably in the loss of chondrocytes which is frequently accompanied by a decrease in cartilage metachromasia.[183] Since the latter is a histochemical measurement of proteoglycans, it would be appropriate to conclude that maintenance of cartilage proteoglycan depends upon joint motion. Thus, decreases in joint motion frequently lead to cartilage degeneration. Additional studies provide evidence that effects of immobilization on chondrocyte proteoglycans are reversible[184] and point to a probable mechanochemical sensing system within cartilage which alters chondrocyte metabolism. Extensive immobilization in addition to resulting in alterations in proteoglycan content also results in loss of cartilage thickness, chondrocyte degeneration (i.e., fibrillation and erosion) with the suggestion that, at some point, effects of inhibition of joint movement may be irreversible.[163] Changes in the nutritional pathway via synovial fluid are compromised in compressive immobilization models of cartilage degeneration.[163]

The signal transduction pathways altered by mechanical forces are not known. Regulation of chondrocyte metabolism in normal synovial joints is, in all likelihood, regulated by cross-talk between several intracellular signaling pathways. Hydrostatic pressure (80 g/cm^2) stimulated DNA synthesis when applied to chondrocytes derived from embryonic growth plate.[185] Increases in the level of cyclic $3',5'$-adenosine monophosphate (cAMP) are induced by applying mechanical tension to chondrocytes in culture[186] or by applying intermittant compressive forces to cartilage explants or chondrocytes.[187] Intermittant compressive forces increase chondrocyte proteoglycan synthesis and the ability of proteoglycans to form aggregates,[188] but other studies using bovine or human cartilage indicate that extracellular matrix and RNA synthesis responses are a function of the magnitude of the applied force and the presence of serum.[189] A decrease in adenylate cyclase has been associated with increased hydrostatic pressure and increased calcium uptake.[190] An increase in proteoglycan synthesis[191-193] in chondrocytes occurred when cells are stimulated by various agonists which led to an increase in accumulation of cAMP. By contrast, canine chondrocyte expression of the ED-A form of fibronectin was decreased by dibutyryl cAMP.[194] HA synthesis by canine cartilage slices was increased by this cAMP analog.[195] Effects of cAMP on collagen synthesis generally result in depressed levels,[196] but is often dependent on cell type.[197] A decrease in cAMP does not necessarily occur due to inhibition of adenylate cyclase, and other possible effects such as stimulation of phosphodiesterase must also be considered. The mechanical sensing "receptor(s)" that result in changes in adenylate cyclase activity may be central to the apparent dichotomy between chondrocyte proliferation and biosynthesis of extracellular matrix macromolecules. In the OA joint, changes in mechanical forces early in disease pathogenesis may govern chondrocyte attempts at repair which is likely to involve these signaling pathways.

REFERENCES

1. **Sokoloff, L.,** Ageing and degenerative diseases affecting cartilage, in *Cartilage,* Vol. 3, Hall, B. K., Ed., Academic Press, New York, 1983, chap. 4.
2. **Radin, E. L., Parker, H. G., Pugh, J. W., Steinberg, R. S., Paul, I. G., and Rose, R. M.,** Responses of joints to impact loading. III. Relationship between trabecular microfractures and cartilage degeneration, *J. Biomechanics,* 6, 51, 1973.
3. **Malemud, C. J. and Shuckett, R.,** Impact loading and lower extremity disease, in *Clinical Concepts in Regional Musculoskeletal Illness,* Hadler, N. M., Ed., Grune and Stratton, Orlando, FL, 1987, chap. 7.
4. **Eyre, D. R., Wu, J.-J., and Apone, S.,** A growing family of collagens in articular cartilage: identification of 5 genetically distinct types, *J. Rheumatol.,* 14 (Suppl. 14), 25, 1987.
5. **Baldwin, C. T., Reginato, A. M., Smith, C., Jimenez, S. A., and Prockop, D. J.,** Structure of cDNA clones coding for human type II procollagen. The alpha 1(II) chain is more similar to the alpha 1(I) chain than two other alpha chains of fibrillar collagens, *Biochem. J.,* 262, 521, 1989.
6. **Meyers, J. C., Loidl, H. R., Stolle, C. A., and Seyer, J. M.,** Partial covalent structure of the human alpha 2 type V collagen chain, *J. Biol. Chem.,* 260, 5533, 1985.
7. **Chu, M.-L., Mann, K., Deutzmann, R., Pribula-Conway, D., Hsu-Chen, C.-C., Bernard, M. P., and Timpl, R.,** Characterization of three constituent chains of collagen type VI by peptide sequences and cDNA clones, *Eur. J. Biochem.,* 168, 309, 1987.
8. **Ninomiya, Y., van der Rest, M., Mayne, R., Lozano, G., and Olsen, B. R.,** Construction and characterization of cDNA encoding the alpha 2 chain of chicken type IX collagen, *Biochemistry,* 24, 4223, 1985.
9. **Bernard, M., Yoshioka, H., Rodriguez, E., van der Rest, M., Kimura, T., Ninomiya, Y., Olsen, B. R., and Ramirez, F.,** Cloning and sequencing of pro-α1(XI) collagen cDNA demonstrates that type XI belongs to the fibrillar class of collagens and reveals that the expression of the gene is not restricted to cartilagenous tissue, *J. Biol. Chem.,* 263, 17159, 1988.
10. **Wu, J. J. and Eyre, D. R.,** Type IX collagen: interfibrillar covalent adhesion protein of the cartilage matrix, *Trans. Orthop. Res. Soc.,* 15, 62, 1989.
11. **Mendler, M., Eich-Bender, S. G., Vaughan, L., Winterhalter, K. H., and Bruckner, P.,** Cartilage contains mixed fibrils of collagen types II, IX, and XI, *J. Cell Biol.,* 108, 191, 1989.
12. **Wu, J. J. and Eyre, D. R.,** Covalent interactions of the cartilage collagens, *Trans. Orthop. Res. Soc.,* 16, 27, 1991.
13. **Thom, J. R. and Morris, N. P.,** Biosynthesis and proteolytic processing of type XI collagen in embryonic chick sterna, *J. Biol. Chem.,* 266, 7262, 1991.
14. **Eyre, D., McDevitt, C. A., Billingham, M. E. J., and Muir, H.,** Biosynthesis of collagen and other matrix proteins by articular cartilage in experimental osteoarthritis, *Biochem. J.,* 188, 823, 1980.

15. **Poole, A. R., Rizkalla, G., Ionescu, M., Rosenberg, L. C., and Bogach, E.,** Increased content of the C-propeptide of type II collagen in osteoarthritic human articular cartilage, *Trans. Orthop. Res. Soc.,* 16, 343, 1991.

16. **Malemud, C. J.,** The biology of cartilage and synovium in animal models of osteoarthritis, in *Handbook of Animal Models for the Rheumatic Diseases,* Vol. 2, Greenwald, R. A. and Diamond, H. S., Eds., CRC Press, Boca Raton, FL, 1988, 3.

17. **Nimni, M. and Deshmukh, K.,** Differences in collagen metabolism between normal and osteoarthritic human articular cartilage, *Science,* 181, 751, 1973.

18. **Jimenez, S. A.,** Defects in the collagen II gene and osteoarthritis, in *Proc. XXXII Eur. Cong. Rheumatol.,* Hodinka, L., Ed., 1991, 23.

19. **Knowlton, R. G., Katzenstein, P. L., Moskowitz, R. W., Weaver, E. J., Malemud, C. J., Pathria, M. N., Jimenez, S. A., and Prockop, D. J.,** Genetic linkage of a polymorphism in the type II procollagen gene (COL2A1) to primary osteoarthritis associated with mild chondrodysplasia, *N. Engl. J. Med.,* 322, 526, 1990.

20. **Palotie, A., Ott, J., Elima, K., Cheah, K., Vaisanen, P., Ryhanen, L., Vikkula, M., Vuoria, M., Vuorio, E., and Peltonen, L.,** Predisposition to familial osteoarthrosis linked to type II collagen gene, *Lancet,* April 29, 924, 1989.

21. **Ala-Kokko, L., Baldwin, C. T., Moskowitz, R. W., and Prockop, D. J.,** Single base mutation in the type II procollagen gene (COL2A1) as a cause of primary osteoarthritis associated with a mild chondrodysplasia, *Proc. Natl. Acad. Sci. U.S.A.,* 87, 6565, 1990.

22. **Eyre, D. R., Weis, M. A., and Moskowitz, R. W.,** Cartilage expression of a type II collagen mutation in an inherited form of osteoarthritis associated with a mild chondrodysplasia, *J. Clin. Invest.,* 87, 357, 1991.

23. **Sandy, J., Adams, M. E., Billingham, M. E., Plaas, A. H. K., and Muir, H.,** In vivo and in vitro stimulation of chondrocyte biosynthetic activity in early experimental osteoarthritis, *Arthritis Rheum.,* 27, 388, 1984.

24. **Hering, T. M. and Sandell, L. J.,** Biosynthesis and processing of bovine cartilage link proteins, *J. Biol Chem.,* 265, 2375, 1990.

25. **Poole, A. R.,** Proteoglycans in health and disease: structures and functions, *Biochem. J.,* 236, 1, 1986.

26. **Bayliss, M. T.,** Proteoglycan structure and metabolism during maturation and ageing of human articular cartilage, *Biochem. Soc. Trans.,* 18, 799, 1990.

27. **Roughley, P. J.,** Changes in cartilage proteoglycan structure during ageing: origin and effects — a review, *Agents Actions,* 18 (Suppl.), 19, 1986.

28. **Morales, T. I. and Hascall, V. C.,** Factors involved in the regulation of proteoglycan metabolism in articular cartilage, *Arthritis Rheum.,* 32, 1197, 1989.

29. **Hascall, V. C., Handley, C. J., McQuillan, D. J., Hascall, G. K., Robinson, H. C., and Lowther, D. A.,** The effect of serum on biosynthesis of proteoglycans by bovine articular cartilage in culture, *Arch. Biochem. Biophys.,* 224, 206, 1983.

30. **McQuillan, D. J., Handley, C. J., Campbell, M. A., Bolis, S., Milway, V. E., and Herington, A. C.,** Stimulation of proteoglycan biosynthesis by serum and insulin-like growth factor-1 in cultured bovine articular cartilage, *Biochem. J.,* 240, 423, 1986.

31. **Luyten, F. P., Hascall, V. C., Nissley, S. P., Morales, T. I., and Reddi, A. H.,** Insulin- like growth factors maintain steady-state metabolism of proteoglycans in bovine articular cartilage explants, *Arch. Biochem. Biophys.,* 267, 416, 1988.

32. **Morales, T. I. and Roberts, A. B.,** Transforming growth factor ß regulates the metabolism of proteoglycans in bovine cartilage organ cultures, *J. Biol. Chem.,* 263, 12828, 1988.

33. **Benton, H. P. and Tyler, J. A.,** Inhibition of cartilage proteoglycan synthesis by interleukin 1, *Biochem. Biophys. Res. Commun.,* 154, 421, 1988.

34. **Morales, T. I. and Hascall, V. C.,** Specificity of action of lipopolysaccharide on bovine articular cartilage metabolism: a comparison with interleukin-1, *Connect. Tissue Res.,* 19, 255, 1989.

35. **Nietfeld, J. J., Wilbrink, B., Helle, M., van Roy, J. L. A. M., den Otter, W., Swaak, A. J. G., and Huber-Bruning, O.,** Interleukin-1-induced interleukin-6 is required for the inhibition of proteoglycan synthesis by interleukin-1 in human articular cartilage, *Arthritis Rheum.,* 33, 1695, 1990.

36. **Kimura, J. H., Thonar, E. J.-M. A., Hascall, V. D., Reiner, A., and Poole, A. R.,** Identification of core protein, an intermediate in proteoglycan biosynthesis in cultured chondrocytes from the Swarm rat chondrosarcoma, *J. Biol. Chem.,* 256, 7890, 1981.

37. **Bayliss, M. T., Ridgway, G. D., and Ali, S. Y.,** Differences in the rates of aggregation of proteoglycans from human articular cartilage and chondrosarcoma, *Biochem. J.,* 215, 705, 1983.

38. **Oegema, T. R.,** Delayed formation of proteoglycan aggregate structures in human articular cartilage disease states, *Nature (London),* 288, 583, 1980.

39. **Plaas, A. H. K. and Sandy, J. D.,** The affinity of newly synthesized proteoglycan for hyaluronic acid can be enhanced by exposure to mild alkali, *Biochem. J.,* 234, 221, 1986.

40. **Sah, R. L.-Y., Grodzinsky, A. J., Plaas, A. H. K., and Sandy, J. D.,** Effects of tissue compression on the hyaluronate-binding properties of newly synthesized proteoglycans in cartilage explants, *Biochem. J.,* 267, 803, 1990.

41. **Mort, J. S., Poole, A. R., and Roughley, P. J.,** Age-related changes in the structure of proteoglycan link proteins present in norman human articular cartilage, *Biochem. J.,* 214, 269, 1983.

42. **Plaas, A. H. K. and Sandy, J. D.,** Age-related decrease in the link-stability of proteoglycan aggregates formed by articular chondrocytes, *Biochem. J.,* 220, 337, 1984.

43. **Perricone, E., Palmoski, M. J., and Brandt, K. D.,** Failure of proteoglycans to form aggregates in morphologically normal aged human hip cartilage, *Arthritis Rheum.,* 20, 1372, 1977.

44. **Bayliss, M. T. and Venn, M.,** Chemistry of human articular cartilage, in *Studies in Joint Diseases,* Vol. 1, Maroudas, A. and Holborow, E. J., Ed., Pitman Medical, London, 1980, 2.

45. **Rostand, K. S., Baker, J. R., Caterson, B., and Christner, J. E.,** Articular cartilage proteoglycans from normal and osteoarthritic mice, *Arthritis Rheum.,* 29, 95, 1986.

46. **Ryu, J. and Treadwell, B. V.,** Characterization of human articular cartilage link proteins from normal and osteoarthritic cartilage, *Ann. Rheum. Dis.,* 41, 164, 1982.

47. **Nguyen, Q., Murphy, G., Roughley, P. J., and Mort, J. S.,** Degradation of proteoglycan aggregate by a cartilage metalloproteinase, *Biochem. J.,* 259, 61, 1989.

48. **Vilamitjana-Amedee, J. and Harmand, M.-F.,** Osteoarthritis-related changes in human articular cartilage. An immunologic study, *Arthritis Rheum.,* 33, 219, 1990.

49. **Campbell, M. A., Handley, C. J., Hascall, V. C., Campbell, R. A., and Lowther, D. A.,** Turnover of proteoglycans in cultures of bovine articular cartilage, *Arch. Biochem. Biophys.,* 234, 275, 1984.

50. **Roughley, P. J. and White, R. J.,** Dermatan sulfate proteoglycans of human articular cartilage, *Biochem. J.,* 262, 823, 1989.

51. **Rosenberg, L. C., Choi, H. U., Tang, L.-H., Johnson, T. L., Pal, S., Webber, C., Reiner, A., and Poole, A. R.,** Isolation of dermatan sulfate proteoglycans from mature bovine articular cartilages, *J. Biol. Chem.,* 260, 6304, 1985.

52. **Oldberg, A., Antonsson, P., Lindblom, K., and Heinegard, D.,** A collagen-binding 59-kd protein (fibromodulin) is structurally related to the small interstitial proteoglycans PG-S1 and PG-S2 (decorin), *EMBO J.,* 8, 2601, 1989.

53. **Melching, L. I. and Roughley, P. J.,** The synthesis of dermatan sulphate proteoglycans by fetal and adult human articular cartilage, *J. Biol. Chem.,* 261, 501, 1989.

54. **Heinegard, D. and Oldberg, A.,** Structure and biology of cartilage and bone matrix noncollagenous macromolecules, *FASEB J.,* 3, 2042, 1989.

55. **Visser, N. A., VanKampen, G. P. J., Van de Stadt, R. J., and Van der Korst, J. K.,** Modulation of the synthesis of dermatan sulfate proteoglycans I and II in intact articular cartilage loaded in vitro, *Trans. Orthop. Res. Soc.,* 16, 364, 1991.

56. **Redini, F., Galera, P., Mauviel, A., Loyau, G., and Pujol, J.-P.,** Transforming growth factor ß stimulates collagen and glycosaminoglycan biosynthesis in cultured rabbit articular chondrocytes, *FEBS Lett.,* 234, 172, 1988.

57. **Malemud, C. J., Killeen, W., Hering, T. M., and Purchio, A. F.,** Enhanced sulfated-proteoglycan core protein synthesis by incubation of rabbit chondrocytes with recombinant transforming growth factor ß₁, *J. Cell. Physiol.,* 149, 152, 1991.

58. **Venn, G., Hardingham, T. E., Lauder, R., and Muir, H.,** TGFß stimulates the synthesis of small CS/DS proteoglycan in canine articular cartilage and its tissue content is increased in experimental OA, *Trans. Orthop. Res. Soc.,* 16, 380, 1991.

59. **Bartholomew, J. S., Handley, C. J., and Lowther, D. A.,** The effects of trypsin treatment on proteoglycan biosynthesis by bovine articular cartilage, *Biochem. J.,* 227, 429, 1985.

60. **Krusius, T. and Ruoslahti, E.,** Primary structure of an extracellular matrix proteoglycan core protein deduced from cloned cDNA, *Proc. Natl. Acad. Sci. U.S.A.,* 83, 7683, 1986.

61. **Chopra, R. K., Pearson, C. H., Pringle, G. A., Fackre, D. S., and Scott, P. G.,** Dermatan sulphate is located on serine-4 of bovine skin proteodermatan sulphate, *Biochem. J.,* 232, 277, 1985.

62. **Sawhney, R. S., Hering, T. M., and Sandell, L. J.,** Biosynthesis of small proteoglycan II (decorin) by chondrocytes and evidence for a procore protein, *J. Biol. Chem.,* 266, 9231, 1991.

63. **Scott, P. G. and Dodd, C. M.,** Self-aggregation of bovine skin proteodermatan sulphate promoted by removal of the three N-linked oligosaccharides, *Connect. Tissue Res.,* 24, 225, 1990.

64. **Lewandowska, K., Choi, H. U., Rosenberg, L. C., Zardi, L., and Culp, L. A.,** Fibronectin-mediated adhesion of fibroblasts: inhibition by dermatan sulfate proteoglycan and evidence for a cryptic glycosaminoglycan-binding domain, *J. Cell Biol.,* 105, 1443, 1987.

65. **Vogel, K. G., Paulsson, M., and Heinegard, D.,** Specific inhibition of type I and type II collagen fibrillogenesis by the small proteoglycan of tendon, *Biochem. J.,* 223, 587, 1984.

66. **Yamaguchi, Y., Mann, D. M., and Ruoslahti, E.,** Negative regulation of transforming growth factor-ß by the proteoglycan decorin, *Nature (London),* 346, 281, 1990.

67. **Schmidt, G., Robenek, H., Harrach, B., Glossl, J., Nolte, V., Hormann, H., Richter, H., and Kress, H.,** Interaction of small dermatan sulfate proteoglycan from fibroblasts with fibronectin, *J. Cell Biol.,* 104, 1683, 1987.

68. **Yamaguchi, Y. and Ruoslahti, E.,** Expression of human proteoglycan in Chinese hamster ovary cells inhibits cell proliferation, *Nature (London),* 336, 244, 1988.

69. **Kresse, H., Rosthoj, S., Quentin, E., Hollmann, J., Glossl, J., Okada, S., and Tonnesen, T.,** Glycosaminoglycan-free small proteoglycan core protein is secreted by fibroblasts from a patient with a syndrome resembling progeroid, *Am. J. Hum. Genet.,* 41, 436, 1987.

70. **Quentin, E., Gladen, A., Roden, L., and Kresse, H.,** A genetic defect in the biosynthesis of dermatan sulfate proteoglycan: galactosyltransferase I deficiency in fibroblasts from a patient with a progeroid syndrome, *Proc. Natl. Acad. Sci. U.S.A.,* 87, 1342, 1990.

71. **Stanescu, V. and Stanescu, R.,** The distribution of proteoglycans of high electrophoretic mobility in cartilages from different species and of different ages, *Biochim. Biophys. Acta,* 757, 377, 1983.

72. **Peyron, J. G., Stanescu, R., Stanescu, V., and Maroteaux, P.,** Distribution electrophoretique particuliere des populations de proteoglycanes dans les zones de regeneration du cartilage arthrosique, *Rev. Rheum. Mal. Osteoartic.,* 45, 569, 1978.

73. **Shuckett, R. and Malemud, C. J.,** Proteoglycans synthesized by chondrocytes of human nonarthritic and osteoarthritic cartilage, *Proc. Soc. Exp. Biol. Med.,* 190, 275, 1989.

74. **Hirotani, H., Iwata, A., Maniwa, S., Kawabe, N., and Tanaka, O.,** Immunohistochemical localization of dermatan sulfate proteoglycans in human articular cartilage, *Trans. Orthop. Res. Soc.,* 15, 302, 1990.

75. **Fisher, L. W., Termine, J. D., and Young, M. F.,** Deduced protein sequence of bone small proteoglycan I (biglycan) shows homology with proteoglycan II (decorin) and several nonconnective tissue proteins in a variety of species, *J. Biol Chem.,* 264, 4571, 1989.

76. **Day, A. A., McQuillan, C. I., Termine, J. D., and Young, M. R.,** Molecular cloning and sequence analysis of the cDNA for small proteoglycan II of bovine bone, *Biochem. J.,* 248, 801, 1987.

77. **Heinegard, D., Larsson, T., Sommarin, Y., Franzen, A., Paulsson, M., and Hedbom, E.,** Two novel matrix proteins isolated from articular cartilage show wide distributions among connective tissues, *J. Biol. Chem.,* 261, 13866, 1986.

78. **Hedbom, E. and Heinegard, D.,** Interactions of a 59 kDa connective tissue matrix protein with collagen I and collagen II, *J. Biol. Chem.,* 264, 6898, 1989.

79. **Von der Mark, K., Mollenhauer, J., Pfaffle, M., van Menxel, M., and Muller, P. K.,** Role of anchorin CII in the interaction of chondrocytes with extracellular collagen, in *Articular Cartilage Biochemistry,* Kuettner, K. E., Schleyerbach, R., and Hascall, V. C., Eds., Raven Press, New York, 1986, 125.

80. **Hewitt, A. T., Varner, H. H., Silver, M. H., Dessau, W., Wilkes, C. M., and Martin, G. R.,** The isolation and partial characterization of chondronectin, an attachment factor for chondrocytes, *J. Biol. Chem.,* 257, 2330, 1982.

81. **Glant, T., Hadhazy, C., Mikecz, K., and Sipos, A.,** Appearance and persistence of fibronectin in cartilage, *Histochemistry,* 82, 149, 1985.

82. **Kosher, R. A., Walker, K. H., and Ledger, P. W.,** Temporal and spatial distribution of fibronectin during development of the embryonic chick limb bud, *Cell Differ.,* 11, 217, 1982.

83. **Melnick, M., Jaskoll, T., Brownell, A. G., MacDougall, M., Bessem, C., and Slavkin, H. C.,** Spatiotemporal patterns of fibronectin distribution during embryonic development. I. Chick limbs, *J. Embryol. Exp. Morphol.,* 63, 193, 1981.

84. **Wurster, N. B. and Lust, G.,** Fibronectin in osteoarthritic canine articular cartilage, *Biochem. Biophys. Res. Commun.,* 109, 1094, 1982.

85. **Kubo, T., Towle, C., Mankin, H. J., and Treadwell, B. V.,** Stress-induced proteins in chondrocytes from patients with osteoarthritis, *Arthritis Rheum.,* 28, 1140, 1985.

86. **Mason, R.,** Recent advances in the biochemistry of hyaluronic acid in cartilage, in *Connective Tissue Research: Chemistry, Biology and Physiology,* Deyl, Z. and Adam, M., Eds., Alan R. Liss, New York, 1981, 87.

87. **Gillard, G. C., Caterson, B., and Lowther, D.,** The synthesis of hyaluronic acid by sheep and rabbit articular cartilage in vitro, *Biochem. J.,* 145, 209, 1973.

88. **Srivastava, V. M. L., Malemud, C. J., and Sokoloff, L.,** Chondroid expression by lapine articular chondrocytes in spinner culture following monolayer growth, *Connect. Tissue Res.,* 2, 127, 1974.

89. **Malemud, C. J., Goldberg, V. M., and Miller, L. M.,** Synthesis of a "free" form of hyaluronic acid by articular chondrocytes in monolayer culture, *Biochem. Int.,* 14, 987, 1987.

90. **Holmes, M. W. A., Bayliss, M. T., and Muir, H.,** Hyaluronic acid in human articular cartilage, *Biochem. J.,* 250, 435, 1988.

91. **Hardingham, T. and Bayliss, M.,** Proteoglycans of articular cartilage: changes in ageing and in joint disease, *Semin. Arthritis Rheum.,* 20(Suppl. 1), 12, 1990.

92. **Manicourt, D., Howell, D. S., Moskowitz, R. W., Goldberg, V., Malemud, C. J., and Pita, J. C.,** Application of new techniques to separation of proteoglycan aggregates from normal and destabilized rabbit articular cartilage, *Acta Biol. Hung.,* 35, 137, 1984.

93. **Manicourt, D., Howell, D. S., Pita, J. C., Moskowitz, R. W., Goldberg, V., and Malemud, C.,** Studies of cartilage proteoglycans in an osteoarthritic rabbit model, in *Osteoarthritis — Current Clinical and Fundamental Problems,* Peyron, J. G., Ed., CIBA-Geigy, Paris, 1985, 192.

94. **Moskowitz, R. W., Davis, W., Sammarco, J., Marteus, M., Baker, J., Mayor, M., Burstein, A. H., and Frankel, V. C.,** Experimentally induced degenerative joint lesions following partial meniscectomy in the rabbit, *Arthritis Rheum.,* 16, 397, 1973.

95. **Bornstein, P. and Sage, H.,** Regulation of collagen gene expression, *Prog. Nucleic Acid Res. Mol. Biol.,* 37, 66, 1989.

96. **Hakala, B. E., Mort, J. S., and Recklies, A. D.,** Amplification of specific messenger RNA species isolated directly from human articular cartilage, *Trans. Orthop. Res. Soc.,* 16, 310, 1991.

97. **Livne, E., von der Mark, K., and Silbermann, M.,** Morphologic and cytochemical changes in maturing and osteoarthritic articular cartilage in the temporomandibular joint of mice, *Arthritis Rheum.,* 28, 1027, 1985.

98. **Floman, Y., Eyre, D. R., and Glimcher, M. J.,** Induction of osteoarthrosis in the rabbit knee joint: biochemical studies on the articular cartilage, *Clin. Orthop. Relat. Res.,* 147, 278, 1980.

99. **Furukawa, T., Eyre, D. R., Koide, S., and Glimcher, M. J.,** Biochemical studies on repair cartilage resurfacing experimental defects in the rabbit knee, *J. Bone Joint Surg. (Am.),* 62, 79, 1980.

100. **Bennett, V. D. and Adams, S. L.,** Identification of a cartilage-specific promoter within intron 2 of the chick α2(I) collagen gene, *J. Biol. Chem.,* 265, 2223, 1990.

101. **Ryan, M. C., Sieraski, M., and Sandell, L. J.,** The human type II procollagen gene: identification of an additional protein-coding domain and location of potential regulatory sequences in the promoter and first intron, *Genomics,* 8, 41, 1990.

102. **Savagner, P., Miiyashita, T., and Yamada, Y.,** Two silencers regulate the tissue-specific expression of the collagen II gene, *J. Biol. Chem.,* 265, 6669, 1990.

103. **Chandrasekhar, S., Harvey, A. K., Higginbotham, J. D., and Horton, W. E.,** Interleukin-1-induced suppression of type II collagen gene transcripton involves DNA regulatory elements, *Exp. Cell. Res.,* 191, 105, 1990.

104. **Ryan, M. C. and Sandell, L. J.,** Differential expression of a cysteine-rich domain in the amino-terminal propeptide of type II (cartilage) procollagen by alternative splicing of mRNA, *J. Biol. Chem.,* 265, 10334, 1990.

105. **Sandell, L. J., Morris, N., Robbins, J. R., and Goldring, M. B.,** Alternatively spliced type II procollagen mRNAs define distinct populations of cells during vertebral development: differential expression of the amino-propeptide, *J. Cell Biol.,* 114, 1307, 1991.

106. **Ninomiya, Y., Gordon, M., van der Rest, M., Schmid, T., Linsenmeyer, T., and Olsen, B. R.,** The developmentally regulated type X collagen gene contains a long open reading frame without introns, *J. Biol. Chem.,* 261, 5041, 1986.

107. **Walker, G., Fischer, M., Thompson, R. C., and Oegema, T. R.,** The expression of type X collagen in osteoarthritis, *Trans. Orthop. Res. Soc.,* 16, 340, 1991.

108. **Friedman, J. B., Winfield, S., Robbins, J., Lanzer, W. L., Sandell, L. J., and Eyre, D. R.,** Metabolic effects on collagen types II, IX and XI of articular cartilage in a rabbit meniscectomy model of osteoarthrosis, *Trans. Orthop. Res. Soc.,* 15, 63, 1990.

109. **Nishomura, I., Muragaki, Y., and Olsen, B. R.,** Tissue-specific forms of type IX collagen-proteoglycan arise from the use of two widely separated promoters, *J. Biol. Chem.,* 264, 20033, 1989.

110. **Brewton, R. G., Wright, D. W., and Mayne, R.,** Structural and functional comparison of type IX collagen-proteoglycan from chicken cartilage and vitreous humor, *J. Biol. Chem.,* 266, 4752, 1991.

111. **Baldwin, C. T., Reginato, A. M., and Prockop, D. J.,** A new epidermal growth factor-like domain in the human core protein for the large cartilage-specific proteoglycan, *J. Biol. Chem.,* 264, 15747, 1989.

112. **Doege, K. J., Sasaki, M., Kimura, T., and Yamada, Y.,** Complete coding sequence and deduced primary structure of the human cartilage large aggregating proteoglycan, aggrecan, *J. Biol. Chem.,* 266, 894, 1991.

113. **Bennett, V. D., Pallante, K. M., and Adams, S. L.,** The splicing pattern of fibronectin mRNA changes during chondrogenesis resulting in an unusual form of the mRNA in cartilage, *J. Biol. Chem.,* 266, 5918, 1991.

114. **Woessner, J. F., Jr.,** Matrix metalloproteinase and their inhibitors in connective tissue remodeling, *FASEB J.,* 5, 2145, 1991.

115. **DiPasquale, G., Caccese, R., Pasternak, R., Conaty, J., Hubbs, S., and Perry, K.,** Proteoglycan- and collagen-degrading enzymes from human interleukin-1-stimulated chondrocytes from several species: proteoglycanase and collagenase inhibitors as potentially new disease-modifying antiarthritic agents, *Proc. Soc. Exp. Biol. Med.,* 183, 262, 1986.

116. **Chengshi, H., Wilhelm, S. M., Pentland, A. P., Marmer, B. L., Grant, G. A., Eisen, A. Z., and Goldberg, G. I.,** Tissue cooperation in a proteolytic cascade activating human interstitial collagenase, *Proc. Natl. Acad. Sci. U.S.A.,* 86, 2632, 1989.

117. **Harris, E. D., Jr., Welgus, H. G., and Krane, S. M.,** Regulation of mammalian collagenases, *Collagen Relat. Res.,* 4, 493, 1984.

118. **Schmid, T. M., Mayne, R., Jeffrey, J. J., and Linsenmeyer, T. F.,** Type X collagen contains two cleavage sites for a vertebrate collagenase, *J. Biol. Chem.,* 261, 4184, 1986.

119. **Murphy, G., Cockett, M. I., Stephens, P. E., Smith, B. J., and Docherty, A. J. P.,** Purification and characterization of a bone metalloproteinase that degrades gelatin and types IV and V collagen, *Biochim. Biophys. Acta,* 831, 49, 1987.

120. **Hibbs, M. S., Hasty, K. A., Seyer, J. M., Kang, A. H., and Mainardi, C. L.,** Biochemical and immunological characterization of the secreted forms of human neutrophil gelatinase, *J. Biol. Chem.,* 260, 2493, 1985.

121. **Okada, Y., Nagase, H., and Harris, E. D., Jr.,** A metalloprotease from human rheumatoid synovial fibroblasts that digests connective tissue matrix components, *J. Biol. Chem.,* 261, 14245, 1986.

122. **Wu, J.-J., Lark, M. W., Chun, L. E., and Eyre, D. R.,** Sites of stromelysin cleavage in collagen types II, IX, X, and XI of cartilage, *J. Biol. Chem.,* 266, 5625, 1991.

123. **Murphy, G., Cockett, M. I., Stephens, P. E., Smith, B. J., and Docherty, A. J. P.,** Stromelysin is an activator of procollagenase, *Biochem. J.,* 248, 265, 1987.

124. **Martel-Pelletier, J., Cloutier, J.-M., and Pelletier, J.-P.,** Role of plasminogen activators and inhibitor in the enzymatic degradation of human osteoarthritic cartilage, *Trans. Orthop. Res. Soc.,* 16, 319, 1991.

125. **Martel-Pelletier, J., Pelletier, J. P., Cloutier, J. M., Howell, D. S., Ghandur-Mnaymneh, L., and Woessner, J. F., Jr.,** Neutral proteases capable of proteoglycan digesting activity in osteoarthritic and normal human articular cartilage, *Arthritis Rheum.,* 27, 305, 1984.

126. **Pelletier, J.-P., Martel-Pelletier, J., Howell, D. S., Ghandur-Mnaymneh, L., Enis, J. E., and Woessner, J. F., Jr.,** Collagenase and collagenolytic activity in human osteoarthritic cartilage, *Arthritis Rheum.,* 26, 63, 1983.

127. **Nojima, T., Towle, C. A., Mankin, H. J., and Treadwell, B. V.,** Secretion of higher levels of active proteoglycanases from human osteoarthritic chondrocytes, *Arthritis Rheum.,* 29, 292, 1986.

128. **Dean, D. D., Martel-Pelletier, J., Pelletier, J. P., Howell, S. S., and Woessner, J. F.,** Evidence for metalloproteinase and metalloproteinase inhibitor imbalance in human osteoarthritic cartilage, *J. Clin. Invest.,* 84, 678, 1989.

129. **Ehrlich, M. G., Houle, P. A., Vigliani, G., and Mankin, H. J.,** Correlation between articular cartilage collagenase activity and osteoarthritis, *Arthritis Rheum.,* 21, 761, 1978.

130. **Kresina, T. F. and Malemud, C. J.,** Susceptibility of interstitial rabbit collagens to rabbit articular chondrocyte collagenase, *Collagen Relat. Res.,* 4, 453, 1984.

131. **Frisch, S. M. and Ruley, H. E.,** Transcription from the stromelysin promoter is induced by interleukin-1 and repressed by dexamethasone, *J. Biol. Chem.,* 262, 16300, 1987.

132. **Hruby, P. S., Harvey, A. K., and Chandrasekhar, S.,** Differential regulation of stromelysin mRNA expression by FGF and IL-1 on rabbit articular chondrocytes, *Trans. Orthop. Res. Soc.,* 16, 318, 1991.

133. **Wood, D. D., Ihrie, I. E., Dinarello, C., and Cohen, D. L.,** Isolation of interleukin-1-like factor from human joint effusions, *Arthritis Rheum.,* 26, 975, 1983.

134. **Pelletier, J. P. and Martel-Pelletier, J.,** Evidence for the involvement of interleukin-1 in human osteoartritic cartilage degradation: protective effect of NSAIDs, *J. Rheumatol.,* 16 (Suppl. 18), 19, 1989.

135. **Pelletier, J.-P., McCollum, R., and Martel-Pelletier, J.,** Regulation of human chondrocytes IL-1 receptor expression by cytokines and antirheumatic drugs, *Arthritis Rheum.*, 33, S13, 1990.

136. **Chin, J. E. and Horuk, R.,** Interleukin-1 receptors on rabbit articular chondrocytes: relationship between biological activity and receptor binding kinetics, *FASEB J.*, 4, 1481, 1990.

137. **Ganu, V. S., Hu, S.-I., Arsenis, C., Melton, R., Winter, C., Goldberg, V., and Malemud, C. J.,** Biochemical and molecular characterization of stromelysin secreted by human chondrocytes in culture stimulated with rhIL-1α and rhIL-1ß, *Arthritis Rheum.*, 33, S12, 1990.

138. **Ganu, V. S. and Malemud, C. J.,** unpublished data, 1991.

139. **Dibattista, J. A., Martel-Pelletier, J., Wosu, L. O., Sandor, T. S., Antakly, T., and Pelletier, J.-P.,** Glucocorticoid receptor mediated inhibition of interleukin-1 stimulated neutral metalloprotease synthesis in normal human chondrocytes, *J. Clin. Endocrinol. Metab.*, 72, 316, 1991.

140. **Mort, J. S., Recklies, A. D., and Poole, A. R.,** Extracellular presence of the lysosomal proteinase cathepsin B in rheumatoid synovium and its activity at neutral pH, *Biochem. J.*, 27, 509, 1984.

141. **Bayliss, M. T. and Ali, S. Y.,** Studies on cathepsin B in human articular cartilage, *Biochem. J.*, 171, 149, 1978.

142. **Roughley, P. J. and Barrett, A. J.,** The degradation of cartilage proteoglycans by tissue proteinases, *Biochem. J.*, 167, 629, 1977.

143. **Etherington, D. J.,** The purification of bovine cathepsin B1 and its mode of action on bovine collagens, *Biochem. J.*, 137, 547, 1974.

144. **Burleigh, M. C., Barrett, A. J., and Lazarus, G.,** Cathepsin B1, a lysosomal enzyme that degrades native collagen, *Biochem. J.*, 137, 387, 1974.

145. **Mason, R. W., Johnson, D. A., Barrett, A. J., and Chapman, H. A.,** Elastinolytic activity of human cathepsin L, *Biochem. J.*, 233, 925, 1986.

146. **Isemura, M., Yosizawa, Z., Takahashi, K., Kosaka, H., Kojima, N., and Ono, T.,** Characterization of porcine plasma fibronectin and its fragmentation by porcine liver cathepsin B1, *J. Biochem. (Tokyo)*, 90, 1, 1981.

147. **Nguyen, Q., Mort, J. S., and Roughley, P. J.,** Cartilage proteoglycan aggregate is degraded more extensively by cathepsin L than by cathepsin B, *Biochem. J.*, 266, 569, 1990.

148. **Barrett, A. J. and Kirschke, H.,** Cathepsin B, cathepsin H, and cathepsin L, *Methods Enzymol.*, 80, 535, 1981.

149. **Sachs, B. L., Goldberg, V. M., Getzy, L., Moskowitz, R. W., and Malemud, C. J.,** A histopathologic differentiation of tissue types in human osteoarthritic cartilage, *J. Rheumatol.*, 9, 210, 1982.

150. **Pelletier, J.-P., Martel-Pelletier, J., Cloutier, J. F., and Woessner, J. F., Jr.,** Acid metalloproteases in human osteoarthritic (OA) cartilage: the effects of intraarticular steroids, *Arthritis Rheum.*, 30, 541, 1987.

151. **Ayad, S. and Weiss, J. B.,** A new look at vitreous-humour collagen, *Biochem. J.*, 218, 835, 1984.

152. **Ayad, S., Evans, H., Weiss, J. B., and Hoh, L.,** Type VI collagen but not type V collagen is present in cartilage, *Collagen Relat. Res.*, 4, 165, 1984.

153. **Miller, R. R. and McDevitt, C. A.,** Thrombospondin is present in articular cartilage and is synthesized by articular chondrocytes, *Biochem. Biophys. Res. Commun.*, 153, 708, 1988.

154. **Poole, C. A., Ayad, S., and Schofield, J. R.,** Chondrons from articular cartilage. I. Immunolocalization of type VI collagen in the pericellular capsule of isolated canine tibial chondrons, *J. Cell Sci.*, 90, 635, 1988.

155. **Keene, D. R., Engvall, E., and Glanville, R. W.,** Ultrastructure of type VI collagen in human skin and cartilage suggests an anchoring function for this filamentous network, *J. Cell Biol.*, 107, 1995, 1988.

156. **Mankin, H. J. and Lippiello, L.,** Biochemical and metabolic abnormalities in articular cartilage from osteoarthritic human hips, *J. Bone Joint Surg., (Am.)*, 52, 424, 1970.

157. **Mankin, H. J., Dorfman, H., Lippiello, L., and Zarins, A.,** Biochemical and metabolic abnormalities in articular cartilage from osteoarthritic human hips. II. Correlation of morphology with biochemical and metabolic data, *J. Bone Joint Surg. (Am.),* 53, 523, 1971.

158. **Malemud, C. J.,** Proteolytic degradation of cartilage sulfated-proteoglycans and its consequence in the osteoarthritic process, in *Proc. XVI Int. Congr. Rheumatology,* Brooks, P. M. and York, J. R., Eds., Elsevier Science, Amsterdam, 1985, 85.

159. **Hirotani, H. and Ito, T.,** Chondrocyte mitosis in the articular cartilage of femoral heads with various diseases, *Acta Orthop. Scand.,* 48, 979, 1975.

160. **Havdrup, T. and Telhag, H.,** Mitosis of chondrocytes in normal adult cartilage, *Clin. Orthop. Relat. Res.,* 153, 248, 1980.

161. **Malemud, C. J. and Sokoloff, L.,** Some biological characteristics of a pituitary growth factor (CGF) for cultured lapine articular chondrocytes, *J. Cell. Physiol.,* 84, 171, 1974.

162. **Malemud, C. J., Goldberg, V. M., and Moskowitz, R. W.,** DNA synthesis and cell proliferation by topographically distinct components of human osteoarthritic cartilage, *Arthritis Rheum.,* 22, 637, 1979.

163. **Stockwell, R. A.,** *Biology of Cartilage Cells,* Cambridge University Press, Cambridge, 1979, 174.

164. **Woods, C. G., Greenwald, A. S., and Haynes, D. W.,** Subchondral vascularity in the human femoral head, *Ann. Rheum. Dis.,* 29, 138, 1970.

165. **Maroudas, A.,** Physicochemical properties of cartilage in the light of nonexchange theory, *Biophys. J.,* 8, 575, 1968.

166. **Shuckett, R. and Malemud, C. J.,** Distinct cartilage proteoglycan chromatographic elution patterns in advanced human hip osteoarthritis: correlations with histologic analysis, *J. Rheumatol.,* 17, 357, 1990.

167. **Malemud, C. J.,** Changes in proteoglycans in osteoarthritis: biochemistry, ultrastructure and biosynthetic processing, *J. Rheumatol.,* 18 (Suppl. 27), 60, 1991.

168. **Sandy, J. D., Flannery, C. R., and Plaas, A. H. K.,** Structural studies on proteoglycan catabolism in rabbit articular cartilage explant cultures, *Biochim. Biophys. Acta,* 931, 255, 1987.

169. **Pelletier, J.-P., Martel-Pelletier, J., Mehraban, F., and Malemud, C. J.,** Immunological analysis of proteoglycan structural changes in the early stage of experimental osteoarthritic canine cartilage lesions, *Arthritis Rheum.,* 34, 562, 1991.

170. **Martel-Pelletier, J., Pelletier, J. P., and Malemud, C. J.,** Activation of neutral metalloprotease in human osteoarthritic knee cartilage: evidence for degradation in the core protein of sulphated proteoglycan, *Ann. Rheum. Dis.,* 47, 808, 1988.

171. **Cruz, T. F., Malcolm, A. J., and Adams, M. E.,** The effect of maturation and anterior cruciate ligament transection on the level of keratan sulfate in the serum of dogs, *J. Rheumatol.,* 16, 1345, 1989.

172. **Calandruccio, R. A. and Gilmer, W. S., Jr.,** Proliferation, regeneration and repair of articular cartilage of immature animals, *J. Bone Joint Surg. (Am.),* 44, 431, 1962.

173. **Mitchell, N. and Shepard, N.,** The resurfacing of adult articular cartilage by multiple perforations through the subchondral bone, *J. Bone Joint Surg. (Am.),* 58, 230, 1976.

174. **Mankin, H. J.,** The response of articular cartilage to mechanical injury, *J. Bone Joint Surg. (Am.),* 64, 460, 1982.

175. **Thompson, R. C., Jr.,** An experimental study of surface injury to articular cartilage and enzyme responses within the joint, *Clin. Orthop. Rel. Res.,* 107, 239, 1975.

176. **Cheung, H. S., Cottrell, W. H., Stephenson, K., and Nimni, M. E.,** In vitro collagen biosynthesis in healing and normal rabbit articular cartilage, *J. Bone Joint Surg. (Am.),* 60, 107, 1978.

177. **Cheung, H. S., Lynch, K. L., Johnson, R. P., and Brewster, B. J.,** In vitro synthesis of tissue-specific type II collagen by healing cartilage. I. Short-term repair of cartilage by mature rabbits, *Arthritis Rheum.,* 23, 211, 1980.

178. **Buckwalter, J. A., Rosenberg, L. C., and Hunziker, E. B.,** Articular cartilage: composition, structure, response to injury and methods of facilitating repair, in *Articular Cartilage and Knee Joint Function: Basic Science and Arthroscopy,* Ewing, J. W., Ed., Raven Press, New York, 1990, chap. 2.

179. **Moskowitz, R. W. and Goldberg, V. M.,** Osteophyte evolution: studies in an experimental partial meniscectomy model, *J. Rheumatol.,* 14 (Suppl. 14), 116, 1987.

180. **Moskowitz, R. W. and Goldberg, V. M.,** Studies of osteophyte pathogenesis in experimentally induced osteoarthritis, *J. Rheumatol.,* 14, 311, 1987.

181. **Stockwell, R. A.,** The interrelationship of cell density and cartilage thickness in mammalian articular cartilage, *J. Anat.,* 109, 411, 1971.

182. **Malemud, C. J., Papay, R. S., Killeen, W., Mehraban, F., Cloutier, J.-M., Pelletier, J.-P., and Martel-Pelletier, J.,** Connective tissue responses to altered joint mechanics: proteoglycans from human knee cartilage differ according to topographical location, in *Proc. First World Congr. Biomech.,* 1990, 23.

183. **Akeson, W. H., Woo, S. L. Y., Amiel, D., Coutts, R. D., and Daniel, D.,** The connective tissue response to immobility: biochemical changes in periarticular connective tissue of the immobilized rabbit knee, *Clin. Orthop. Relat. Res.,* 93, 356, 1973.

184. **Palmoski, M., Perricone, E., and Brandt, K. D.,** Development and reversal of a proteoglycan aggregation defect in normal canine knee cartilage after immobilization, *Arthritis Rheum.,* 22, 508, 1979.

185. **Rodan, G. A., Mensi, T., and Harvey, A.,** A quantitative method for the application of compressive forces to bone in tissue culture, *Calcif. Tissue Res.,* 18, 125, 1975.

186. **DeWitt, M. T., Handley, C. R., Oakes, B. W., and Lowther, D. A.,** In vitro response of chondrocytes to mechanical loading. The effect of short term mechanical tension, *Connect. Tissue Res.,* 12, 97, 1984.

187. **Veldhuijzen, J. P., Bourret, L. A., and Rodan, G. A.,** In vitro studies of the effect of intermittant compressive forces on cartilage cell proliferation, *J. Cell. Physiol.,* 98, 299, 1979.

188. **VanKampen, G. P. J., Veldhuijzen, J. P., Kuijer, R., van de Stadt, R. J., and Schipper, C. A.,** Cartilage response to mechanical force in high-density chondrocyte cultures, *Arthritis Rheum.,* 28, 419, 1985.

189. **Lippiello, L., Kaye, C., Neumata, T., and Mankin, H. J.,** In vitro metabolic response of articular cartilage segments to low levels of hydrostatic pressure, *Connect. Tissue Res.,* 13, 99, 1985.

190. **Bouret, L. A. and Rodan, G. A.,** The role of calcium in the inhibition of cAMP accumulation in epiphyseal cartilage cells exposed to physiological pressures, *J. Cell. Physiol.,* 88, 353, 1976.

191. **Malemud, C. J. and Papay, R. S.,** Stimulation of cyclic AMP in chondrocyte cultures: effects on sulfated-proteoglycan synthesis, *FEBS Lett.,* 167, 343, 1984.

192. **Malemud, C. J., Mills, T. M., Shuckett, R., and Papay, R. S.,** Stimulation of sulfated proteoglycan synthesis by forskolin in monolayer cultures of rabbit articular chondrocytes, *J. Cell. Physiol.,* 129, 51, 1986.

193. **Vasan, N. S. and Lamb, K. M.,** Stage-related changes in extracellular matrix and intracellular events during cytodifferentiation and morphogenesis of limb development, *J. Cell Biol.,* 101, (Abstr.) 335, 1985.

194. **Burton-Wurster, N., Leipold, H. R., and Lust, G.,** Dibutyryl cyclic AMP decreases expression of ED-A fibronectin by canine chondrocytes, *Biochem. Biophys. Res. Commun.,* 154, 1088, 1988.

195. **Stack, M. T. and Brandt, K. D.,** Dibutyryl cyclic AMP affects hyaluronate synthesis and macromolecular organization in normal adult articular cartilage in vitro, *Biochim. Biophys. Acta,* 631, 264, 1980.

196. **Chao, W.-T. H. and Walkenabach, R. J.,** Effect of cyclic AMP on collagen production by corneal fibroblasts, *Curr. Eye Res.,* 5, 177, 1986.

197. **Manner, G. and Kuleba, M.,** Effect of dibutyryl-cyclic AMP on collagen and noncollagen protein synthesis in cultured human cells, *Connect. Tissue Res.,* 2, 167, 1974.

Chapter 13

REGULATION OF TUMOR-DERIVED AND IMMORTALIZED CHONDROCYTES

Motomi I. Enomoto and Masaharu Takigawa

TABLE OF CONTENTS

I. INTRODUCTION

Numerous cell lines have been established which have brought great progress to the fields of cell and molecular biology. It would have been impossible to investigate the mechanism of the regulation of both cell proliferation and differentiation without the use of cell lines. Furthermore, established cell lines have played an important role in the investigation and determination of cancer therapy. The manner of establishment of cell lines can be classified into three groups: (1) spontaneous immortalization of normal cells, (2) high-frequency immortalization of cells derived from malignancies, and (3) immortalization by transformation, either by chemical carcinogens or oncogenic viruses. The introduction of oncogenes into cells has become a popular method of obtaining immortalized cell lines because the immortalization mechanism is clear.

Isolated chondrocytes can form a solid tissue when implanted subcutaneously into nude mice[1] and grow in agarose gels[2] or in suspension culture.[3] These characteristics have been associated with transformed cells; therefore, chondrocytes seem to have tumor cell-like properties. However, after several passages in culture, they easily lose their differentiated phenotype and their ability to proliferate.[1,2] This suggests that there is a unique control of the proliferation and differentiation in chondrocytes. Many investigators have tried to understand this uniquely controlled system by using normal chondrocytes in culture. However, the comparative study of chondrosarcoma-derived cells and transformed chondrocytes with normal chondrocytes would provide a more comprehensive understanding of the mechanism involved in the regulation of proliferation and differentiation. To perform these studies easily, efficiently, and practically, attempts have been made to establish clonal chondrocyte cell lines. Very recently, our research group has established two clonal cell lines which retain the chondrocyte phenotype.[4-9] These cell lines, along with several other immortalized chondrocyte cell lines, are critical to these studies. In this chapter, we discuss the regulation of proliferation and the expression of the differentiated phenotype of chondrocytes derived from chondrosarcoma or infected with oncogenic viruses, as compared with those in normal chondrocytes.

II. CHONDROCYTES DERIVED FROM CHONDROSARCOMA

A. HUMAN CHONDROSARCOMA (HCS) CELL LINES

Several investigators have tried to establish chondrogenic cell lines from human and rodent chondrosarcomas. However, only two clonal cell lines, HCS-2/8 and HCS-2/A, which we established from a human chondrosarcoma have retained the chondrocyte phenotype during many serial passages.[4-9] The phenotype of other cell lines has been fibroblastic or fibrosarcoma-like. For example, Miller et al.[10] tried to culture cells from a human chondrosarcoma, but the cells did not produce a metachromatic matrix, even in primary culture. Thien and Lotan[11] reported on the isolation of chondrosarcoma cell lines, but their cell lines were fiboblast-like in morphology and there was no determination of the expression of chondrogenic markers. The immortalized cell lines which maintain chondrogenic properties provide an important tool for the investigation of the regulation of chondrocytes in malignancy. Therefore, we will concentrate on the HCS-2/8 and 2/A cell lines in this section.

1. HCS-2/8 Cells

HCS cells were isolated from a human chondrosarcoma of the proximal part of the humeral bone of a 72-year-old Japanese male. This tumor was identified as a well-differentiated type of chondrosarcoma.[4] The cells were cloned at the 22nd passage in culture, and the HCS-2/8 clone was selected using proteoglycan synthesis as a marker.

a. Growth Properties and Morphology[4,5]

HCS-2/8 cells proliferate with a doubling time of 3.5 d in MEM containing 20% fetal bovine serum. After reaching confluence, the cells continue to slowly proliferate to a saturation density of about $2.5 \times 10^5/cm^2$. In sparse cultures, HCS-2/8 cells have a slightly elongated polygonal shape (Plate 1, part A). As they become confluent, they become polygonal and then spherical, forming nodules which show strong metachromasia when stained with toluidine blue (Plate 1, parts B and C). The nodules are a three-dimensional structure; the cells are multilayered in the surface regions, overlying a thick layer of extracellular matrix which shows metachromasia (Plate 1, part D). By electron microscopic analysis, it was found that the cells within the nodules, which resemble chondrocytes *in vivo,* are surrounded by an extracellular matrix consisting of thin collagen-like fibrils with numerous fine granules, presumably composed of proteoglycans (Figures 1A and B). In other words, HCS-2/8 cells produce a cartilage-like matrix.

b. Characterization of Proteoglycans[4,5]

HCS-2/8 cells produce proteoglycan aggregates composed of proteoglycan monomers, hyaluronic acid, and a link protein. Using glycerol gradient centrifugation and size exclusion chromatography, the proteoglycan monomers synthesized by the cells are identified as cartilage-specific, large proteoglycan monomers. The major glycosaminoglycans contained within the proteoglycan monomers is about 38 kDa when determined by Superose 12 FPLC. The glycosaminoglycans are mainly composed of chondroitin sulfate, especially chondroitin-6-sulfate. The proteoglycan monomers also contain keratan sulfate. The cells synthesize a single link protein with a molecular weight of 42 kDa. These findings indicate that HCS-2/8 cells synthesize proteoglycans of similar composition and structure to those in normal cartilage.

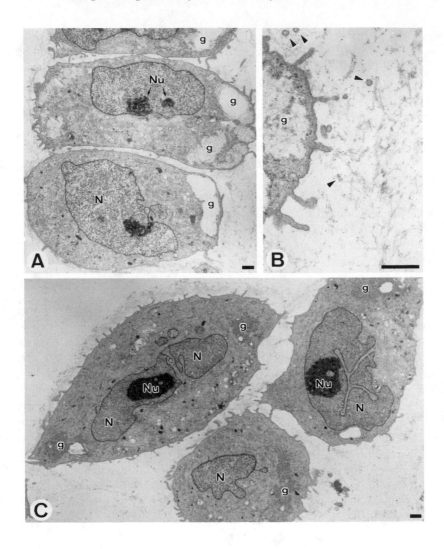

FIGURE 1. Electron micrographs of clonal HCS cells. Upper panels: HCS-2/8 cells; Left, typical HCS-2/8 cells located in the nodule. N, nucleus; Nu, nucleolus; g, glycogen (Some glycogen was washed away during histological processing). Right, a clonal cell at higher magnification, showing numerous microprojections. Note the presence of fine granules associated with fibrillar structures in the extracellular matrix. Arrowheads indicate transverse sections of microprojections. g, glycogen. Lower panel: HCS-2/A cells located in the nodule. Each cell has a highly irregular shaped nucleus (N) showing evenly distributed chromatin and a distinct nucleolus (Nu).

The incorporation of [^{35}S]sulfate into proteoglycans in HCS-2/8 cells is 0.7 times the rate of incorporation in primary cultures of rabbit costal chondrocytes, 2.5 times that of primary cultures of mouse costal chondrocytes, and 9 to 24 times that of cells derived from other tissues such as NRK, Swiss 3T3, and

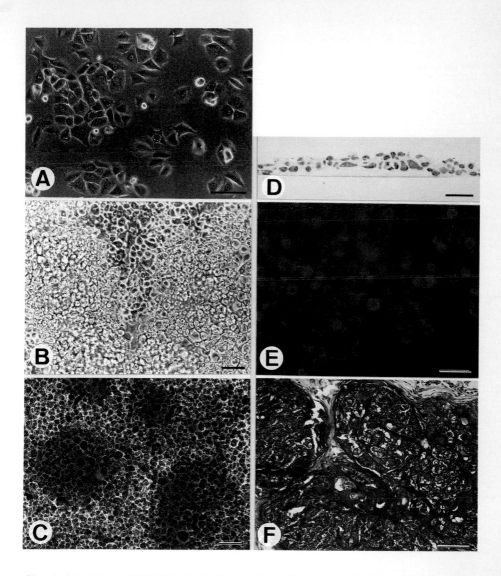

Plate 1. Morphology of HCS-2/8 cells. A: Sparse cultures. Phase-contrast. B: Over-confluent cultures. Phase-contrast. C: The cells in culture B were stained with Toluidine Blue. D: Light micrographs of sections of nodules of HCS-2/8 cells with toluidine blue staining. E: Light micrographs of sections of chondrosarcomas formed by HCS-2/8 cells in nude mice with toluidine blue staining. Bars = 100 μm. (From Takigawa, M., Tajima, K., Pan, H.-O., Enomoto, M., Kinoshita, A., Suzuki, F., Takano, Y., and Mori, Y., *Cancer Res.,* 49, 3996, 1989. With permission.)

mouse fibroblasts (Figure 2). In contrast to other cartilage- or chondrosarcoma-derived cells, HCS-2/8 cells have retained the ability to synthesize proteoglycans for more than 70 passages (Figure 2).

c. Characterization of Collagens[4,5,7,8]

HCS-2/8 cells stained by indirect immunofluorescence are positive for type II collagen (Figure 1E), but are negative for type I collagen. The production of collagen type II, but not type I by the cells has been confirmed by biochemical analyses such as SDS-PAGE and Western blotting. These cells also synthesize two minor bands of molecular weights corresponding to the $\alpha 1$ and $\alpha 2$ bands of type XI collagen.

The level of collagen synthesis in HCS-2/8 cells, determined by [^{3}H]proline incorporation into collagenase digestible protein, is lower than in primary cultures of rabbit costal chondrocytes from young rabbits, but higher than human fibroblasts which were isolated from a 60-year-old Japanese male.

d. Tumorigenicity[4,5,8]

Inoculation of 0.5 to 1×10^7 viable HCS-2/8 cells into nude mice results in development of palpable s.c. tumors which grow very slowly. They grow to 5 to 10 mm in diameter within a month after inoculation and 15 to 25 mm in diameter within 6 months. Grossly, the tumors are poorly vascularized and are solid at the periphery, with central necrotic and cystic degeneration. They are histologically well-differentiated chondrosarcomas retaining the characters of the original tumor. They show strong metachromasia when stained with toluidine blue (Plate 1, part F).

e. Autocrine Regulation of Expression of the Differentiated Phenotype of Chondrocytes by Insulin-Like Growth Factors (IGFs)[9]

HCS-2/8 cells have maintained a cartilage phenotype such as the ability to synthesize proteoglycans during many serial passages. Since IGFs have been shown to be involved in the expression of the differentiated phenotype of chondrocytes, the role of IGFs in retention of the cartilage phenotype of HCS-2/8 cells has been studied. Both IGF-I and II increase the synthesis of cartilage-specific proteoglycan in HCS-2/8 cells. In contrast to stimulatory effects of IGFs on costal chondrocytes of the young rabbit, the stimulatory effect of IGF-II on proteoglycan synthesis in HCS-2/8 cells is more potent than that of IGF-I.[8]

[^{125}I]IGF-I binds to HCS-2/8 cells. This binding can be competed by low concentrations of unlabeled IGF-I, higher concentrations of IGF-II, and much higher concentrations of insulin. [^{125}I]IGF-II also binds to the cells and its binding is competed for by IGF-II and high concentrations of IGF-I. After cross-linking the IGF receptors on the HCS-2/8 cells to [^{125}I]IGF-I, the proteins were separated by SDS-PAGE and subjected to autoradiography. Two major bands at 260 and 130 kDa, which correspond to the IGF type II receptor, and

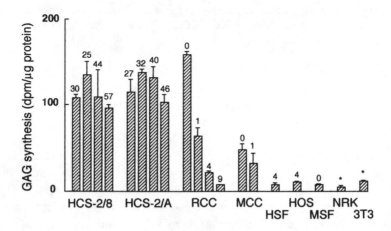

FIGURE 2. Proteoglycan synthesis by clonal HCS cells and cells derived from other tissues. RCC, rabbit costal chondrocytes; MCC, mouse costal chondrocytes; HSF, human skin fibroblasts; HOS, human osteosarcoma cells; MSF, mouse skin fibroblasts; NRK, normal rat kidney cells; and 3T3, Swiss 3T3 fibroblasts. Numbers above parentheses include passage numbers from before cloning. *, >30 passages.

the subunit of the IGF type I receptor are observed. These findings indicate that HCS-2/8 cells have both IGF-I and IGF-II receptors.

When confluent cultures of HCS-2/8 cells are maintained in serum-free medium for 4 d, proteoglycan synthesis does not decrease unless the medium is repeatedly replaced. Moreover, conditioned medium of HCS-2/8 cells stimulates them to synthesize proteoglycans. The cells produce IGF-II which stimulates their own proteoglycan synthesis. The cells also produce IGF-I, but its concentration is not enough to stimulate their proteoglycan synthesis. These findings suggest that IGFs, especially IGF-II, act on HCS-2/8 cells as autocrine differentiation factors.

Although IGFs stimulate DNA synthesis in normal chondrocytes cultured in serum-free medium, neither IGF-I nor IGF-II stimulate DNA synthesis or proliferation of HCS-2/8 cells. However, in contrast to normal chondrocytes, the DNA synthesis in HCS-2/8 cells in confluent cultures does not decrease dramatically, even if they are placed in serum-free medium. Since the stimulation of DNA synthesis by IGFs in normal chondrocytes is only slight, the high basal level of DNA synthesis in HCS-2/8 cells might mask the effect of IGF.

f. Regulation of Proliferation and Differentiation of HCS-2/8 Cells by Other Factors

Like normal chondrocytes, proteoglycan synthesis in HCS-2/8 cells is stimulated by insulin[12] and transforming growth factor(TGF)-ß.[13] Basic fibroblast growth factor (bFGF) and retinoic acid which decrease proteoglycan synthesis in normal chondrocytes,[14-16] also decrease proteoglycan synthesis in HCS-2/8 cells.

Factors such as TGF-ß, bFGF, and retinoic acid, which stimulate DNA synthesis or proliferation of normal chondrocytes,[13,16,17] stimulate the DNA synthesis or proliferation of HCS-2/8 cells. However, there are some subtle differences between the responsiveness of HCS-2/8 cells and that of normal rabbit chondrocytes to these factors. For instance, the stimulatory effect of bFGF on DNA synthesis in HCS-2/8 cells is observed only in sparse cultures. This may be due to the production of bFGF by the HCS-2/8 cells, which reaches a level in confluent cultures of two- to threefold that of normal rabbit chondrocytes. bFGF which has been accumulated in the extracellular matrix of HCS-2/8 cells might mask its cellular receptors. Although TGF-ß has been shown to stimulate DNA synthesis in normal chondrocytes cultured in medium containing serum, but inhibited DNA synthesis in the absence of serum,[13] it was found to stimulate DNA synthesis in HCS-2/8 cells in serum-free media. Since TGF-ß and FGF act synergistically to stimulate DNA synthesis in normal chondrocytes,[13] the overproduction of bFGF by the HCS-2/8 cells might relate to the ability of TGF-ß to DNA synthesis in these cells in serum-free medium.

Epidermal growth factor (EGF) has been shown to stimulate DNA synthesis in rabbit costal chondrocytes,[17] but have no effect on HCS-2/8 cells. Since the number of EGF receptors on chondrocytes decreases as they differentiate, this finding might indicate that HCS-2/8 cells are highly differentiated.

Although parathyroid hormone (PTH) increases proteoglycan synthesis, ornithine decarboxylase activity, and the level of cyclic AMP in rabbit costal chondrocytes,[18,19] HCS-2/8 does not respond to parathyroid hormone. Since growth cartilage cells have a larger number of PTH receptors than resting cartilage cells,[20] this might indicate that HCS-2/8 cell phenotype resembles resting cartilage cells in some ways. Inasmuch as interferons are species-specific cytokines and, until recently, only those of human origin had been available, their effects on chondrocytes have not been reported. However, it is difficult to obtain primary cultures of human chondrocytes. The establishment of HCS-2/8 cells, a permanent chondrocyte cell line, has made it possible to investigate the effect of interferons on chondrocyte metabolism. The cytokines, especially interferons α and ß, dramatically inhibited proteoglycan synthesis in HCS-2/8 cells.[21] The interferons also markedly inhibit DNA synthesis and proliferation of HCS-2/8 cells.[21]

g. Regulation of Calcification[8]

HCS-2/8 cells do not calcify *in vitro* in their normal growth medium, Eagle's minimum essential medium containing 20% fetal bovine serum. Alkaline phosphatase, which is used as a marker of calcification, is very low under normal growth conditions. Ascorbic acid (5–10 µg/ml) increases the alkaline phosphatase activity in HCS-2/8 cells to two times that of control at 2 weeks after its addition. This is also observed in primary cultures of rabbit costal chondrocytes. Moreover, when HCS-2/8 cells are subcultured once a week in the presence of ascorbic acid, alkaline phosphatase activity markedly increases, reaching 20-fold that of control after 20 weeks. Ascorbic acid stimulates the proliferation of HCS-2/8 cells in an early stage of serial passages, stimulates collagen and proteoglycan syntheses in a middle stage, and finally suppresses their proliferation and changes the shape of cells from polygonal to spherical, large cells which have high alkaline phosphatase activity. Therefore, ascorbic acid may stimulate the differentiation of HCS-2/8 cells to hypertrophic cells along the process of endochondral ossification.

2. HCS-2/A Cells

HCS-2/8 cells express the cartilage phenotype very well, but their tumorigenicity is low. To understand better the regulation of chondrocytes in chondrosarcomas, another immortal cell line, HCS-2/A, was established from the same human chondrosarcoma as HCS-2/8.[7]

a. General Characteristics[7]

HCS-2/A cells proliferate with a doubling time of 3.5 d in Eagle's MEM containing 20% fetal bovine serum. This growth rate is comparable to that of HCS-2/8 cells. However, in the presence of 2 to 10% fetal bovine serum HCS-2/A cells proliferate more rapidly than HCS-2/8 cells. Like HCS-2/8 cells, HCS-2/A cells have a polygonal morphology in sparse culture and become spherical as they become confluent. After reaching confluence, they form nodules composed of multilayers of cells and deposit a large quantity of extracellular matrix, as shown by strong metachromasia. The nodules formed by HCS-2/A cells are thicker and also larger in diameter than those formed by HCS-2/8 cells. Upon electron microscopical analysis, the cells in the nodules have been shown to resemble chondrocytes *in vivo,* but the nuclei of these cells are irregularly shaped, a characteristic of tumor cells (Figure 1C). The cells actively synthesize cartilage-specific, large proteoglycans. The level of proteoglycan synthesis in the cells is comparable to that of HCS-2/8 and has been retained more than 46 passages. Immunostaining reveals that HCS-2/A cells synthesized type II collagen, but not type I collagen. However, the level of collagen synthesis of HCS-2/A cells is lower than that of HCS-2/8 cells. Inoculation of HCS-2/A cells into athymic mice results in the formation of well-differentiated types of chondrosarcomas which are histologically similar to those formed by HCS-2/8 cells, but grow faster than HCS-2/8 tumors. HCS-2/A cells are more tumorigenic than HCS-2/8 cells; HCS-2/A tumors, but not HCS-2/8 tumors, are transplantable *in vivo.*

b. Regulation of Proliferation and Differentiation

Since malignant tumor cells generally require less serum in culture, the difference between growth rates of HCS-2/8 and 2/A cells cultured in medium containing low concentrations of serum might reflect the difference between their growth rates *in vivo.* Insulin, which stimulates proteoglycan and DNA synthesis in cultured normal chondrocytes,[12] only slightly increases proteoglycan synthesis in HCS-2/8 cells and has no effect on their DNA synthesis. However, the hormone markedly increases proteoglycan and DNA synthesis in HCS-2/A cells. The lower serum requirement for HCS-2/A cell growth may be related to its higher responsiveness to insulin. The growth stimulatory effect of insulin on cartilage and chondrocytes in *in vitro* systems has been demonstrable only using pharmacological doses,[12,22] while stimulation by insulin of proteoglycan synthesis in primary cultures of Swarm rat chondrosarcoma cells is observed at physiological concentrations.[23] The responsiveness of HCS-2/A cells to insulin is higher than that of normal chondrocytes and lower than that of Swarm rat chondrosarcoma cells. In other words, the HCS-2/A cell insulin responsiveness falls between rabbit normal chondrocytes and rat chondrosarcoma cells. Moreover, IGF-I stimulates proteoglycan synthesis by HCS-2/A cells *in vitro* and the responsiveness is higher than that of HCS-2/8 cells. Therefore, like Swarm rat chondrosarcoma,[24,25] HCS-2/A cells might form insulin-and somatomedin-dependent tumors in nude mice, and the high responsiveness of HCS-2/A cells to insulin and IGF-I might contribute to the faster development of HCS-2/A tumors. However, it is noteworthy that these findings also indicate that both HCS-2/8 and HCS-2/A cells retain one important characteristic; that is, insulin and somatomedin stimulation of proteoglycan synthesis.

The characteristics of the HCS-2/A line should make these cells useful in basic studies on the diagnosis, treatment, and etiology of human chondrosarcomas. Moreover, further comparative studies on biological, biochemical, and endocrinological differences between the HCS-2/A and HCS-2/8 cell lines should be helpful in understanding the relationship between tumorigenicity and altered regulation of cancerous chondrocytes of human origin.

B. SWARM RAT CHONDROSARCOMA

An *in vivo* transplantable rat chondrosarcoma has been established by Choi et al.[26] Originally, this tumor arose spontaneously in a Sprague-Dawley rat and has been maintained by subcutaneous implantation.[26] The histological appearance of the tumor shows a well-differentiated chondrosarcoma with chondrocytes occupying well-developed lacunae and prominent vascular channels and some of chondrocytes having multiple nuclei.[27] The tumor cells synthesize cartilage-specific proteoglycans which are recognized with an antibody against normal cartilage proteoglycans.[26] However, their mucopolysaccharide consists of 97.8% heterogeneous chondroitin 4-sulfate and 1.2% hyaluronic acid,[26] while normal cartilage is composed of chondroitin 6-sulfate and keratan sulfate.[28] Type II and type IX collagens were also produced by these tumor cells.[29,30] This chondrosarcoma provides us with a good source for the purification of cartilage-specific products and the determination of their amino acid and DNA sequences,[30-33] as well as to study their metabolism in chondrocytes.[34]

In contrast to HCS-2/8 and 2/A cells, it is difficult to culture the cells isolated from this tumor *in vitro* because the expression of the differentiated phenotype is unstable unless the cells were cultured in agarose.[35,36] However, several *in vitro* studies suggest that the tumor cells maintain some normal chondrocyte characteristics. The cells isolated from this tumor respond to multiplication-stimulating activity (MSA), IGFII, and insulin although their insulin receptor turnover is reported to be different from normal chondrocytes.[37] These factors promote the rate of the synthesis of proteoglycans in these cells,[23] as has also been reported in primary rabbit normal chondrocytes.[12,14] In contrast, the effect of MSA on the hydrodynamics size of proteoglycan shows a difference between the tumor cells and normal chondrocytes. MSA and insulin induce a slightly larger hydrodynamic size of monomeric proteoglycan in the tumor cells,[23] whereas MSA does not affect the size in normal rabbit chondrocytes.[14] Interestingly, the mediation of growth hormone is required for the growth of this tumor. When this tumor is implanted into hypophysectomized rats, the growth is remarkably inhibited and the tumor begins to atrophy.[38] Treatment with growth hormone or cortisone restores growth of the tumor and each of these factors show a synergistic effect.[38] Furthermore, anti-FGF antibodies suppress the growth of this rat transplantable tumor.[39] It is likely that this tumor is under the control of humoral factors such as somatomedin-like factors, cortisone, and FGF, as are normal cartilage cells.[12,14,40-42] The mechanism of the tumorigenesis of this transplantable rat chondrosarcoma has not been clear. It has been reported that the interaction of this tumor with host-derived chondrocytes induced by morphogenetic bone matrix cause chondrolysis and calcification of the tumor.[43] Furthermore, exogenous collagen and proteoglycan aggregates isolated from normal chondrocytes inhibit the synthesis of proteoglycan and collagen by these tumor cells *in vitro*.[44] These findings indicate that some regulatory factors, soluble factors, or extracellular matrices produced by normal chondrocytes can modulate the phenotype of these tumor cells and induce the organogenesis and terminal differentiation of the tumor. The lack of the production of these factors might cause the transformation of these tumor cells.

III. CHONDROCYTES TRANSFORMED WITH ONCOGENIC VIRUSES

Oncogenes are known to cause transformation of cells and affect the expression of their differentiated phenotypes. The regulation of the expression of protooncogenes has been correlated with development changes of many cell types. Therefore, studies on the effect of oncogenes in differentiated cells is very important in order to understand the regulation of proliferation and differentiation of these cells. In the case of chondrocytes, both the src oncogene, which has tyrosine kinase activity and is associated with cell membranes, and the myc oncogene, which has DNA binding activity and localizes to the nucleus, have been introduced into these cells, and their effects on differentiation have been examined.

A. src
When chicken embryo vertebral chondrocytes are infected with a tempera-

ture-sensitive src viral mutant, many phenotypical changes are demonstrated at the permissive temperature for src transformation; (1) a change in cell morphology from a polygonal shape to a bipolar shape,[45] (2) a dramatic decrease in the rate of synthesis of type II collagen at both the RNA and protein levels,[46] (3) the initiation of synthesis of fibroblast-type mRNA for type I collagen RNA,[46] (4) the inhibition of the synthesis of cartilage-specific proteoglycan[45,47] and the stimulation of the synthesis of fibroblast-specific proteoglycans,[47] (5) the increase in fibronectin mRNA and its protein products,[46,48] and (6) the increased synthesis of the cytoskeletal proteins such as actin, vimentin, and tubulin.[49] Furthermore, it has been reported that chick embryo chondrocytes transformed by Rous sarcoma virus, which carries the src gene, no longer synthesize a 64-kDa collagenase-sensitive protein, presumably type X collagen, as well as type II collagen and cartilage-specific proteoglycans.[50] This is accompanied by rapid growth and the initiation of the synthesis of type I collagen and fibronectin.[50] Although the src gene product, pp60[v-src], seriously affects expression of the extracellular matrix in chondrocytes, the effect is thought to be indirect. In chick fibroblasts, pp60[v-src] has the reverse effect, causing a reduction in the synthesis of type I collagen and fibronectin.[51-53] Fibroblasts transformed by pp60[v-src] have a disrupted cytoskeleton, reduced fibronectin matrix, and reduced substrate attachment,[54,55] all which may be directly related to phosphorylation by the src kinase.[56-58] By comparison, it has been concluded that pp60[v-src] does not directly regulate the expression of the individual genes or proteins of the cytoskeleton or extracellular matrix in chondrocytes, but that these changes stem from the inhibition by pp60[v-src] of the expression of the differentiated phenotype of the chondrocytes. In addition, pp60[v-src] has been shown to block the responsiveness of the chondrocytes to both soluble factors and the extracellular matrix. In normal chondrocytes, the synthesis of both fibronectin and vimentin are decreased by the transfer of these cells from attached culture to suspension culture.[49] In contrast, in src-transformed chondrocytes, the rate of the synthesis of fibronectin and cytoskeletal proteins does not change when the cells are transferred to suspension culture.[49] Furthermore, rabbit chondrocytes transfected with the src oncogene lose the ability to respond to FGF, which is a strong growth promoter for normal chondrocytes.[43]

B. myc

In contrast to src, the myc oncogene does not block the differentiation of chondrocytes. The expression of myc in chicken vertebral chondrocytes by the infection of the cells with MC29 (an avian myelocytosis virus) or HB1 virus promotes the proliferation of the cells, but does not induce a morphological change[59] as does src.[45] The myc-transformed cells are polygonal and organized into pavement-packed colonies similar to normal chondrocytes except that the individual chondrocytes are smaller.[59] Furthermore, the rate of synthesis of both type II collagen and a core protein of the cartilage-specific proteoglycan does not change in myc-transformed cells.[59] Similar findings have also been reported in MC29-infected quail embryo chondrocytes.[60] Furthermore, Horton et al.[61] suggest that myc expression in fetal rat chondrocytes induces rapid cell division, but does not block the expression of the differentiated phenotype, although the synthesis of type II collagen is reduced and the transformed chondrocytes maintain their responsiveness to retinoic acid as a suppressor of their differentiation.[18] The process of differentiation of chondrocytes is separated into three stages: proliferation, maturation, and calcification.[62] It is likely that the myc oncogene promotes the growth of chondrocytes without altering their phenotype during maturation. However, myc may block calcification because the synthesis of type X collagen, which is closely related to endochondral calcification, is inhibited by myc expression.[60] Therefore, myc might inhibit the terminal differentiation of chondrocytes as it does in myoblasts[63,64] and erythroblasts.[65]

C. OTHERS

It is reported that mil oncogene product, a cytosolic protein, inhibits the differentiated phenotype in chondrocytes. MH2 virus, which carries both the myc and mil oncogenes, induces the suppression of cartilage-specific proteoglycan and type II collagen and promotes the cell proliferation in chicken embryo chondrocytes.[59] Since myc does not affect the synthesis of either cartilage-specific proteoglycans or type II collagen in chondrocytes as described before, it is presumed that mil is responsible for the inhibition of differentiation of chondrocytes seen with transformation by the MH2 virus.

Although the transformation of chondrocytes by c-fos has not yet been reported, c-fos is a notable oncogene in the understanding of the regulation of chondrocyte differentiation. High expression of c-fos is observed in the zone of hypertrophic chondrocytes in mouse mandibular condyles organ culture[66] and in epiphysial cartilage of the developing human long bone, especially in the chondrocytes bordering the joint space.[67] This suggests that the c-fos may be involved in the regulation of calcification in chondrocytes.

IV. CHEMICALLY TRANSFORMED CHONDROCYTES

Katoh and Takayama[68] have reported the transformation of chondrocytes by treatment with 4-nitroquinoline-1-oxide. They have established two clonal chondrogenic cell lines from chemically transformed hamster sternal chondrocytes.[69] The cells have maintained the typical ultrastructural appearance of chondrocytes and synthesized a metachromatic matrix.[69] However, it has not been determined whether the cells synthesize cartilage-specific proteoglycan and collagen type II. Moreover, the cells do not form tumors *in vivo*, although they are transformed cell lines.[69]

V. SPONTANEOUSLY IMMORTALIZED CELL LINES

No spontaneously immortalized cell lines expressing the cartilage pheno-type have been established thus far. Only three clonal cell lines with differences in osteoblastic phenotype and with the ability to produce an endothelial cell growth inhibitor(s) were established by our group from growth cartilage of mouse ribs.[70] The proteoglycan synthesis of these cells is lower than in primary cultures of mouse growth cartilage cells. Unlike mouse growth cartilage cells, these cell lines synthesize type I collagen. However, they have some interesting characteristics. The three clonal cell lines, designated MGC/T1.4, MGC/T1.17, and MGC/T1.18, respond to parathyroid hormone determined by increase in c-AMP level, which is a characteristic of osteoblastic cells[70] and growth cartilage cells.[18-20] The PTH-stimulated level of c-AMP in MGC/T1.18 cells is 200 times that of control. The three clonal cell lines, especially MGC/T1.17, possess a high alkaline phosphatase activity which can indicate both growth cartilage cells and osteoblastic cells. MGC/T1.17 also forms nodules which calcify in the presence of ß-glycerophosphate. The three clonal lines, especially MGC/T1.4, secrete an endothelial cell growth inhibitor which is known to be present in cartilage[71] and produced by normal chondrocytes[72] and HCS-2/8 cells in culture.[73] Regulation of the proliferation and differentiation of these cell lines has not been extensively investigated, but it has been shown that EGF and TGF-ß stimulate the proliferation of MGC/T1.17 cells. Due to their different properties, these cell lines should be useful for studies on endochondral ossification, the actions of PTH on skeletal cells, and antiangiogenesis factors.

VI. CONCLUSION

In this chapter we have focused on the comparison of the characteristics of immortalized chondrocyte cell lines with those of normal chondrocytes in culture. Among them, HCS-2/8 should be useful for studies on the differentiated phenotype of human chondrocyte because it has important markers of differentiated chondrocytes and is of human origin. On the other hand, the HCS-2/A cell line should be a useful experimental model for studies on human chondrosarcomas. Like *in vivo* transplantable rat chondrosarcoma, these cell lines, however, do not calcify in their normal growth medium or *in vivo*. Calcification is the final stage of differentiation of chondrocytes in the process of endochondral ossification and is associated with cessation of proliferation. Therefore, if a means of induction of calcification of these cells can be determined, it might be a key step in establishing control over their tumorigenic state. Furthermore, the determination of the expression of oncogenes by these cells is necessary to understand the mechanism of their tumorigenicity.

The src and myc oncogene products inhibit different stages of differentiation of chondrocytes. This suggests the possibility of obtaining immortalized chondrocytes which retain the phenotype of a specific stage of differentiation by using src, myc, or other oncogenes to transform the cells and then investigate specific gene expression at each stage of chondrocyte differentiation.

ACKNOWLEDGMENTS

This work was supported in part by Grants-in-Aid for Scientific Research from the Ministry of Education and Culture of Japan (Motomi Enomoto and Masaharu Takigawa) and grants from the Kudo Scientific Foundation, the Kowa Life Science Foundation, and the Naito Foundation (Masaharu Takigawa). We thank Dr. David Boettiger for giving us the opportunity to write this chapter, and Dr. Sue Menko for critical reading and assistance in its preparation. We also thank Dr. Pan Hai-Ou, Dr. Koji Tajima, Dr. Akihiro Kinoshita, and Dr. Akira Sakawa for their industrious work in establishing and characterizing HCS cell lines; and to Miss Yoshie Endo for technical assistance.

REFERENCES

1. **Takigawa, M., Shirai, E., Fukuo, K., Tajima, K., Mori, Y., and Suzuki, F.,** Chondrocytes dedifferentiated by serial monolayer culture from cartilage nodules in nude mice, *Bone Miner.,* 2, 449, 1987.

2. **Benya, P. D. and Shaffer, J. D.,** Dedifferentiated chondrocytes reexpress the differentiated collagen phenotype when cultured in agarose gels, *Cell,* 30, 215, 1982.

3. **Castagnola, P., Moro, G., Descalzi-Cancedda, F., and Cancedda, R.,** Type X collagen synthesis during in vitro development of chick embryo chondrocytes, *J. Cell Biol.,* 102, 2310, 1986.

4. **Takigawa, M., Tajima, K., Pan, H.-O., Enomoto, M., Kinoshita, A., Suzuki, F., Takano, Y., and Mori, Y.,** Establishment of a clonal human chondrosarcoma cell line with cartilage phenotypes, *Cancer Res.,* 49, 3996, 1989.

5. **Takigawa, M.,** Establishment of clonal cartilage cell lines, *Soshiki-baiyou (Cell Culture),* 15, 165, 1989.

6. **Takigawa, M.,** HCS-2/8, *Protein, Nucleic Acid, Enzyme,* 34, 898, 1989.

7. **Takigawa, M., Pan, H.-O., Kinoshita, A., Tajima, K., and Takano, Y.,** Establishment from a human chondrosarcoma of a new immortal cell line with abilities to form proteoglycan-rich cartilage-like nodules and to respond to insulin in vitro and high tumorigenicity in vivo, *Int. J. Cancer,* 48, 717, 1991.

8. **Takigawa, M.,** A permanent cartilage cell line established from a human chondrosarcoma, *The Bone,* 12, 37, 1990, in Japanese.

9. **Takigawa, M.,** A clonal cartilage cell line established from a human chondrosarcoma, *Soshiki-baiyou Kenkyuu (Cell Culture Research),* 9, 68, 1991.

10. **Miller, D. R., Treadwell, B. V., and Mankin, H. J.,** *De novo* protein synthesis by human chondrosarcoma in cell and organ culture: evidence of unusually high collagen production by a neoplastic tissue, *Connect. Tissue Res.,* 8, 9, 1980.

11. **Thein, R. and Lotan, R.,** Sensitivity of cultured human osteosarcoma and chondrosarcoma cells to retinoic acid, *Cancer Res.,* 42, 4771, 1982.

12. **Kato, Y., Nasu, N., Takase, T., Daikuhara, Y., and Suzuki, F.,** A serum-free medium supplemented with multiplication-stimulating activity (MSA) supports both proliferation and differentiation of chondrocytes in primary culture, *Exp. Cell Res.,* 125, 167, 1986.

13. **Hiraki, Y., Inoue, H., Hirai, R., Kato, Y., and Suzuki, F.,** Effect of transforming growth factor-beta on cell proliferation and glycosaminoglycan synthesis by rabbit growth-plate chondrocytes in culture, *Biochim. Biophys. Acta,* 969, 91, 1988.

14. **Hiraki, Y., Yutani, Y., Takigawa, M., Kato, Y., and Suzuki, F.,** Differential effects of parathyroid hormone and somatomedin-like growth factors on the size of proteoglycan monomers and their synthesis in rabbit costal chondrocytes in culture, *Biochim. Biophys. Acta,* 845, 445, 1985.

15. **Takigawa, M., Takano, T., and Suzuki, F.,** Restoration by cyclic AMP of the differentiated phenotype of chondrocytes from de-differentiated cells pretreated with retinoids, *Mol. Cell. Biochem.,* 42, 145, 1982.

16. **Enomoto, M., Pan, H.-O., Suzuki, F., and Takigawa, M.,** Physiological role of vitamin A in growth cartilage cells: low concentrations of retinoic acid strongly promote the proliferation of rabbit costal growth cartilage cells in culture, *J. Biochem.,* 107, 743, 1990.

17. **Hiraki, Y., Kato, Y., Inoue, H., and Suzuki, F.,** Stimulation of DNA synthesis in quiescent rabbit chondrocytes in culture by limited exposure to somatomedin-like growth factor, *Eur. J. Biochem.,* 158, 333, 1986.

18. **Takigawa, M., Ishida, H., Takano, T., and Suzuki, F.,** Polyamine and differentiation: induction of ornithine decarboxylase by parathyroid hormone is a good marker of differentiated chondrocytes, *Proc. Natl. Acad. Sci. U.S.A.,* 77, 1481, 1980.

19. **Takigawa, M., Takano, T., and Suzuki, F.,** Effects of parathyroid hormone and cyclic AMP analogues on the activity of ornithine decarboxylase and expression of the differentiated phenotype of chondrocytes in culture, *J. Cell. Physiol.,* 106, 259, 1981.

20. **Enomoto, M., Kinoshita, A., Pan, H.-O., Yamamoto, I, Suzuki, F., and Takigawa, M.,** Demonstration of receptors for parathyroid hormone on cultured rabbit costal chondrocytes, *Biochem. Biophys. Res. Commun.,* 162, 1222, 1989.

21. **Sakawa, A., Yutani, Y., Asada, K., Shimazu, A., and Takigawa, M.,** Effect of interferon on a human chondrosarcoma cell line (HCS-2/8), *Cent. Jpn. J. Orthop. Traumat.,* 34, 348, 1991.

22. **Levovitz, H. E. and Eisenbarth, G. S.,** Hormonal regulation of cartilage growth and metabolism, *Vitam. Horm.,* 33, 575, 1975.

23. **Stevens, R. L., Nissley, S. P., Kimura, J. H., Rechler, M. M., Caplan, A. I., and Hascall, V. C.,** Effects of insulin and multiplication-stimulating activity on proteoglycan biosynthesis in chondrocytes from the Swarm rat chondrosarcoma, *J. Biol. Chem.,* 256, 2045, 1981.

24. **Salomon, D. S., Paglia, L. M. and Verbruggen, L.,** Hormone-dependent growth of a rat chondrosarcoma in vivo, *Cancer Res.,* 39, 4387, 1979.

25. **McCumbee, W. D., Lebovitz, H. E. and McCarty, K. S.,** Hormone responsiveness of a transplantable rat chondrosarcoma. III. Ultrastructural evidence of in vivo hormone dependence, *Am. J. Pathol.,* 103, 56, 1981.

26. **Choi, H. U., Meyer, K., and Swarm, R.,** Mucopolysaccharide and protein-polysaccharide of a transplantable rat chondrosarcoma, *Proc. Natl. Acad. Sci. U.S.A.,* 68, 877, 1971.

27. **Breitkreutz, D., deLeon, L. D., Paglia, L., Gay, S., Swarm, R. L., and Stern, R.,** Histological and biochemical studies of a transplantable rat chondrosarcoma, *Cancer Res.,* 39, 5093, 1979.

28. **Kaplan, D. and Meyer, K.,** Ageing of human cartilage, *Nature (London),* 183, 1267, 1959.

29. **Poole, C. A., Wotton, S. F., and Duance, V. C.,** Localization of type IX collagen in chondrons isolated from porcine articular cartilage and rat chondrosarcoma, *Histochem. J.,* 20, 567, 1988.

30. **Furuto, D. K., Bhown, A. S., and Miller, E. J.,** Characterization of mammalian type IX collagen fragments from limited pepsin digest of a transplantable swarm rat chondrosarcoma, *Matrix,* 9, 353, 1989.

31. **Neame, P. J., Christner, J. E., and Baker, J. R.,** Cartilage proteoglycan aggregates. The link protein and proteoglycan amino-terminal globular domains have similar structures, *J. Biol. Chem.,* 262, 17768, 1987.

32. **Doege, K., Sasaki, M., Horigan, E., Hassell, J. R., and Yamada, Y.,** Complete primary structure of the rat cartilage proteoglycan core protein deduced from cDNA clones, *J. Biol. Chem.,* 262, 17757, 1987.

33. **Doege, K., Hassell, J. R., Caterson, B., and Yamada, Y.,** Link protein cDNA sequence reveals a tandemly repeated protein structure, *Proc. Natl. Acad. Sci. U.S.A.,* 83, 3761, 1986.

34. **Kimura, J. H., Thonar, E. J.-M., Hascall, V. C., Reiner, A., and Poole, A. R.,** Identification of core protein an intermediate in proteoglycan biosynthesis in cultured chondrocytes from the Swarm rat chondrosarcoma, *J. Biol. Chem.,* 256, 7890, 1981.

35. **Sun, D., Aydelotta, M. B., Maldonado, B., Kuettner, K. E., and Kimura, J. H.,** Clonal analysis of the population of chondrocytes from the Swarm rat chondrosarcoma in agarose culture, *J. Orthop. Res.,* 4, 427, 1986.

36. **Bansal, M. K., Ward, H., and Mason, R. M.,** Proteoglycan synthesis in suspension cultures of Swarm rat chondrosarcoma chondrocytes and inhibition by exogenous hyaluronate, *Arch. Biochem. Biophys.,* 246, 602, 1986.

37. **Otsu, K., Geary, E. S., and Stevens, R. L.,** Aberrant regulation of the metabolism of the insulin receptor in Swarm rat chondrosarcoma chondrocytes, *Biochem. J.,* 254, 203, 1988.

38. **McCumbee, W. D., McCarty, K. S., Jr., and Lebovitz, H. E.,** Hormone responsiveness of a transplantable rat chondrosarcoma. II. Evidence for in vivo hormone dependence, *Endocrinology,* 106, 1930, 1980.

39. **Baird, A., Mormede, P., and Bohlen, P.,** Immunoreactive fibroblast growth factor (FGF) in a transplantable chondrosarcoma: inhibition of tumor growth by antibodies to FGF, *J. Cell Biochem.,* 30, 78, 1986.

40. **Takano, T., Takigawa, M., and Suzuki, F.,** Stimulation by glucocorticoids of the differentiated phenotype of chondrocytes and the proliferation of rabbit costal chondrocytes in culture, *J. Biochem.,* 97, 1093, 1985.

41. **Kato, Y. and Gospodarowicz, D.,** Stimulation by glucocorticoid of the synthesis of cartilage-matrix proteoglycans produced by rabbit costal chondrocytes in vitro, *J. Biol. Chem.,* 260, 2364, 1985.

42. **Kato, Y., Iwamoto, M., and Koike, T.,** Fibroblast growth factor stimulates colony formation of differentiated chondrocytes in soft agar, *J. Cell. Physiol.,* 133, 491, 1987.

43. **Miller, G. J., Hamburg, R. J., and Ferrara, J. A.,** Phenotypic modulation of the Swarm rat chondrosarcoma induced by morphogenetic bone matrix, *Cancer Res.,* 47, 3589, 1987.

44. **Lucas, P. A. and Dziewiatkowski, D. D.,** Feedback control of selected biosynthetic activities of chondrocytes in culture, *Connect. Tissue Res.,* 16, 323, 1987.

45. **Pacifici, M., Boettiger, D., Roby, K., and Holtzer, H.,** Transformation of chondroblasts by Rous sarcoma virus and synthesis of the sulfated proteoglycan matrix, *Cell,* 11, 891, 1977.

46. **Allebach, E. S., Boettiger, D., Pacifici, M., and Adams, S. L.,** Control of type I and II collagen and fibronectin gene expression in chondrocytes delineated by viral transformation, *Mol. Cell. Biol.,* 5, 1002, 1985.

47. **Shanley, D. J., Cossu, G., Boettiger, D., Holtzer, H., and Pacifici, M.,** Transformation by Rous sarcoma virus induces similar patterns of glycosaminoglycan synthesis in chick embryo skin fibroblasts and vertebral chondroblasts, *J. Biol. Chem.,* 258, 810, 1983.

48. **Adams, S. L., Pacifici, M., Boettiger, D., and Pallante, K. M.,** Modulation of fibronectin gene expression in chondrocytes by viral transformation and substrate attachment, *J. Cell Biol.,* 105, 483, 1987.

49. **Boettiger, D. and Menko, S. A.,** Effect of oncogenes on chondrogenesis, *Pathol. Immunopathol. Res.,* 7, 32, 1988.

50. **Gionti, E., Capasso, O., and Cancedda, R.,** The culture of chick embryo chondrocytes and the control of their differentiated functions in vitro, *J. Biol. Chem.,* 258, 7190, 1983.

51. **Delclos, K. B. and Blumberg, P. M.,** Decrease in collagen production in normal and Rous sarcoma virus-transformed chick embryo fibroblasts induced by phorbol myristate acetate, *Cancer Res.,* 39, 1667, 1979.

52. **Kamine, J. and Rubin, H.,** Coordinate control of collagen synthesis and cell growth in chick embryo fibroblasts and the effect of viral transformation on collagen synthesis, *J. Cell. Physiol.,* 92, 1, 1977.

53. **Olden, K. and Yamada, K. M.,** Mechanism of the decrease in the major cell surface protein of chick embryo fibroblasts after transformation, *Cell,* 11, 957, 1977.

54. **Parsons, J. T. and Weber, M. J.,** Genetics of src: structures and functional organization of a protein tyrosine kinase, *Curr. Top. Microbiol. Immunol.,* 137, 79, 1989.

55. **Boschek, C. B., Jockush, B. M., Friis, R.R., Back, R., Gundmann, E., and Bauer, H.,** Early changes in the distribution and organization of microfilament proteins during cell transformation, *Cell,* 24, 175, 1981.

56. **Menko, A. S., Croop, J., Toyama, Y., Holtzer, H., and Boettiger D.,** The response of chicken embryo dermal fibroblasts to cytochalasin B is altered by Rous Sarcoma Virus-induced cell transformation, *Mol. Cell. Biol.,* 2, 320, 1982.

57. **Weber, M. J.,** Inhibition of protease activity in cultures of Rous sarcoma virus-transformed cells: effect on the transformed phenotype, *Cell,* 5, 253, 1975.

58. **Chen, W.-T., Wang, J., Hasegawa, S., Yamada, S., and Yacroda, K. M.,** Regulation of fibronectin receptor distribution by transformation, exogenous fibronectin, and synthetic peptides, *J. Cell Biol.,* 103, 1649, 1986.

59. **Alema, S., Tato, F., and Boettiger, D.,** myc and src oncogene have complementary effects on cell proliferation and expression of specific extracellular matrix components in definitive chondroblasts, *Mol. Cell. Biol.,* 5, 538, 1985.

60. **Gionti E., Pontarelli, G., and Cancedda, R.,** Avian myelocytomatosis virus immortalizes differentiated quail chondrocytes, *Proc. Natl. Acad. Sci. U.S.A.,* 82, 2756, 1985.

61. **Horton, W. E., Jr., Cleveland, J., Rapp, U., Nemuth, G., Bolander, M., Doege, K., Yamada, Y., and Hassell, J. R.,** An established rat cell line expressing chondrocytes properties, *Exp. Cell Res.,* 178, 457, 1988.

62. **Kato, Y., Iwamoto, M., Koike, T., Suzuki, F., and Takano, Y.,** Terminal differentiation and calcification of rabbit chondrocytes cultures grown in centrifuge tubes: regulation by transforming growth factor beta and serum factors, *Proc. Natl. Acad. Sci. U.S.A.,* 85, 9552, 1988

63. **Endo, T. and Nadal-Ginard, B.,** Transcription and posttranscriptional control of c-myc myogenesis: its mRNA remain inducible in differentiated cells and does not suppress the differentiated phenotype, *Mol. Cell. Biol.,* 6, 1412, 1986.

64. **Falcone, G., Boettiger, D., Tato, F., and Alema, S.,** Role of cell division in differentiation of myoblasts infected with a temperature-sensitive mutant of Rous sarcoma virus, *EMBO J.,* 3, 1327, 1984.

65. **Coppola, J. A. and Cole, M. D.,** Constitutive c-myc oncogene expression blocks mouse erythroleukemia cell differentiation but not commitment, *Nature (London),* 320, 760, 1986.

66. **Closs, E. I., Murray, A. B., Schmidt, J., Schon, A., Erfle, V., and Strauss, P. G.,** c-fos expression precedes osteogenic differentiation of cartilage cells in vitro, *J. Cell Biol.,* 111, 1313, 1990.

67. **Sanberg, M., Vuorio, T., Hirvonen, H., Alitalo, K., and Vuorio, E.,** Enhanced expression of TGF-beta and c-fos mRNAs in the growth plates of developing human long bones, *Development,* 192, 461, 1988.

68. **Katoh, T. and Takayama, S.,** In vitro transformation of hamster chondrocytes by 4-nitroquinoline 1-oxide, *Cancer Lett.,* 2, 31, 1976.

69. **Katoh, Y. and Takayama, S.,** Establishment of clonal cell lines of hamster chondrocytes maintaining their phenotypic traits, *Exp. Cell Res.,* 106, 285, 1977

70. **Takigawa, M., Shirai, E., Enomoto, M., Kinoshita, A., Pan, H.-O., and Suzuki, F.,** Establishment from mouse growth cartilage of clonal cell lines with responsiveness to parathyroid hormone, alkaline phosphatase activity, and ability to produce an endothelial cell growth inhibitor, *Calcif. Tissue Int.,* 45, 305, 1989.

71. **Takigawa, M., Shirai, E., Enomoto, M., Hiraki, Y., Suzuki, F., Shiio, T., and Yugari, Y.,** Cartilage-derived anti-tumor factor (CATF): partial purification and correlation of inhibitory activity against tumor growth with anti-angiogenic activity, *J. Bone Miner. Metab.,* 6, 83, 1988.

72. **Takigawa, M., Shirai, E., Enomoto, M., Pan, H.-O., Suzuki, F., Shiio, T., and Yugari, Y.,** A factor in conditioned medium of rabbit costal chondrocytes inhibits the proliferation of cultured endothelial cells and angiogenesis induced by B16 melanoma: its relation with cartilage-derived anti-tumor factor (CATF), *Biochem. Int.,* 14, 357, 1987.

73. **Takigawa, M., Pan, H.-O., Enomoto, M., Kinoshita, A., Nishida, Y., Suzuki, F., and Tajima, K.,** A clonal human chondrosarcoma cell line produces an anti-angiogenic antitumor factor, *Anticancer Res.,* 10, 311, 1990.

INDEX